Musculoskeletal Pain and Disability

Musculoskeletal Pain and Disability

Paul E. Kaplan, MD
Professor and Chairman, Department of Physical Medicine and Rehabilitation,
University of Missouri—Columbia School of Medicine
Medical Director—Rusk Rehabilitation Center, Columbia, Missouri

Ellen D. Tanner, RPT
Private Practice: Bensalem, Pennsylvania
Guest Lecturer

APPLETON & LANGE
Norwalk, Connecticut/San Mateo, California

ISBN 0-8385-6560-3

89 90 91 92 93 / 10 9 8 7 6 5 4 3 2 1

Prentice Hall International (UK) Limited, *London*
Prentice Hall of Australia Pty. Limited, *Sydney*
Prentice Hall Canada, Inc., *Toronto*
Prentice Hall Hispanoamericana, S.A., *Mexico*
Prentice Hall of India Private Limited, *New Delhi*
Prentice Hall of Japan, Inc., *Tokyo*
Simon & Schuster Asia Pte. Ltd., *Singapore*
Editora Prentice Hall do Brasil Ltda., *Rio de Janeiro*
Prentice Hall, *Englewood Cliffs, New Jersey*

Library of Congress Cataloging-in-Publication Data

Kaplan, Paul E., 1940–
    Musculoskeletal pain and disability / Paul E. Kaplan, Ellen D.
Tanner.
        p. cm.
    ISBN 0-8385-6560-3
    1. Musculoskeletal system—Diseases.   2. Pain—Treatment.
  I. Tanner, Ellen D.   II. Title
    [DNLM: 1. Bone Diseases.   2. Muscular Diseases.
  3. Musculoskeletal System—physiopathology.   4. Pain.   WE 140 K17m]
  RC925.5.K36   1989
  616.7—dc19
  DNLM/DLC
  for Library of Congress   88–38495
                                                                  CIP

Acquisitions Editor: Stephany S. Scott
Production Editor: Laura K. Giesman
Designer: Michael J. Kelly

PRINTED IN THE UNITED STATES OF AMERICA

Dedication

We dedicate this book to our families. They provided vital support and encouragement during the long time it took to prepare, revise, and revise this manuscript yet again. We would also like to especially acknowledge Dr. Shephard of Temple University, the manufacturers who provided us with illustrations, and also the many models and contributors who worked so very hard on this project.

Ellen D. Tanner
Paul E. Kaplan

Contents

Contributors

Susan P. Buckelew, PhD
Assistant Professor of Physical Medicine and Rehabilitation, Adjunct Assistant
Professor of Psychology, and Director, Departmental Pain and Industrial Care
Program–Department of Physical Medicine and Rehabilitation,
University of Missouri–Columbia, Columbia, Missouri

Kip Burkman, MD
Staff Physiatrist, Immanuel Rehabilitation Center, Omaha, Nebraska
Consulting Physiatrist, Memorial Hospital of Dodge City, Fremont, Nebraska

Robert R. Conway, MD
Assistant Professor of Clinical Physical Medicine and Rehabilitation,
Staff Physician–Rusk Rehabilitation Center,
Columbia, Missouri
Director, Medical Student Education, and Director, Electromyography Lab–
Department of Physical Medicine and Rehabilitation,
University of Missouri–Columbia, Columbia, Missouri

Robert G. Frank, PhD
Associate Professor and Associate Chairman—Department of Physical Medicine and
Rehabilitation, Associate Professor of Psychiatry (Medical Psychology), Adjunct
Associate Professor of Psychology and Director, Psychology Division—Department
of Physical Medicine and Rehabilitation,
University of Missouri—Columbia, Columbia, Missouri

Gerard J. Jablonowski, PT, MS
Director, Physical Medicine and Rehabilitation, The Lower Bucks Hospital, Bristol,
Pennsylvania
Clinical Assistant Professor, Physical Therapy, Philadelphia College of Pharmacy and
Science, Philadelphia, Pennsylvania.

Joseph Kahn, PT, PhD
Clinical Assistant Professor, State University of New York at Stony Brook, New York
Adjunct Associate Professor, Touro College, Huntington, New York
Clinical Associate, New York University, New York, New York

Adjunct Clinical Professor, Daemen College, Amherst, New York
Private Practice: Plainview, New York
Author: *Principles & Practice of Electrotherapy*, Churchill Livingstone Inc.,
   New York, New York 1987

Richard T. Katz, MD

Assistant Professor of Rehabilitation Medicine, Northwestern University School of
   Medicine, Chicago, Illinois
Associate Director of Rehabilitation Institute of Chicago Brain Trauma Program,
Associate Director of Rehabilitation Institute of EMG and Clinical Neurophysiology
   Labatories, Director of Rehabilitation Institute of Chicago Multidisciplinary Task
   Force on Spasticity, Chicago, Illinois

Christina A. Marciniak, MD

Associate in Clinical Rehabilitation Medicine, Northwestern University Medical
   School, Chicago, Illinois
Director of Northwestern Memorial Hospital–Rehabilitation Consult Service,
   Chicago, Illinois
Staff Physician of Rehabilitation Institute of Chicago, Chicago, Illinois

Saroj Shah, MD

Assistant Professor of Clinical Physical Medicine and Rehabilitation Director, Outpa-
   tient Clinics—Department of Physical Medicine and Rehabilitation, University of
   Missouri—Columbia, Columbia, Missouri
Attending Physician, Rusk Rehabilitation Center, Columbia, Missouri

Mark Walsh, PT, MS

Private Practice: Hand & Orthopedic Rehabilitation Services, Levittown,
   Pennsylvania
Assistant Adjunct Professor, Philadelphia College of Pharmacy and Science,
   Philadelphia, Pennsylvania

Gary M. Yarkony, MD

Director, Rehabilitation, Midwest Regional Spinal Cord Injury Care System and
Attending Physician, Rehabilitation Institute of Chicago, Chicago, Illinois
Assistant Professor, Northwestern Medical School, Chicago, Illinois
Adjunct Assistant Professor, Pritzker Institute for Medical Engineering–Illinois
   Institute of Technology, Chicago, Illinois

Preface

This textbook represents a new and original effort at interdisciplinary communication. It was not easy, and our families will confirm the intermittent frustration and fatigue.

Through clinical research, knowledge about the rehabilitation of pain and disability from musculoskeletal syndromes has exploded in the past decade. We have endeavored to produce a book reflecting the perspectives of the many health disciplines working in this field. After reviewing the most commonly seen clinical problems, we have closely examined issues surrounding musculoskeletal rehabilitation in each major anatomic region. Specialists in the field offer new considerations in the evaluation and treatment of each area and give specific guidelines for professionals working with these patients.

This was a different type of project, and we hope it presents new aspects of medical or therapeutic perspectives. This book does, however, try to reproduce the type of clinical interaction that is helpful, friendly, and positive. We have endeavored to be accurate but not critical of these observations regarding musculoskeletal pain and disability.

We could at this point produce many statistics emphasizing the importance of this area. You, the reader, know that these disorders produce millions of lost hours of fruitful function. You also know the anguish and anxiety involved. Indeed, as part of this book, we have presented implications of the technology and the outlook of the eighties so that it will also be an effective update of our interests in this active field.

Treating people who are hurt and anxious is difficult. We hope that this book will yield relevant tips and push the memory for reminders of how to treat these patients. Most of all we would be happy if it facilitated more interdisciplinary cooperation. Everyone benefits from an inclusive, thorough, hard-hitting treatment plan.

Paul E. Kaplan, MD
Columbia, Missouri

Ellen D. Tanner, RPT
Bensalem, Pennsylvania

Acknowledgments

We would like to acknowledge the assistance of: Nancy Edden, *illustrations*; Chattanooga Corporation, Camp International, Henley International, *photographs*; Frank J. Malone & Sons, Inc., *model assistance*; and Christopher Carcia, Jud Davidson, John Tanner, *models*.

1

Tendinitis, Bursitis, and Fibrositis

Paul E. Kaplan and
Ellen D. Tanner

OVERVIEW

Soft tissue disorders can baffle even the most intrepid observers. The clinical trend of bursitis, tendinitis, and fibrositis contributes more than its fair share of confusion. The usual response to this challenge is at first confident: try this method and it will help. Nonetheless, as weeks pass, the patient continues to return with symptoms intact. Anxiety, anger, and resentment grow. To prevent failure, one must review the diagnosis and management in a consistent and orderly manner.

Tendinitis, bursitis, and fibrositis are diagnoses that are not mutually exclusive. Since they all are painful conditions that exist in soft tissue, they also overlap. Tendinitis can occur without bursitis, but bursitis often requires an existing tendinitis. This combination is seen with patterns of relative overuse[13]: normal structures undergoing heavy strain, slightly weakened structures subjected to repetitive forces, and greatly weakened structures required to handle otherwise normal stress (Fig. 1-1).

Bursitis and tendinitis are often present in industrial work settings.[4,5,13] They are frequently accompanied by myalgae, produced by decreased blood flow,[33,40,45] and hypoxemia.[51] Pain, however, will also present itself after the tendon has been chronically compressed and worn.[20]

Bursitis and secondary fibrositis will often be observed in patients who have had long-standing arthritic disorders.[46] Arthritis directly weakens the bony structure. It also indirectly weakens soft tissue through inflammation of surrounding muscles, tendons, and bursae. Subsequently, if these areas are subjected to even weak or repetitive forces, pain will be experienced. The symptomatology of the tendinitis will vary throughout the day,[71] and will commonly not be as apparent. Any resultant failure of soft tissue function is, therefore, always multifactorial (Fig. 1-2).

Soft tissue disorders are often associated with an unbalanced or unprepared response to force vectors. Inflammation associated with arthritis can weaken any residual counter-

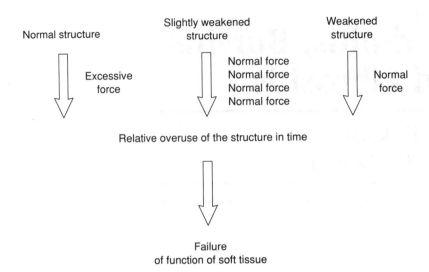

Figure 1-1. Generation of shoulder or hip girdle dysfunction.

force. Those conditions that further reduce blood flow to muscles (vasculitis, atherosclerosis), generate hypoxemia, pain, and disability. The multifactorial nature of soft tissue dysfunction has been highlighted by a hypothesis suggesting that fibrositis is caused by any process interrupting stage IV sleep.[15] Failure of function does not exist in a vacuum: the stage or milieu has been prepared by fatigue, exhaustion, depression, and anxiety. Unless the full dimensions of the crisis are evaluated and managed, any limited treatment effort will not succeed in relieving the pain over a long period of time. In a few cases, the dimensions of the problem are so great that complete pain relief is not likely. But even in these situations, some modification of pain behavior or pain response is possible.

TREATMENT OBJECTIVES

Rehabilitation therapy is not very effective if it is not planned and directed. There is a 50% chance of any novel procedure alleviating symptomatology. But with familiarity, the

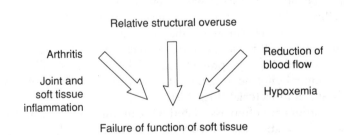

Figure 1-2. Generation of soft tissue dysfunction.

helpfulness of even this effect rapidly dissipates. Achieving an accurate diagnosis is a first step and must be followed by a reasonable and realistic series of goals.

Long-term and short-term treatment objectives commonly direct both inpatient and outpatient care. Short-term objectives frequently cover a 2-week span. Both short-term and long-term goals can have serious disadvantages. Short-term goals can be too specific or too short. For example:

1. Problem: To mobilize a contracted, painful shoulder.
2. Short-term objective: To reduce pain by 12%.

Long-term goals, on the other hand, can be too broad and unfocused. For example:

1. Problem: To mobilize a contracted, painful shoulder.
2. Long-term objective: To eliminate pain.

It is difficult to be against objectives that sound so good—it's like being against apple pie. In theory, short-term objectives should lead to the realization of the long-term objective. The reality is that objectives must generate a successful milieu. Both short- and long-term objectives must be appropriate and efficient so that the patient has a reasonably good chance of achieving them. The multifactorial nature of the processes themselves supplies guidance on how to fashion these objectives. Patients are asked to significantly modify their lifestyles. Therefore, these treatment objectives are as educational as they are therapeutic, and the patient embarks on a guided or modified self-directed learning (SDL) program. Much attention has been given to SDL programs, mainly as part of a curriculum for continuing medical education.[7,42,70] This type of teaching is also relevant to the instruction given patients enrolled in a rehabilitative program. In fact, the most effective objectives are educational objectives. A few specific characteristics of these educational objectives will help illustrate the appropriate and efficient nature of applying educational objectives to clinical problems.

Each objective should involve performance. Ideally, the action should be something that can be measured. This principle is important whether long- or short-term objectives are being constructed. "Reduction of pain" or "reduction of spasticity" are inappropriate objectives because it is unclear how the reduction will be measured. These objectives are qualitative symptoms. One observes and can measure the patient's behavior. But centering a treatment objective around an abstract idea of a medical condition is like fighting with one disabled arm. The result vastly increases the chance for failure. Rather than emphasize subjective and emotional issues, attention could be shifted toward objectives expressing functional performance. Increasing range of motion, lifting a weight, ambulating, and performing knee bends are examples of actions that can be observed and measured.

The conditions under which the patient is to perform have to be precisely and thoroughly stated. One condition is whether the patient will be independent, need assistance, or be dependent. Another is whether the range of motion is active or passive. A third is qualifying a strength grade with a plus or a minus. All three are relevant, appropriate conditions to be applied to the objectives. In fact, they make it possible to progress in definite, transient stages (short-term objectives) toward the end point of the treatment plan (long-term objectives). The overall performance has not changed in the serial progression from short- to long-term objectives, but the conditions have been modified. For

TABLE 1-1. FORMULATING TREATMENT OBJECTIVES

Process	Application
Performance	Directs therapy toward functional behavior that can be observed, measured, influenced
Conditions	Helps structure a measured, reasonable, appropriate plan of action toward the successful resolution of a problem
Criterion	Helps determine whether the effort is successful and standards are met

example, during the progression toward independent ambulation (long-term objective), the patient must first achieve success shifting body weight and then walking between the parallel bars. Part of the information given for one short-term objective is the achievement of the prior short-term objective. But the conditions must represent a logical progression, must be appropriate to the functional impairment, and must present—when achieved—a significant improvement. Only then is the atmosphere set for success.

Objectives are also related to acceptance: What are the patient's criteria for success? What are the therapist's criteria for success? The surface manifestations have to do with the speed, accuracy, and smoothness of the functional performance. The acceptance of the quality of the function is heavily dependent upon the original contract between the patient and the therapist. What were their original expectations? What was the agreement with regard to the patient's reliability in follow up performance and the therapist's reliability in providing personalized instruction? What provision was made if follow up home instructions were not carried out or were ineffective? Many of the answers to these questions are assumed, not negotiated. Yet circumstances change, particularly with disorders that are managed by outpatient therapy. Specific performance, however, cannot be evaluated without clear answers to these questions. The full process of defining treatment objectives is outlined in Table 1-1.

TENOSYNOVITIS

Tenosynovitis is a more appropriate name than tendinitis since the tendon is not at all a vascular tissue.[18] The inflammation is usually not located within the tendon itself but occurs within the tendon sheath. During the period of acute inflammation, effusion can accumulate.[58] The origin of the inflammation is frequently traumatic, involving workers' shoulders, secretaries' wrists or hands, and dancers' ankles. Arthritic disorders as osteoarthritis often associated with tendosynovitis.[18,54,58] But any form of acute or chronic inflammation may be present at any time.[54] With chronic inflammation, annular fibrosis of the sheath that extends to the surface of the tendon[18] will limit the range of motion of the tendon. Distal to the ring of fibrous tissue, the tendon can, in time, become thickened, further limiting the range of its motion.[54] When adhesions and calcific tendinitis are present, crepitus will often be palpated over the tender tendon as the patient attempts to move it.[18,54,58] There is some opinion that fibrous adhesions can be stretched to restore function,[58] presumably with use. Figure 1-3 shows the pathogenetic flow chart.

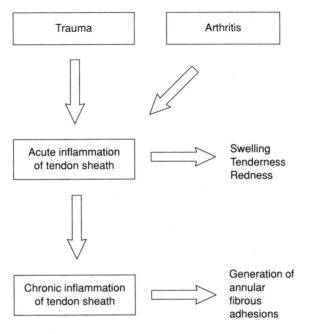

Figure 1-3. Pathogenesis of tenosynovitis.

Acute

Treatment objectives begin with management of the acute inflammation.[1] Aspirin and other nonsteroidal agents can be prescribed orally to help make the patient more comfortable and reduce the inflammation. Corticosteroid agents are usually used only after nonsteroidal therapy, and given by injection with a local anesthetic. Oral corticosteroid intake is the last line of defense and is not frequently required.

Medical management in the acute stage is often effective in resolving the inflammation and thereby the pain. Used alone or in conjunction with rest and local ice application, the treatment regimen often resolves the problem. The patient can still profit, however, from clinical care during the acute phase of tendinitis.

The application of ice to an acutely inflamed area is a widely accepted method of treatment.[13,27,28,34,43,48,49,56,59,64] The effects of cold include reduction of edema via vasoconstriction, which reduces hyperemia; reduction of pain by affecting the activity of pain nerve fibers and receptors; and alleviation of painful muscle spasm.[34,35,47,64]

Ice can be applied many ways, and a number of commercial products make application easy. Gel packs, ice massage, ice and ice water packs in various forms that are easily made or obtained are advocated by multiple sources. The form to be utilized can best be determined by its availability, ease of application, and the preference of the patient and therapist. Care should be taken, however, to avoid damage to the skin. A layer of wet toweling is placed between the patient's body and any stationary ice modality to prevent freezing the skin to the ice.

Here, as with any treatment modality, it is important to explain the procedure to the patient. In the shoulder, for example, pain is often referred to the area where the deltoid inserts on the humerus. If the rationale is not thoroughly explained, patients wonder why the painful area is not being treated. Confidence in the clinician is necessary so that treatment regimens will be followed. The patient's trust may be lost if the problem for which he seeks care does not seem to be given appropriate attention by the clinician.

The patient should be taught about the application of ice and instructed how to gently exercise in order to prevent stiffness or contracture while healing takes place. Follow up and reinstruction are usually necessary to be sure that the treatment program is being done correctly. Visits should continue as required during the healing process.

Application of resting splints or slings may be necessary in the presence of severe pain. The "resting position" of a joint should be maintained to allow healing to occur without further stress, and to reduce pain. Splinting at night can be helpful in maintaining the affected limb in the position of function. Corrective orthoses should not be used, as added stress could be applied across the involved area.

Too much rest, however, can lead to reduced mobility and compromised function. Histological evidence of fibrosis can be detected almost immediately after injury, and gross evidence of restriction of movement can be shown in as little as 4 days.[74] Therefore, maintenance of range of motion is an important goal while healing proceeds. Once past the acute stage, when any movement is painful, careful, gentle motion within the limits of pain should be started.

The inflammatory process, which is present during the acute phase, is usually precipitated by trauma and accompanied by edema. Reduced circulation might also be present. If immobilization is then added, there will then be four factors that promote the formation of dense connective fibrosis.[74] Maintenance of mobility during this time is essential in order to prevent the most common sequela associated with acute trauma around a joint—restriction of function.

Some movement of the affected joint is indicated, whether it is "active rest" (i.e., muscle contraction without load),[48] gentle passive or active assistive range of motion,[25,32,59] gentle stretching,[47] or active range of motion exercises.[13,25,28,48,59] Strengthening exercises in the pain-free ranges are also recommended to restore strength and function as soon as healing permits.[28,44] The choice should be determined by the joint involved, the personality of the patient, and the clinician's evaluation, since careful instruction is essential. The patient should be taught to recognize the difference between the pain associated with increased trauma and that which accompanies gentle therapeutic stretching.

The joint most often restricted after tendinitis is usually the shoulder. "Frozen" shoulder can inadvertently be induced by prolonged immobilization. Thus, simply instructing a patient to raise the arm two or three times daily, for example, might not prevent an imbalance in the normal scapulohumeral rhythm. Since pain restricts movement, favoring and substitution can occur. Fibrosis and restriction of motion can then result at either the glenohumeral joint or the scapulothoracic juncture. Restriction of movement can be prevented by carefully following a patient for the first critical week or two, until the inflammation subsides.

Further discussion of treatment of the individual joints affected will accompany chapters specific to the part involved.

As mentioned previously, tendinitis is technically **tenosynovitis** and is usually initiated by trauma, either acute or as a result of overuse. When associated with arthritis, tendinitis may be caused by the arthritis. Treatment of the tendinitis then requires concomitant management of the accompanying arthritis, as discussed in Chapter 2.

Chronic

The problem with constructing objectives for chronic tenosynovitis is that it is only occasionally isolated. Chronic tenosynovitis is usually accompanied by acute and chronic bursitis or adhesive capsulitis. The capsulitis and bursitis can become more difficult to treat than the original tenosynovitis, and thus are more obvious. But capsulitis and bursitis originally generated by tenosynovitis will not completely subside until the original tenosynovitis has become quiescent. The application of modalities are useful when combined with exercises. This treatment is most effective when frequently repeated during the day. The management of adhesive capsulitis can be much more intensive, involving traction, passive range of motion exercises, and transcutaneous electrical nerve stimulation.[53] Unfortunately, chronic tenosynovitis can involve not only contracture formation, but also calcification. Usually the calcification occurs at the hip and shoulder, but it can happen retropharyngeally or at the elbow, ankle, knee, foot, or hand.[53,57] Under these circumstances, passive range of motion, traction, or manipulation could lead to hemorrhage or tendon rupture. Treating the contracture as above or by careful serial casting procedures that are, in fact, corrective, are probably safer protocols. Figure 1-4 shows the management flow chart.

The decision as to when the acute stage ends may be difficult, rendering the patient more liable to complications. Treatment of simple, chronic tendinitis begins after care during the acute stage has not proven effective enough in reducing the pain associated

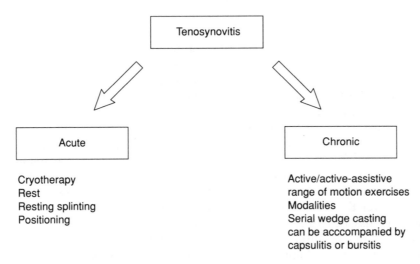

Figure 1-4. Management of tenosynovitis.

with this problem. The acute phase can be as short as 2 days[13] or until the pain resolves.[48]

The goals of treatment are to

1. Reduce or eliminate pain,
2. Prevent restrictive scar formation and reduced range of motion,
3. Maintain strength.

There will always be those patients who have been noncompliant with the treatment of the acute stage and who have not continued therapy. They then return to the clinic with chronic pain weeks or months later, often with complications, including stiff, or "frozen" joints, and calcific tendinitis.

The first goal, reduction or elimination of pain, is often achieved with pain medication. Medication, however, tends to mask the symptom of pain when the patient inappropriately stresses the affected tendon. This can be a particular problem with patients who do not limit themselves to the recommended activities. As long as the pain is not incapacitating, the use of modalities to control pain, after the acute phase, is frequently more functional for these patients.

Complications are less likely to occur in patients who are compliant with the treatment program. For these patients, the use of modalities and careful guidance during the healing process are often sufficient. Pain medication is frequently unnecessary for these individuals, since therapy controls most of the pain associated with chronic tendinitis. Of course, medication is a problem for some patients because of allergy or drug interactions. For these patients, the clinician can offer significant pain reduction with modalities. These techniques reduce pain because the **cause** of the pain is being dealt with. This applies to spasm, ischemia, or fibrosis formation.

The most commonly used modality, which has multiple benefits, is heat. Heat has a reflex effect that is valuable in pain reduction via the gate theory of pain of Melzack and Wall.[34,64,67] Local heat is effective in reducing muscle spasm and in increasing circulation, thereby reducing ischemia.[34,64] Heat, specifically deep heat, increases tissue distensibility and allows stretching to occur with much less pain and trauma.[35,36,64] Heat also reduces joint stiffness[34,64] and allows freer range of motion.

The only modality that can be truly considered to heat deeply is ultrasound.[37,64] Ultrasound speeds the resolution of inflammatory exudates through increased blood flow, which aids in the healing process.[28,35] Ultrasound is absorbed at a higher rate by tissues that have a higher protein content, e.g., muscle, joint, capsule, tendon, and extracapsular ligament. Ultrasound has therefore proven of benefit in increasing the extensibility of tight capsular tissue when combined with exercise[49] and is widely advocated in the treatment of tendinitis.[27,28,35,48,49,64] The major drawback to ultrasound is the amount of clinician time required. Phonophoresis, the use of ultrasound to drive medication into the skin, has also proved useful in the treatment of tendinitis. The medication is locally applied directly over the tissue affected, rather than taken systemically, and mechanically pushed into the tissues by the action of the sound waves. Hydrocortisone is the medication of choice in this condition.[26,49,69] Ultrasound is the only "diathermy" safe to use in the presence of metallic implants, which is not true of shortwave and microwave diathermy[34,35,38] (see Fig. 1-5).

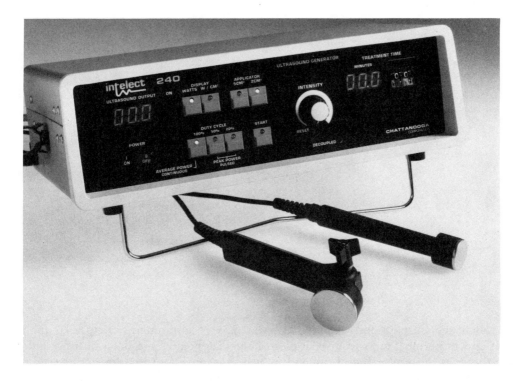

Figure 1-5. Ultrasound equipment.

Although shortwave and microwave diathermy in various forms have been thought to be deep-heating modalities when operating under specific frequencies, their effects are considered to be relatively superficial.[37,64] Many units have not met approved standards. Diathermy has been shown to heat superficially,[37] and the benefits of the diathermies have been described as doubtful.[48] Microwave diathermy, however, does not demand inordinate clinician time to set up and operate and has been popular in the past. Newer units do have approval, but clinicians are no longer being routinely trained in the use of diathermy. These newer units uniformly deliver superficial heat.

Fluidotherapy, a fairly recent development, is also claimed to be a deep-heating modality. Fluidotherapy raises the in vivo temperature as measured in the hand and foot more significantly than other modalities.[11,12,72] Unfortunately, its application is limited because of the mechanical design. At present, the mechanics of the units restricts their use to the more distal joints of the extremities and to the back (Fig. 1-6).

Although moist heat packs have no deep-heating effect, they are often used in combination with ultrasound to promote the psychologic and reflex effects for which they are so valuable. Ultrasound, when properly applied, is not perceived by the patient as heat. Without the sensation of heating, the patient could believe that little is being accomplished, so the effects of ultrasound should be explained (Fig. 1-7).

Figure 1-6. Fluidotherapy unit.

Some therapists prefer to apply hot packs prior to ultrasound to help relax the patient and to permit exercises to take place immediately after the deep heating. Others like the moist heat after the ultrasound to avoid the shock of the cold coupling gel and ultrasound head after the skin has been comfortably heated. With the advent of gel heaters, this problem is easily avoided.

If the ultrasound is to be followed by mobilization, i.e., stretching toward full range of motion, the ultrasound should be applied as closely as possible to the time of the exercise. Otherwise the body dispels the heat, and its beneficial effects on deep fibrotic tissue are lost or diminished, especially if the 20 minutes of moist heat utilization follows the application of ultrasound.

Transcutaneous electronic nerve stimulation (TENS) is another modality designed specifically for the control of pain. TENS can be invaluable in reducing or eliminating pain when properly applied. There are very few contraindications to its use, and TENS

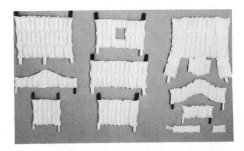 

Figure 1-7. Hot packs and hot pack covers.

can be used for almost every part of the body. This device can replace pain medication when other modalities alone do not suffice. TENS can also be a valuable adjunct in the restoration of motion and function when joint restriction has occurred by reducing pain, which allows freer movement while remobilization is occurring. Injudicious use permits stress of healing tissues, so careful supervision during application of this modality and thorough explanation to the patient are important. The danger, however, may be self-limiting, since TENS does not completely block pain. The "break-through" pain during exercise warns patients that they have reached the safe limits of range (Fig. 1-8).

The helium-neon laser, or "cold laser," is another useful device in the control of pain.[26] Cold laser provides temporary pain relief, which then allows easier restoration of mobility. The laser is also effective in accelerating the healing process. It should not be used as a panacea but as part of the armamentarium of physical agents used in the rehabilitation process.

The second goal is prevention of restrictive scar formation and the resultant reduction of range of motion. This is accomplished by one or both of two basic techniques, stretching and friction massage. Gentle stretching in some form is recommended by most sources as an important part of the treatment of tendinitis.[13,25,28,35,47,48,49,59] The purpose is to prevent fibrosis and scarring from restricting mobility and causing loss of function.

Once the acute phase has passed, gentle active assistive exercises should be added to the patient's program. Again, the patient should know how to differentiate between the pulling sensation that is necessary for successful restoration of range and the pain associated with further trauma to already damaged tissues. Patients should be carefully instructed as to the difference between the two types of pain. The "no pain, no gain" axiom does not apply here. Active movement within the pain-free range should also be encouraged to further aid in overall rehabilitation. Weakening of the affected muscle while healing occurs can thus be minimized.

Stretching has been shown to be most effective in producing a lasting length increase when loading is accompanied by heat and maintained during cooling.[36] Loading of

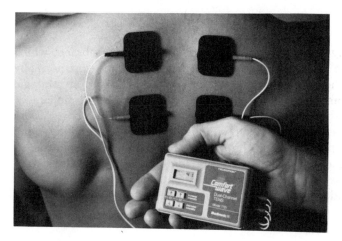

Figure 1-8. Transcutaneous electronic nerve stimulation (TENS) unit.

the joint should begin as soon as possible after heating has taken place to maximize the benefit of the heat. Optimum results occur when heat and load are applied simultaneously, but this is usually impossible.

Various manual therapy techniques are effective in restoring the normal joint relationships when treatment is carried out by experienced practitioners. Since there are benefits and drawbacks to most every style, no particular technique is exclusively advocated. Rather, the practitioner should be eclectic in the use of the available techniques. Each is most beneficial when competently practiced in appropriate situations. This must be determined by the individual clinician and is dependent upon personal skill levels. Clinicians should try to be familiar with the applications of the many techniques currently utilized and the practitioners who use them effectively.

Proprioceptive neuromuscular facilitation (PNF) techniques offer several methods that can promote relaxation.[30] These are valuable in gaining lost range of motion once fibrosis and stiffness are imposed upon a joint, especially after the original insult has healed. They involve strong voluntary contraction, which is normally painless.

Friction massage is helpful in reducing or preventing restriction by fibrosis, especially in chronic soft tissue conditions.[19,28,73] The deep, transverse massage across a tendon stretches or ruptures the small adhesions that form during healing. This technique is not supposed to be painful and should be used judiciously. Friction massage may be used in conjunction with one or more of the above-mentioned modalities to help control the pain. If possible, patients should be taught to give themselves friction massage as part of their home program. Thus they can determine their own tolerance to the associated discomfort and self-treat accordingly. This might prevent them from missing treatment sessions because of disproportionately high expectations of pain.

The maintenance of strength is the third goal. With diminished strength come alterations in the biomechanics of a joint. These changes could lead to continued or increased trauma. As soon as pain permits, active exercise should be initiated. Progression to resisted exercise should be added when active movement is painless, but should be stopped if pain returns.

BURSITIS

A bursa, the Latin word meaning "purse," is a sac lined by a synovial membrane, which looks like a small bag-type purse. There is some opinion that is does not exist normally but is formed in response to the friction generated by connective tissue at strategic points.[18,54,58] When acutely inflamed, the sac is distended. When chronically inflamed, fibrous scar tissue is formed in the walls and across the inner space of the bag. As the synovial membrane is both vascular and well supplied with nerve endings, any inflammation will be painful—especially when the bursa is under additional mechanical stress. The source of the inflammation is usually traumatic tenosynovitis with extension to the nearby bursa. As with tenosynovitis, bursitis is frequently found in patients with arthritic disorders. The larger and weight-bearing joints are often affected—the shoulder, hip, knee and elbow (Fig. 1-9).

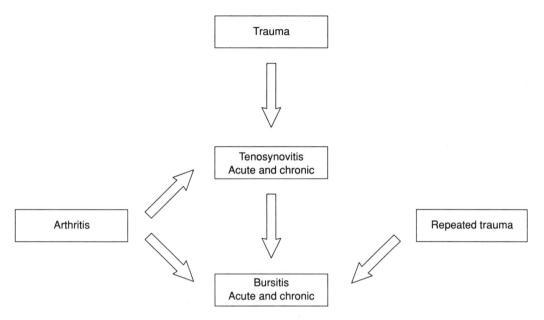

Figure 1-9. Pathogenesis of bursitis.

When bursitis is added to tenosynovitis, the area of the bursal inflammation is slightly distended and very tender. Pain is produced when stress is applied across the bursa so that the synovial membrane is placed under increased tension. For example, if a patient with shoulder pain is positioned prone on an examining table with the arm dangling over the edge, moving the extended arm in rotary directions produces pain caused by shoulder abduction without external rotation of the humerus, which in turn causes impingement of the acromion process upon the biceps or supraspinatus tendons. This pain is intermittent, whereas pain produced by the bursitis will be constant.[50,53] Impingement syndrome itself is a clinical manifestation of tenosynovitis, associated with the sensitivity of the supra- and infraspinatus muscles to the weight of hand tools used during heavy manual labor.[57] As noted above, it can clinically progress and generate an additional bursitis.[25] Calcific bursitis could be another cause of chronic shoulder pain.[21] Additionally, bursitis might also be a cause of hip or back pain and can become evident by tenderness at the inferior edge of the greater trochanter.[24] But bursitis can also be found at the knee, elbow, wrist, hand, or foot—any place tenosynovitis can be found[25] (Table 1-2).

While bursitis and tenosynovitis are commonly associated, and might require concomitant treatment, distinction should be made between the two entities. Differentiating between the tissues affected is possible through careful manual muscle testing.[19] If a resisted, isometric contraction of a muscle causes pain, the contractile unit, the muscle or tendon or both, is probably involved. Clinicians who are familiar with manual muscle testing run less risk of mislabeling the specific problem. Although there is little differ-

TABLE 1-2. CLINICAL PRESENTATION OF SHOULDER TENOSYNOVITIS AND BURSITIS

Pathogenic Process	Clinical Manifestation
Tenosynovitis	Pain on attempted humeral abduction. As external rotation is limited, scapular elevation (shrugging) is substituted for humeral abduction; forward flexion of the shoulder is not affected.
	Crepitus of the tendon and tendon tenderness.
	Positive Yergason's sign is consistent with bicepts tendinitis.[27] Shoulder pain is generated by an abducted and fixed arm resisting supination and external rotation.
Bursitis	Constant shoulder pain when patient is in a prone position, dangling the arm and moving it in a rotary position.
	Tenderness and distension at the area over the bursitis.
	Often associated with partial or full supraspinatus tendon rupture. Shoulder abduction is therefore limited and weak.

ence in the basics of the treatment regimen, appropriate therapy requires prior specific identification. The clinician needs to know what tissue is affected, its location and depth, and what modalities are most appropriate. If there is a tendinitis in the shoulder, for example, the tendon that is inflamed, the location of the tendon in relation to the superficial anatomy, and the position required to expose the tendon to treatment must be known, as well as the stage of the inflammation and the modalities indicated at that stage. Some tissues require rest, while others should be stretched. These questions and many more can be answered only by accurate evaluation. If treatment is not specific and appropriate, results will be disappointing (Fig. 1-10).

When treating bursitis or tendinitis, the clinician requires a good working knowledge of the anatomical relationships of the surrounding tissues. Ultrasound is given at a lower intensity for superficial structures and at a higher intensity for deeper structures. Higher doses are required to penetrate thick, overlying fat. Moist heat might reach superficial tendons, but deeper bursae might not be affected, except by the reflex effects discussed above (Fig. 1-11).

The pain of bursitis, like tendinitis, frequently radiates, creating difficulty in pinpointing its source, and causing problems for the patient. Which structure is involved, where the affected tissue is located, and how the pain is referred should be explained to the patient when treatment begins. Pain felt at the insertion of the deltoid is common in shoulder injury.[59] Patients might wonder why the area under the acromion, for example, is being treated when the pain is perceived further down the arm (Fig. 1-12).

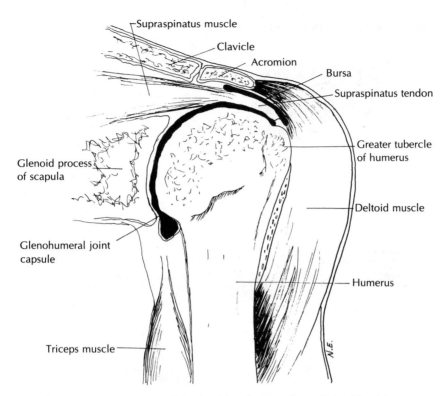

Figure 1-10. Cross-section of the shoulder, showing close relationship of tissues.

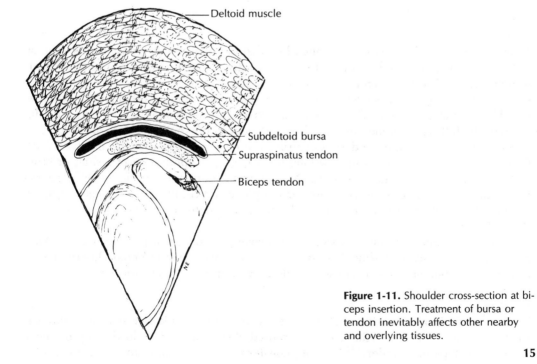

Figure 1-11. Shoulder cross-section at biceps insertion. Treatment of bursa or tendon inevitably affects other nearby and overlying tissues.

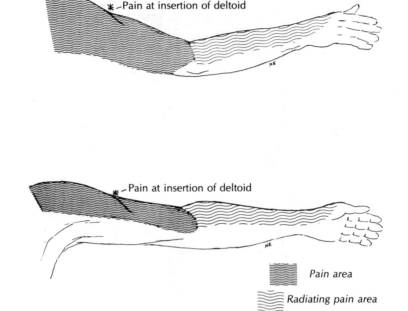

Figure 1-12. Pain of supraspinatus tendonitis and subacromial bursitis with keys for pain area and radiating pain area.

Acute

Treatment of acute bursitis is very much like that outlined under tenosynovitis. The affected joint must be rested, but as with tenosynovitis, range of motion is preserved while healing occurs. The most commonly recommended protocol, however, is ice in the early stage. Cryotherapy reduces edema and augmented muscle tone,[15] although it may be resisted by the patient at first. Gentle active-assistive movement one to three times per day through the full range of motion should be sufficient to prevent adhesive capsulitis. Posterior night-resting splints and slings during the day can help rest the involved area. Carefully instruct patients as to how long to rest, proper application of a sling or other resting device, and wearing schedules. Noncompliance or complications such as restriction of functional range or disuse muscular atrophy could result from prolonged immobilization. With effective rest and immobilization, the acute phase should not last more than a few days.

Noncompliance with prescribed resting, however, can slow or prevent healing. Most patients would discard a sling that is causing neck pain, for example. Instruction in proper application and judicious use of padding will prevent noncompliance.

Chronic

Treatment of chronic bursitis is also similar to that for chronic tenosynovitis. Physical treatments can be very imaginative; e.g., pulsed electromagnetic field therapy has been used in this type of disorder.[3,10,22] While nonsteroidal anti-inflammatory oral medication has been effective,[6] complications generated by corticosteroid injection have been frequent enough to limit its usefulness.[3,8,11,22] Active and active-assistive exercise and local heat in the form of moist heat and ultrasound remain the most efficient combination, effective in increasing local circulation, reducing pain and muscle spasm, and speeding healing.

If arthrography reveals a tendon rupture, surgical repair is often indicated. One difficulty has been that it is hard to estimate how large the tear is by arthroscopy. Arthropneumotography might be one new variation of arthrography that could provide some data as to the size of the rupture and the thickness of the cuff remainder[10,29] (Fig. 1-13).

Good results have been reported with the application of iontophoresis for bursitis and tenosynovitis.[26,72] Mild, surging sine wave (electrical stimulation) to the muscle overlying the involved bursa is also recommended.[26] This treatment is often helpful in increasing circulation and removing inflammatory irritants and debris.

The reason for the trauma that initiated the inflammation should be identified, if possible, during the rehabilitation of a patient after bursitis or tenosynovitis. For example, it might be determined that a homemaker's pain started after washing all the windows in one day. The trauma probably was caused by impingement of the acromion. The patient might be convinced that the overhead swinging movement of the arm was the cause of the discomfort and might be happy to learn of alternative movements to avoid repeating the trauma. For a manual laborer, the causal stimulus is sometimes difficult to identify. Surveillance of the work environment could determine the cause of the trauma

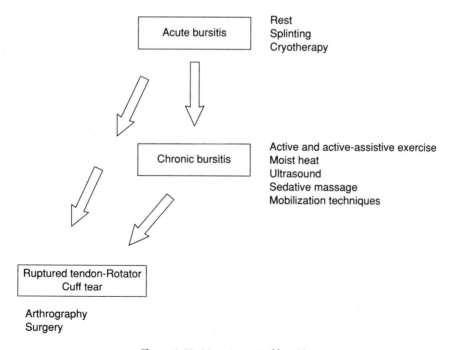

Figure 1-13. Management of bursitis.

and arrangements made for the faulty movements to be avoided. Work-hardening programs should be added to the patient's care when indicated. In the athlete, assessment of gait mechanics, running patterns, posture, flexibility, and even the shoes being used may be necessary to determine the problem.[13] Computer-assisted analysis of movements can aid in the identification of a faulty pattern if other methods are not effective. In all patients, the offending movement must be avoided or corrected to prevent patterns of chronic injury, which are often concomitant with loss of function. Some patients are even deemed unsuited for their jobs if intervention is not successful.

Complications

The most common complication of tenosynovitis and bursitis is restriction of movement at the joint involved, as discussed above. The most effective treatment for this potentially serious problem is **prevention**. Once range of motion is restricted, the affected joint and the surrounding structures must be aggressively treated in order to restore functional movement.

Calcific tendinitis or bursitis is another potentially troublesome complication that frequently appears some time after the acute phase of either problem. Ultrasound has been used for these cases, although there is some controversy as to whether it is effective in speeding the disappearance of the calcium deposits.[35] Whether the calcium is actually "broken up," ultrasound is effective in increasing the local circulation, which may improve reabsorption of the deposits. Again, maintenance of joint mobility is an important part of the treatment of this sequela.

FIBROSITIS

Of the three conditions discussed in this chapter, this disorder is by far the most common and the least well understood. The pathogenesis of fibrositis is remarkable for the presence of major trauma affecting major joints, arthritis (polymyalgia rheumatica, rheumatoid arthritis, ankylopoetica spondylitis), endocrinologic dysfunction (hypothyroidism), neuritis (radiculitis), or even a preceding viral infection.[2,16,31,60,61] In other words, fibrositis is a response to a variety of underlying clinical conditions, all of which generate major muscular tension in and around the larger, weight-bearing proximal joints—joints intimately associated with type I pennate muscle fibers. As these fibers contract and relax slowly, tension is not quickly dispelled.

The pathologic analysis of the condition does not reveal inflammatory changes at painful sites. Fatty and fibrous nodules are often present, aligned with the posterior or superior iliac spines, but these are usually not the location of painful trigger points.[16,65] These trigger sites are frequently found scattered throughout the shoulder or pelvic girdles or in the calf muscles of the lower extremities.[9,31,61] The clinical entity, therefore, includes painful trigger areas, tense large muscles, joint stiffness, and fibrofatty nodules[60,61] (Fig. 1-14).

If a specific arthritic disorder is present, treatment of the fibrositis includes control

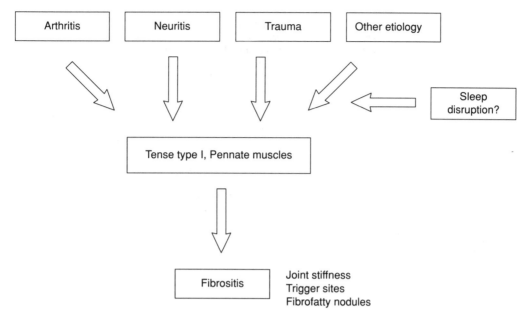

Figure 1-14. The pathogenesis of fibrositis.

of the underlying arthritis.[17] Treatment protocols usually emphasize neutralization of trigger points, applying injection of anesthesia, or freezing techniques.[9,28,52] But this management option modifies only the obvious symptomatology. The same objection can be made to prescription of nonsteroidal anti-inflammatory agents. These options are indeed effective, but should be accompanied by some plan to release the accumulated tension in type I pennate musculature. The most efficient way of accomplishing this goal is to instruct the patient in relaxation exercises.[13] Preventive rehabilitation should also include instruction in postural exercises because a person's anxieties and strains can also be "frozen" in proximal muscles by the way the head and trunk are held. For postural exercises to be effective, they might need to be given along with supportive psychotherapy (Fig. 1-15).

This clinical entity is referred to by many names, including myofascial pain syndrome, myofascitis, and fibromyositis. Patients who experience the symptoms of fibrositis often have "type A" personalities. They are busy, fast-moving people who frequently have difficulty relaxing. The cycle that starts the pathologic process—i.e., retention of abnormal levels of muscle tension—begins with the basic personality structure. These people, who suffer the symptoms of painful nodules or "trigger points," frequently ignore the pain until it becomes severe. Also, since there is not always an easily recognizable clinical entity, these problems might go unrecognized for a long time. The source of pain is unknown, and this adds to the already high anxiety levels. Anxiety, stress, and poor sleeping habits can further exacerbate the predisposing factors.[74] Therefore, the importance of rest, relaxation, and the avoidance of stress and anxiety should be conveyed to patients.

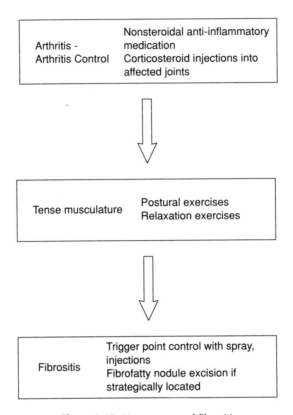

Figure 1-15. Management of fibrositis.

Stress and anxiety can be reduced in a number of ways. Some are more effective than others on an individual basis. Taking a vacation helps some people but increases stress for others. Some individuals have to change jobs to avoid the stress and anxiety associated with their work. Electromyographic biofeedback can help teach patients how to relax.[63,74] Relaxation is an important step in dealing with tension. Psychological support is essential, including counseling to help the patient deal with stress and anxiety. Both the patient and the clinician should understand the connection between the physical symptoms and the psychological precursors. Not every symptom without an obvious source is imaginary, as evidenced by the recent recognition of temporomandibular joint dysfunction as the cause of multiple, ill-defined painful syndromes.

Treatment of the painful trigger points is aimed at increasing local circulation, relaxing muscle spasm, stretching tight musculature, and eliminating the hard, painful nodules at the trigger sites. Restoration of full range of motion of affected joints is another important goal of treatment. Many specific techniques can be helpful in accomplishing these goals. However, the overall program must be directed at improving physical fitness generally. Comprehensive treatment should include use of techniques of muscle stretching and strengthening, relaxation exercises, identifying and reducing the sources of stress, and improving the quality of sleep.[23] Correction of any postural imbalances is essential.[13,68] Weight reduction might also be indicated.[74]

For local treatment of the affected trigger points, many sources advocate a vapocoolant spray with concomitant stretch of the affected muscle to its normal maximal length.[13,34,39,55,62,66-68,75] Most sources agree that most of the modalities available to a physical therapist can and have proven effective in the treatment of this complex problem. Included in this list are moist heat, ultrasound, ice or ice massage, electrical stimulation, massage in various forms, helium-neon laser, acupressure, and iontophoresis.

Deep pressure, kneading, and ischemic compression are terms associated with massage techniques for this problem. They have proven useful clinically and are particularly successful in combination. If the patient tolerates the ischemic compression—a forceful capture of the nodule between the fingers or thumbs of the practitioner and the deeper structures—the compression is effective in quickly reducing the size of the nodule. Pressure must be maintained then increased as the associated pain diminishes. The pressure should be followed by kneading massage to increase local circulation and relax the musculature. For patients who cannot tolerate or should not have this type of massage, gentle, soothing techniques can be valuable in reducing spasm and relaxing the patient (Fig. 1-16).

No matter what technique or combination of techniques are used, the basic goals of treatment remain the same. Local pain relief is necessary in order to allow restoration of normal movement. Normal movement can be restored through a number of techniques, including stretching of tight muscles, mobilization of stiff joints, and strengthening of weakened muscle groups. Normal movement must be maintained through relaxation of tension and postural realignment. Positions and movements that cause prolonged misalignment of body parts should be avoided. The patient must then learn to deal effectively with stress and anxiety to prevent recurrence of the symptoms.

Many of the techniques that can keep a patient free from pain are taught for use at home. These devices include stretching and strengthening exercises, relaxation exercises,

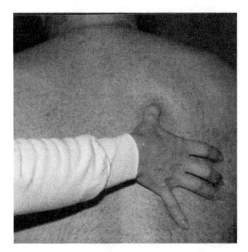

Figure 1-16. Ischemic compression at the upper back, with trigger point nodule captured between clinician's thumb and medial border of scapula.

use of local heat or ice, and mechanical means necessary to maintain good postural alignment. Braces and lifts might be included.[14,41] The patient must be taught to change position frequently at work. Relocation or adjustment of desks, chairs, typewriters, machine heights, etc., have proven excellent means of reducing strain and tension. Effective ways of recognizing and reducing muscular tension must be utilized on a regular basis.

Each of the three clinical entities discussed has its own multifocal pathogenesis and guidelines for management. Management, not treatment, is really the correct word. These disorders are rarely "cured." The pain expressed—the chief complaint for all three—is often an expression of a crisis. Usually, it is not that the pain has suddenly gotten that much worse, but the patient's ability to tolerate the anxiety, the discomfort, and the unease has declined. The pain has destroyed the patient's function. Return of that function frequently is a painful process in itself. Nonetheless, if the patient is successful and function is regained, painful sensations will abate and even if still present, will remain at tolerable levels. If there is any one key to successful management of patients in pain, it is the patients' being able to express their feelings to a sympathetic, yet professional, listener. The crisis specifically related to loss of sensation must be steadily, gradually disassociated from the myriad daily problems, tensions that augment musculoskeletal tension. Commonly, people with these painful disorders will be hard-driving, competitive people without avocational pursuits to relax them. In their concentration on not wasting time, they have transferred outlets for frustration to the wear-and-tear in their own muscles and joints. If this fundamental lifestyle is not modified, one complaint will be replaced with another. Part of rehabilitative therapy is to address these issues in a discreet and effective manner.

REFERENCES

1. Anderson WAD and Kissane JM: *Pathology,* 7th ed. St. Louis, C.V. Mosby Co., 1977.
2. Awod EA: Interstitial myofibrositis: Hypothesis of the mechanism. *Arch Phys Med Rehabil* 1973; 54:449.
3. Binder A, et al: Pulsed electromagnetic field therapy of persistent rotator cuff tendinitis. *Lancet* 1984; 1 (8379):695.
4. Bjelle A, Hagberg M, Michaelson G: Clinical and ergonomic factors in prolonged shoulder pain among industrial workers. *Scand J Work Environ Health* 1979; 5:205.
5. Bjelle A, Hagberg M, Michaelson G: Occupational and individual factors in acute shoulder-neck disorders among industrial workers. *Br J Ind Med* 1981; 38:356.
6. Bland JH, et al: The painful shoulder. *Semin Arthritis Rheum* 1977; 7:21.
7. Bloom BS (editor): *Taxonomy of Educational Objectives: Classification of Educational Goals. Handbook I: Cognitive Domain.* New York, Longmans, 1956.
8. Bonica JJ: Management of myofascial pain syndromes in general practice. *JAMA* 1957; 164:732.
9. Bonica JJ: *The Management of Pain.* Philadelphia, Lea and Febiger, 1953.
10. Bono RF, et al: A multicenter, double-blind comparison of oxaprozin, phenylbutazone and placebo therapy in patients with tendinitis and bursitis. *Clin Therapu* 1983; 6:79.
11. Borrell RM, et al: Comparison of in vivo temperatures produced by hydrotherapy, paraffin wax treatment, and fluidotherapy. *Phys Ther* 1980; 60:1273.
12. Borrell RM, et al: Fluidotherapy: Evaluation of a new heat modality. *Arch Phys Med Rehabil* 1977; 58:69.
13. Calliet R: *Shoulder Pain,* 2nd ed. Philadelphia, F. A. Davis Co., 1981.

14. Calliet R: *Soft Tissue Pain and Disability*. Philadelphia, F. A. Davis Co., 1977.
15. Cooney WP III: Bursitis and tendinitis in the hand, wrist and elbow. *Minn Med* 1983; 66:491.
16. Copeman WSU: Fibro-fatty tissue and its relationship to certain "rheumatic" syndromes. *Br Med J* 1949; 2:191.
17. Coulehan JL: Primary Fibromyalgia. *Am Fam Phys* 1985; 32:170.
18. Currie DM: Self-directed learning and medical education: A comparison. *Arch Phys Med Rehabil* 1985; 66:454.
19. Cyriax J: *Textbook of Orthopedic Medicine*, Vol. 1, 6th ed. Baltimore, Williams & Wilkins Co., 1976.
20. Fassbender HC, Wegner K: Morphologie and pathogenesis des Weichteilrheumatismus. *Z Rheumaforschung* 1973; 32:355.
21. Halvorson GA: Sports-related injuries. In Kaplan PE (editor): *The practice of Physical Medicine*. Springfield, Ill, Chas. C. Thomas, 1984.
22. Hartviksen K: Ice therapy in spasticity. *Acta Neurol Scand* 1962; Suppl 3: 79.
23. Hawley D, Cathey MA: Fighting fibrositis. *Am J Nurs* 1985; 85:404.
24. Herberts P, et al: Shoulder pain and heavy manual labor. *Clin Orthop Rel Reas* 1984; 191:166.
25. Hoffman GS: Tendinitis and bursitis. *Am Fam Phys* 1981; 23:103.
26. Kahn J: *Low Volt Technique*, 4th ed. New York, Joseph Kahn, 1985.
27. Kendall HO, et al: *Posture and Pain*. Huntingdon, N.Y. Drieger Pub. Co. Inc., 1970.
28. Kessler RM, Hertling D: *Management of Common Musculoskeletal Disorders*. Philadelphia, Harper & Row, 1983.
29. Kilcoyne RF, Matsen FA, III: Rotator cuff tear management by arthropneumotography. *Am J Roentg* 1983; 140:315.
30. Knott M, Voss DE: *Proprioceptive Neuromuscular Facilitation*, 2nd ed. New York, Harper & Row, 1968.
31. Kraft GH, et al: The fibrositis syndrome. *Arch Phys Med Rehabil* 1968; 49:155.
32. Krusen FH, et al: *Handbook of Physical Medicine and Rehabilitation*, 2nd ed. Philadelphia, W.B. Saunders, 1971, pp 386–9.
33. Kvamstron S: Occurrences of musculoskeletal disorders in a manufacturing industry with special attention to occupational shoulder disorders. *Scand J Rehabil* 1983; Suppl 8.
34. Leek JC, et al (editors): *Principles of Physical Medicine and Rehabilitation in the Musculoskeletal Diseases*. Orlando, Grune & Stratton, Inc., 1986.
35. Lehmann JF (editor): *Therapeutic Heat and Cold*, 3rd ed. Baltimore, Williams & Wilkins, 1982.
36. Lehmann JF, et al: Clinical evaluation of a new approach in the treatment of contracture associated with hip fracture after internal fixation. *Arch Phys Med Rehabil* 1961; 42:95.
37. Lehmann JF, et al: Effect of therapeutic temperatures on tendon extensibility. *Arch Phys Med Rehabil* 1970; 51:481.
38. Lehmann JF, et al: Heating patterns produced by shortwave diathermy applicators in tissue substitute models. *Arch Phys Med Rehabil* 1983; 64:575.
39. Licht S. (editor): *Arthritis and Physical Medicine*. New Haven, Elizabeth Licht, 1969.
40. Lindblom K: On the pathogenesis of ruptures of the tendon aponeurosis of the shoulder joint. *Acta Radiol* 1929; 20:563.
41. Lowdon A, et al: The effect of heel pads on the treatment of achilles tendinitis: A double blind trial. *Am J Sports Med*, 12:431, 1984.
42. Mager RF: *Preparing Instructional Objectives*, 2nd ed. Belmont, Calif. Fearon Publishers, 1975.
43. Martens M, et al: Patellar tendinitis: pathology and results of treatment. *Acta Orthop Scand* 1982; 53:445.
44. Mens J, VanDerKorst JK: Calcifying supracoracoid bursitis as a cause of chronic shoulder pain. *Ann Rheum Dis* 1984; 43:758.
45. Mosley HF, Goldie I: The arterial pattern of the rotator cuff of the shoulder. *J Bone Joint Surg* 1963; 45B:780.
46. Neer CS II: Impingement lesions. *Clin Orthop* 1983; 173:70.
47. Nevaider RJ: Lesions of the biceps and tendinitis of the shoulder. *Orthop Clin North*

Am 1980; 11:343.

48. Peterson L, Renstrom P: *Sports Injuries— Their Prevention and Treatment*. Chicago, Year Book Medical Publishers, Inc., 1986.

49. Prentice WE: *Therapeutic Modalities in Sports Medicine*. St. Louis, Times Mirror/ Mosby, 1986.

50. Raman D, Haslock I: Trochanteric bursitis—a frequent cause of "hip" pain in rheumatoid arthritis. *Ann Rheum Dis* 1982; 41:602.

51. Rathburn JB, MacNab I: The microvascular pattern of the rotator cuff. *J Bone Joint Surg* 1970; 52B:540.

52. Richards AJ: Carpal tunnel syndrome and subsequent rheumatoid arthritis in "fibrositis" syndrome. *Ann Rheum Dis* 1984; 43:232.

53. Rizle TE, et al: Adhesive capsulitis (frozen shoulder): A new approach to its management. *Arch Phys Med Rehabil* 1983; 64:29.

54. Robbins SL, Cotran RS: *Pathologic Basis of Disease*, 2nd ed. Philadelphia, W.B. Saunders, 1979.

55. Ruskin AP (editor): *Current Therapy in Physiatry*. Philadelphia, W.B. Saunders, 1984.

56. Sarkozi J, Fam AG: Acute calcific retropharyngeal tendinitis: An unusual cause of neck pain. *Arthritis rheum* 1984; 27:708.

57. Selby CL: Acute calcific tendinitis of the hand: An infrequently recognized and frequently misdiagnosed form of periarthritis. *Arthritis Rheum* 1984; 27:337.

58. Sheldon H: *Boyd's Introduction to the Study of Disease*, 9th ed. Philadelphia, Lea & Febiger, 1979.

59. Simkin PA: Tendinitis and bursitis of the shoulder. *Postgrad Med* 1983; 73:177.

60. Simons DG: Muscle pain syndromes—Part I. *Am J Phys Med* 1975; 54:289.

61. Simons DG: Muscle pain syndromes—Part II. *Am J Phys Med* 1976; 55:15.

62. Simons DG, Travell JG: Myofascial origins of low back pain. *Postgrad Med* 1983; 73:66.

63. Swezey RL: *Arthritis*. Philadelphia, W.B. Saunders Co, 1978.

64. Tepperman PS, Devlin M: Therapeutic heat and cold. *Postgrad Med* 1983; 73:69.

65. Travell J: Symposium on mechanism and management of pain syndromes. *Proc Rudolf Virchow Med Soc* 1957; 16:1.

66. Travell JG, Simons DG: *Myofascial Pain and Dysfunction*. Baltimore, Williams & Wilkins, 1983.

67. Wall PD, Melzack R (editors): *Textbook of Pain*. New York, Churchill Livingstone, 1984.

68. Weed ND: When shoulder pain isn't bursitis—the myofascial pain syndrome. *Postgrad Med* 1983; 74:97.

69. Weinstein PS, et al: Long term follow-up of corticosteroid injection for traumatic olecranon bursitis. *Ann Rheum Dis* 1984; 43:44.

70. Wilke WS, Mackenzie AH: Proposed pathogenesis of fibrosis. *Cleve Clin Q* 1985; 52: 147.

71. Wolfe F, Cathey MA: Prevalence of primary and secondary fibrositis. *Cleve Clin Q* 1985; 52:147.

72. Wolfe S (editor): *Electrotherapy*. New York, Churchill Livingstone, Inc., 1981.

73. Woodman RM, Pare L: Evaluation and treatment of soft tissue lesions of the ankle and forefoot using the Cyriax approach. *Phys Ther* 1982; 62:1144.

74. Yunus M, et al: Primary fibromyalgia. *Am Fam Phys* 1982; 25:115.

75. Zohn DA, Mennell JM: *Musculoskeletal Pain*. Boston, Little, Brown & Co., 1976.

2

Arthritis, Radiculitis, and Neuritis

Paul E. Kaplan and
Ellen D. Tanner

As with the conditions discussed in the first chapter, these three diagnoses are not mutually exclusive. Our understanding of their relationship, however, has been altered recently by a series of advances in medical technology. In the past, effective treatment rested on the foundation of a meticulous physical examination and a few basic clinical procedures—spinal tap, arthrocentesis, and electromyography. While these are still important, their meanings have been irreversibly changed by vast improvements in in vitro laboratory investigations and noninvasive diagnostic techniques. Sketchy principles in the relationship of pathogenesis to treatment have been replaced with specific information generated by this technology. Consequently, we no longer view these diseases in quite the same way. The clear and clean-cut differentiation into monolithic diagnostic entities has become much less relevant than the realization that these are highly complex disease processes that can produce similar effects through highly diverse means. As with many medical disorders, the general classification has not changed but the scientific data has.

ARTHRITIS

The specific arthritic disorders and their rehabilitation have been presented in detail elsewhere.[61,86] Table 2-1 summarizes certain aspects of the clinical presentation.

Our evaluation of the mechanisms underlying the onset of arthritis has been modified by immunologic technical advances. For example, rheumatoid arthritis seems to have a close association with delayed monocytic immune reactions.[16] Immunologic assay has been used to identify arthritic patients with anergic peripheral blood monocytes.[60] These patients have synovial biopsies with more intense inflammation.[16] Synovial fibroblasts, when stimulated by monocytes, secreted augmented levels of prostaglandins and collagenase.[66] Cartilage near inflammation has proteoglycan and collagen loss,[36] and monocytes and fibroblasts can help generate bone resorption using prostaglandins.[74]

TABLE 2-1. PRESENTATION OF ARTHRITIC DISEASES

Disease	History/Physical Examination	Laboratory
I. Rheumatoid Arthritis	Arthritis—symmetrical small joints in hand and feet. Characteristic deformity can follow. Insidious onset of morning stiffness, pain, fatigue. Course is variable, but can be progressive. Synovial hypertrophy of affected joints. Subcutaneous nodules, pericarditis, pleuritis, vasculitis are often noted among other systemic abnormalities. Felty's Syndrome—leukopenia, splenomegaly, leg ulceration, multiple bacterial infections.	Anemia—normochromic, normocytic. Elevated sedimentation rate. Positive rheumatoid factor—high titers assay—with systemic disease. Inflamed synovial fluid.
II. Spondylopathies—includes Reiter's disease, arthritis and bowel inflammation, psoriasis, ankylosing spondylitis	Insidious onset; noted especially in males. Reiter's Disease—conjunctivitis, urethritis, mucocutaneous eruptions. Sacroiliac inflammation early. Arthritis may return. In ankylosing spondylitis, the spine will fuse; proximal, larger joints will be affected late in the disease. In psoriasis with arthritis, nails will be pitted and arthritis may closely resemble rheumatoid arthritis. Arthritis of peripheral joints might be asymmetrical.	Radiographs might reveal sacroiliac joint inflammation, spondylitis with fusion, erosive arthritis of peripheral joints, bony spurs on heel.
III. Gout, pseudogout	Recurrent, acute, monarticular inflammation, especially of the first metatarsophalangeal joint. Tophi, especially on extensor surfaces. Uric acid nephrolithiasis. Differentiate from pseudogout with its calcium pyrophosphate crystals, chondrocalcinosis, involvement of proximal large joints, secondary osteoarthritis, association with endocrinologic or metabolic disorders.	Increased serum urate concentration. Intraleukocyte crystal deposition. Urate deposits in tophi. Characteristic radiographic patterns of affected joints. Synovial fluid examination can also reveal crystals. Pyelography or ultrasound examinations for kidney stones. Increased leukocyte count and elevated sedimentation rate.
IV. Osteoarthritis	History of long career as laborer or performing repetitive weight-lifting tasks. Onset of symptoms can be sudden, especially after trauma. Major symptom is pain. Spine and larger proximal joints affected, often asymmetrically. Heberden's nodes.	Radiographs reveal spurs, loss of joint space, erosive changes. Increased leukocyte count and sedimentation rate.

When complement-activating properties of rheumatoid factor were investigated by radioimmunoassay, the levels obtained matched articular and systemic arthritis intensity.[76] Moreover, when C-reactive protein and sedimentation rates are controlled therapeutically, radiologic progression is also decreased.[15] The inviting factor for this immunologic chain of events is, as yet, not fully characterized.[16] Figure 2-1 summarizes some aspects of the pathogenesis of rheumatoid arthritis.

The pain characteristic of arthritis marks a complex interaction between neurons and the affected joints they supply. For example, an intraneuronal substance may contribute to the intensity of arthritic disorders.[54] Moreover, nociception secondary to chronic arthritis can employ spinal cord dymorphin and opioid receptors to transmit appropriate impulses.[64] Finally, somatosensory neurons in the posterior intralaminar area of the thalamus are sensitive to the nociception secondary to chronic arthritis and could play a significant role in the modulation of pain, pain behavior, and pain response.[42] Elements of both the peripheral and central nervous systems are probably mobilized at an early stage in an arthritic disease process, and by the time chronic arthritic pain is predominant, areas of the basal ganglia have already been activated. Since these are also involved in the limbic lobe reverberating circuit, the association between chronic arthritic pain and emotion is bound to be close.

What factors can be used to modify the development of chronic arthritis and arthritic pain? One surprising and controversial factor is climate. A controlled climate chamber demonstrated that falling barometric pressure and rising humidity (a "low" on the weather map) augmented arthritic signs and symptoms.[34] Since then, the effect of

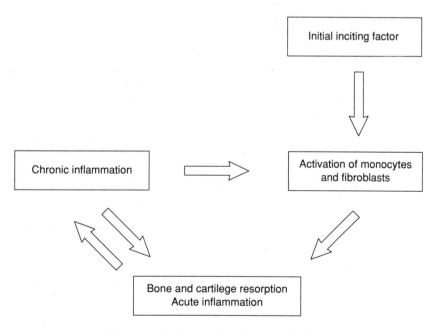

Figure 2-1. Pathogenesis of rheumatoid arthritis.

weather upon arthritic pain has been confirmed.[7,37,77] This is also confirmed by anyone whose aching ankle warned them of an approaching storm. A prospective study demonstrated that arthritic pain varies positively with temperature and negatively with vapor pressure.[69] Another recent controlled study, however, failed to record any significant differences. But this investigation was dependent upon the use of a rather confusing visual analog scale.[79] The balance of the studies indicate that hotter, drier weather at lower altitudes helps ameliorate chronic arthritic pain. Such conditions are found in the southwestern United States in the lower deserts. More work needs to be done to document the effects of climate and to climatologically map the United States.

Since it is difficult to rehabilitate arthritic patients unless they are medically stable, pharmacologic management is important. Nonetheless, it is interesting how little of the chemical armamentarium has been investigated using well-controlled double-blind techniques. In many cases, side effects of the drugs can be alarming—and quite as disabling as the arthritis. Some of the arthritic disorders themselves spontaneously wax and wane. In these instances, how much of the improvement is the medication, and how much is the natural course of the disease? A summary of certain often used medications is presented in Table 2-2.[8,25,43,86]

The past several years have brought technical advances that have changed the traditional presentation and therefore the treatment of the arthritic disorders. Computerized

TABLE 2-2. CHARACTERISTICS OF COMMONLY USED ANTIARTHRITIC MEDICATIONS

Drug	Application	Major Side Effects
Salicylates	Can be used to manage any acute inflammatory arthritis	Gastrointestinal hemorrhage
Nonsteroidal anti-inflammatory drugs	Aspirin substitutes—not as effective	Less chance of gastrointestinal hemorrhage, but augmented chance of abnormal hepatic function or nephritis
Adrenocorticoids	Excellent orally for vasculitic disorders, collagen diseases, but not to be used in rheumatoid arthritis	Cushingoid clinical presentation; if used for a time, adrenal dependency occurs
	By injection into osteoarthritic joints to reduce inflammation, but only a few times a year	Gastrointestinal hemorrhage; bony necrosis if given by injection into a joint
Gold/Penicillamine	Rheumatoid arthritis as second line medications	Stomatitis Aplastic anemia Autoimmune syndrome
Immunosuppressive agents	Rheumatoid arthritis as tertiary medications	Aplastic anemia, hepatitis, cystitis, reticuloendothelial cancer
Antihyperuricemic agents	Gout: (1) allopurinol inhibits xanthine oxidose; (2) probenecid augments urinary excretion of uric acid	(1) Allopurinol—aplastic anemia and hepatisis; should be given cautiously in patients taking thiazides or those with kidney failure (2) Probenecid—salictates may block uricosuric effect

axial tomography (CAT) applies radiographic laminar methodology to radioisotope techniques and, in the process, makes it possible to "see" the affected synovial membrane.[2,77] Consequently, synovial hyperplasia could be accurately quantified. Moreover, small microcystic lesions could be detected earlier in the course of erosive bone disease. Magnetic resonance imaging (MRI) can give the resolution of CAT scanning without contrast materials and has been used to find lesions in cruciate ligaments, ankle ligaments, tumors, and infections.[5] It is now possible to construct accurate clinicopathologic histories using these noninvasive procedures. They should change the way we evaluate and manage arthritis.

Arthrography and arthroscopy are important for patients with arthritis affecting larger and more proximal joints. Though these procedures are much more invasive than CAT or MRI, they have additional therapeutic significance. In knee lesions, for example, arthrography has been accurate in more than eight of ten cases in diagnosing medial and lateral meniscus tears and has been accurate 50% of the time when anterior cruciate ligament tears were present.[14,17,35,68] A 90% diagnostic accuracy rate could be achieved using arthroscopy when anterior cruciate ligament tears were present. When both arthrography and arthroscopy techniques were combined with the initial clinical impressions of patients with one or more of these three lesions, diagnostic accuracy reached almost 95%.[78] Furthermore, arthroscopic meniscectomy methodology has reduced the postoperative rehabilitative time from months (standard operative procedure) to weeks.[72,87,92] Certain influential factors, however, remain unchanged, even if arthroscopic surgery is performed. These factors include the extent of the surgery, additional associated lesions, and individual characteristics.[94] Professional athletes can, unfortunately, be overly ambitious regarding an early return to work.

The rehabilitation of patients with disorders producing acute or chronic inflammation is discussed in Chapter 1. While these principles often apply to the rehabilitation of arthritides with acute and chronic inflammation, the presenting problem most often involves an arthritis with subacute inflammation—an inimitable combination of acute and chronic inflammation. These smoldering inflammations have the potential to evolve to either frank acute or chronic inflammation, but usually they maintain themselves with varying proportions of both varieties.

The treatment of arthritis requires a multifaceted rehabilitation program that encompasses a broad spectrum of options. The goals include helping patients achieve their maximum potential for normal living.[50] As Swezey has stated, "The challenge to all who treat the arthritic patient is to minimize impairment and lessen the burden of the handicap so as to prevent disability."[86] Gloag says that "rehabilitation is prevention—not an end-stage affair, after the event, but used continuously right from the start and merged into treatment."[26]

The most common forms of arthritis are rheumatoid arthritis and osteoarthritis, or degenerative joint disease (DJD). Less frequently seen in physical or occupational therapy departments are other arthritic disorders described in Table 2-1. The treatment of all these entities, however, are based upon the status of the inflammatory process in the affected joints. Inflamed joints are treated differently from those that are not acutely inflamed.

Rheumatoid arthritis affects between 1% and 3% of the adult population of the United States.[83] Its effects can be extremely debilitating, even crippling. The gamut of symptoms includes persistent pain and swelling of joints due to inflammation, multiple joint destruction, and loss of functional abilities. Because of the severity of the disease and the complexity of sequelae, rehabilitation must address not only the physical but also the emotional and psychosocial consequences. General goals of treatment include reduction of inflammation and concomitant pain and swelling, restoration or maintenance of range of motion and strength, prevention of joint contractures and deformities, prevention of further joint destruction, maintenance or restoration of functional abilities, conservation of energy, provision of adaptive and assistive devices, and aid in dealing with the frequent and sometimes devastating emotional and social changes that can accompany this potentially disabling disease.

Osteoarthritis might not strike as many joints of the body, but its effects are frequently extremely deleterious to functional independence. Inflammation and joint destruction are often concomitant processes. The joints that appear to be at risk in order of decreasing frequency are the knees, first metatarsophalangeal joints, distal interphalangeal joints, carpometacarpal joints, hips, shoulders, cervical spine, and lumbar spine.[67] Goals of treatment include pain relief, maintenance of range of motion, maintenance of strength, conservation of energy, provision of adaptive equipment, and improved strategies for dealing with changes in lifestyle.

Evaluation

A treatment plan is formulated after a thorough and comprehensive evaluation. Physical, psychosocial, and vocational aspects of patients' lives are naturally included.[45] Normally, the physical evaluation focuses on the musculoskeletal system with attention to any neurologic changes present. Strength, range of motion, deformities, crepitus, pain, swelling, and heat of individual joints are all assessed. Functional abilities, such as gait, activities of daily living, and vocational skills, require careful review. The necessity of modifications to the environment at work or at home in order to preserve functional independence should be determined as well.

Great strides have been made within the past 10 years in the study of human motion, or kinematics. Computer-assisted evaluation of gait can help, for example, in determining which joints require intervention to reduce or prevent deformity or in finding compensatory gait deviations. When appropriate treatment is rendered, stability, joint preservation, strength, and energy conservation are all enhanced. Although these advanced evaluatory tools are not available in most settings at present, they are often within reasonable distance and should be utilized whenever possible.

Compliance

In order to maximize patient compliance, therapists must be effective and efficient as teachers. The patient needs to understand the nature, effects, and course of the disease and how it will affect function and independence. In a study of 200 patients with arthritis, the educated participants showed significant gains in knowledge and decreases in pain and disability for 20 and 8 months respectively, independent of their age group.[59] Rapport between the physician or therapist and patient also tends to increase compliance.[45] When patients have a clear picture of why each intervention is helpful and how it will change their specific situations, they are more likely to participate and cooperate in their own rehabilitation. Splints and braces, for example, are often discarded after an initial period of use. Adequate instruction in the reasons for the devices and what they are expected to accomplish, along with regular follow up by the physician or therapist, will improve the likelihood of compliance by the patient in using the device.

Written instructions are usually more effective in enhancing compliance. Instructions should describe the application and wearing schedule for a splint or brace, the specific exercise program to be followed, or activities to be carried out. The less complex and more specific the instructions, the more likely they are to be appropriately executed. In addition, the number of exercises or activities should be limited, starting with those most likely to produce results valued by the patient.[45]

Rest

Rest and exercise constitute the fundamental basis of treatment in all forms of arthritis.[55] The debilitating physical and emotional effects of bed rest are now well known. As a result, there has been movement away from long-term complete bed rest and toward specific joint rest. The need for rest periods punctuating the day is widely accepted, however, and especially important in rheumatoid arthritis.[33,53,82,86,88]

Therapeutic rest for inflamed joints has proven effective in reducing inflammation, joint destruction, deformity, and pain.[33,44,45,53,57,73,82] This type of rest usually requires a splint. There are two basic types of splints: dynamic or functional, and static or resting. Resting splints are usually designed to immobilize a joint and stabilize it in a desired position. Air splints can also reduce localized edema, especially in the hands.[63] Functional splints stabilize an involved joint while it is being used. Orthotics are discussed in more depth in Chapter 14.

Exercise

There is a certain dichotomy in the treatment of arthritis—both rest and exercise are essential. Judgments about how much of each is advisable should be made with the patient's participation.[44] Decisions are dependent upon how much pain, swelling, and inflammation is present. The more acutely inflamed a joint is, the more protection and rest it requires, and the less exercise should be performed.

Two cautions apply in the application of exercises:

1. Passive range of motion might cause dislocation of the joint.
2. Excessively applied strengthening exercises could overwork muscles weakened through soft tissue inflammation.

The goals of an exercise program may include maintenance of joint motion, increased mobility, maintenance of strength, increased strength, and generally improved function.[86] Cardiovascular fitness could be another goal.[6] These goals should be formulated according to the individual needs of the patient and the current status of the disease. Goals are re-evaluated as the acuteness of the disease changes, and exercises are modified to meet the specific requirements of the patient during each stage of pathologic development.

Guidelines in the use of exercise for arthritis are often common sense. Patients who tend to overperform need to be restricted and provided with specific "rules" to prevent unnecessary pain and trauma to already inflamed joints. Patients who exercise regularly or remain active show better disease outcomes than those who do not.[81] The combined use of exercise and steroids in rheumatoid arthritis, for example, proved more effective in maintaining activity than a regimen of bed rest and no steroids.[65] Rheumatoid arthritics who exercise as little as 15 minutes three times per week can significantly improve aerobic capacity and reduce joint pain and fatigue.[29] In one 5-year study, rheumatoid arthritic patients who underwent an exercise training program were found to have decreased joint destruction, while the untrained control group showed increases in joint destruction.[21] The conclusion drawn is that "physical training does not increase but rather slows down the rate of destruction of affected joints."[21] Other beneficial results were shown by the exercised group in that study, including hospitalization for half the number of days of the control group and no sick days or missed work, while the days used by the control group increased by 30%.[21]

If the patient is taught to be aware of certain signs and symptoms of increased inflammation and learns how to perform exercises correctly, there will likely be minimal harmful results. If, however, the patient ignores inflammation and excessive fatigue, more harm than good might be done. The patient should also be taught to recognize certain physical changes that might signal developing problems and know when to seek help.[82]

There are three phases of inflammatory activity in rheumatoid arthritis, each of which should be treated differently. The first is the acute phase. Severe joint disease is characterized by hot inflamed joints, effusion, and pain. During this period extreme care must be observed to avoid increasing joint destruction, and more rest is required, both systemic and for the individual joints most severely affected, than during other phases. Goals of local therapy are to maintain range of motion or minimize loss of range of motion and deformity.[86]

The second category is the subacute stage. Although the joint condition appears quiescent, any significant stress can be followed by a flare up of the arthritic symptoms.[55] Care must be taken to prevent a return to the acutely inflamed state.

The third stage, the chronic state, is one of minimally active joint disease with reduced pain, swelling, and inflammation. During this phase, the patient can undergo rigorous rehabilitation and try to regain any function previously lost.

Osteoarthritis is usually seen in a static form, and exercise therapy similar to that applied in the chronic stage of rheumatoid arthritis can be administered.[55] When an osteoarthritic joint is subjected to trauma and suffers an exacerbation in pain, swelling, and effusion, it is similar to an acutely inflamed rheumatoid arthritic joint and should be treated similarly. Likewise, when the acutely inflamed joints of the spondylopathies are treated, the therapy is like that of the acute stage of rheumatoid arthritis. Once the activity in the joints has been controlled and the pain and inflammation reduced, these joints can be treated similarly to those in the chronic stage of rheumatoid arthritis. The more acutely inflamed the joint, however, the more care is required to prevent further damage to the articular surfaces and to prevent deformity.

Exercise should always be performed when a patient is at his or her best, well rested and with minimal discomfort. During the acute stage, exercise is used sparingly and with extreme care so as to avoid an increase in pain and inflammation. Pain, contracture, and deformity should be kept to a minimum, but the muscle wasting that may occur can contribute to deformity. A muscle can atrophy up to 30% in 1 week.[67] A muscle at complete rest will lose function at a rate of 5% per day. Exercise should at least maintain strength, but increased pain and inflammation might actually be caused in isotonic exercise.[53,67,81] Therefore, strength should be maintained with isometric exercise. Isometric exercise can maintain strength when as little as 20% to 35% of maximal tension is generated by a given muscle, once per day.[49] A single isometric contraction maintained for 6 seconds that generates 65% or more of maximal tension can increase strength.[49] Isometrics done infrequently through the day, therefore, can preserve or increase strength without increasing inflammation and pain.

To prevent or reduce stiffness, contracture, and deformity, range of motion exercises must also be performed. Exercise programs are recommended from once to twice per day.[6,44,82,86] If the patient is unable to assist, passive motion (by the therapist or a well-instructed family member) of the joint through its full range, as limited by pain, is done one or two times per day. Preferably, the patient will assist in the movement, making it an active-assistive motion within the limits of pain. The motion can be accomplished with a device, such as an overhead pulley, or with the manual aid of a therapist. Patients must be appraised of the deleterious effects caused by too strenuous a stretch during the acute stage and warned to advise the therapist of discomfort. Painful passive exercises are usually dangerous.[6] Likewise, the therapist should be watchful for signs of too much exercise, including increased heat, swelling, pain, or loss of range. These range of motion exercises should be done in a position of maximal comfort, and with gravity eliminated, if possible. Frequently, modalities to help reduce pain and spasm are valuable adjuncts in maintaining or improving range of motion.

Stretching of tight structures should be done with the utmost care, especially during the acute stage. Vigorous passive range of motion may do more harm than good by overstretching ligamentous structures or even by fracture through the shaft of an osteoporotic bone.[44] Repetitions should be limited to one to three movements per session, one to two sessions per day.[86] They should be designed to cause the least possible pain and be repeated only as frequently as is essential to achieve a goal.[34] Slow, gradual stretches, held at the point of pain, then resumed when the discomfort subsides and repeated until further gains are not possible, are effective in increasing range without causing undue pain. Since pain is a signal of tissue damage, clinicians should avoid causing anything greater than moderate discomfort. "No pain, no gain" does not apply in the treatment of acute arthritis.

During the subacute phase, there is less active joint disease, less inflammation, and less pain. Therefore, more activity is tolerated. Care must still be taken, however, since overindulgence in joint strain or movement is followed by flare-up of the arthritic symptoms.[55] Goals during this phase may be to maintain or to increase mobility and strength. Gravitational and other stresses should be avoided. Exercises should be done in comfortable positions when the patient is rested and pain and stiffness is at a minimum. Modality use precedes exercise, or exercise may be done in a therapeutic pool or Hubbard tank. Isometrics are the exercise of choice to maintain or increase strength. The patient must be carefully monitored and be instructed against overexercising to prevent fatigue, increased pain, and inflammation. Stretching sessions may be increased during this stage but must still be carefully graded. The prone position for gentle stretching or prevention of hip flexion contracture should be encouraged.[86]

During the chronic stage, while the disease is minimally active, the goals are again re-evaluated. Function is the ultimate goal, and all means are directed to that end. Strengthening, stretching, or maintaining strength and range of motion are appropriate at different times. Isotonic exercises may be added but should be done without resistance (active) or with low resistance as tolerated. Again, patients must be aware of the precautions to be taken and the dangers of overexercising. Generally, if there is increased pain for more than 2 hours, if symptoms worsen 24 hours after exercise, or if there is marked fatigue, the intensity of the program must be reduced.[6,82,86]

Recreational exercises that enhance physical fitness for patients with arthritis include walking, swimming, bicycling, light racquet games, croquet, golf, and dancing.[6,26,41,81] Pool therapy or hydrotherapy is the most widely advocated and well received therapeutic/recreational strategy. However, care must be taken that the water temperature is not so high as to allow excessive peripheral vasodilation, which can lead to collapse and injury or to myocardial or cerebral ischemia.[86] Sports to be avoided include those that cause impact or jolting to the joints, including running, jumping, hopping, hardball games, highly competitive team sports, and contact sports.[6] Patients must avoid increasing pain, learn to warm up and cool down, and gradually increase exercise levels within their individual tolerances.

Table 2-3 provides an overview of the three stages of arthritis and the types of exercises usually provided during each.

Modalities

Treatment of the pain, spasm, and stiffness associated with arthritis can be quite a challenge. Frequently, the best palliative is only temporary in its effect and may need to be repeated several times per day. Obviously, it may be impossible to do so on an outpatient basis, but some of the heat and cold devices are often utilized by the patient at home.

During the acute phase, when pain and swelling are at their worst, cold is more effective in reducing pain, muscle spasm, and spasticity, and is frequently used in management of edema.[45,53,67,80,81] Cold is contraindicated if the patient finds it uncomfortable, and specifically in patients with vasculitis, Raynaud's phenomenon, cold hypersensitivity, paroxysmal cold hemoglobinuria, and cryoglobulinemia.[67,80,81] Both heat and cold should be applied with caution where there is scar tissue, hyposensitivity, hypersensitivity, sedation, decreased alertness, skin infection, or circulatory insufficiency.[80,81] Extreme care should be taken with older patients who may have diminished cardiac and respiratory reserves and therefore less tolerance for heat.[80] Specific forms of application of both heat and cold are discussed in Chapter 1.

Clinically, heat is contraindicated in treatment of acutely inflamed joints (see Chapter 1), although it is indicated in the treatment of the complications of inflammation, including joint contracture, tendinitis (subacute), muscle spasm, and stiffness. The results of heat are often temporary in the treatment of arthritis and do not affect the progression of the disease.[45] Most types of heat are superficial. Ultrasound has been shown to reach the joint and heat the surrounding tissues[67] and is considered deep heat. Ultrasound,

TABLE 2-3. THE TREATMENT OF ARTHRITIS DURING ITS THREE STAGES

Category	Acute	Subacute	Chronic
Disease state	Severe joint disease	Moderately active joint disease	Minimally active joint disease
Precautions	Work only to the point of pain; do not increase pain or edema	Reduce intensity of exercises if pain increases > 2 hours, symptoms increase, or there is marked fatigue	Same as for subacute
Goals	Maintain range of motion; prevent loss of range of motion and deformity	Increase or maintain strength and range of motion	Increase or maintain strength and mobility
Treatment technique	Active-assistive or gentle passive range of motion and positioning; gentle isometrics with no pain increase	Active-assistive range of motion, isometric exercises, may use some simple exercise devices	Active or active-assistive to stretch; isometrics and some isotonics to strengthen; may do some gentle aerobics or other recreational activities (e.g., swim)
Positioning	Gravity eliminated, maximal comfort	Same as acute	Comfort—may do antigravity exercise when tolerable
Time of day	After rest, when stiffness is least	Same as acute	Same as acute
Repetitions (per joint or muscle group)	1–3 repetitions 1–2 times/day for stretching and isometrics	3–10 repetitions 1–2 times/day for stretch, 1–3 6-second isometric (maximal)/day for strengthening	Start at level for subacute, increase to tolerance; add one set isotonic exercise/day, increase to tolerance
Pretreatment	Cold preferably, or gentle superficial heat; may use hydrotherapy—goal is analgesia	Cold or superficial heat modality, other required analgesics prior to exercises; may use hydrotherapy	Same as subacute, but may add deep heat (ultrasound) as required
Other	Resting splints, joints must be in functional position, use joint protection and energy conservation techniques; place patient in prone position for total of 2 hr/day to prevent hip flexion contracture	Use working (dynamic) splints during activity; prone position to prevent hip flexion contracture; avoid overfatigue; use swimming for general conditioning	Use daily range of motion for maintenance, strengthening, and postural exercise, joint protection and energy conservation techniques, and splints

however, is specifically contraindicated in the treatment of acute inflammation in arthritis, as increased temperature might accelerate cartilage destruction in the presence of inflammation.[80] Further, animal studies have shown that under certain circumstances, ultrasound can actually increase early and acute joint inflammation before it reduces it.[24] Ultrasound should therefore be used only with chronic or subacute arthritis, not acute arthritis.

The most widely recommended form of superficial heat is moist heat, possibly because of its deeper penetration[81] and enhanced effectiveness in pain relief.[33] The most popular form of moist heat in the treatment of arthritis is hydrotherapy. Its many benefits are results of the thermal and physical properties of water, i.e., buoyancy and viscosity, which ease joint movement and muscular contraction while the patient exercises.[80,81] Recommended temperatures are 90–100 degrees for full body immersion.[50] Prolonged immersion or higher temperatures can cause increased body temperature, increased pulse rate and metabolism, decreased blood pressure and increased oxygen consumption.[80] Local peripheral vasodilation and often reflex vasodilation at a site remote from the warmth may also occur.[45] In a patient whose circulatory reserve is compromised, heat may cause collapse and injury or myocardial or cerebral ischemia,[86] as mentioned above.

Since hydrotherapy rapidly relieves arthritic and myalgic symptomatology, it must be applied precisely. Specific goals must be agreed upon between therapist and patient before therapy. When these have been realized, hydrotherapy should be stopped. Otherwise, the patient will experience such relief that stopping therapy will become difficult. The most effective application of hydrotherapy is the therapeutic pool. Both mat exercises and gait training can be begun after the hydrotherapy. The rate-limiting reaction, however, is the transition between pool and "dry" gait exercises. After hydrotherapy, "dry" gait seems heavier, stiffer, and more painful. One must be prepared for the patient to regress somewhat. Prolonged pool therapy can also interfere with progress on land (see Figure 2-2).

Other forms of moist heat are baths and showers, hot moist packs, and paraffin, which are comforting and help ease morning stiffness.[33,44] Paraffin is applied at temperatures between 126 and 130 degrees F and is a good home treatment procedure once patients have been carefully trained in the application technique.[81] One study found no significant difference between treatment with paraffin and exercise and treatment with ultrasound/faradic current/exercise and ultrasound/exercise programs[30] for rheumatoid arthritis of the hands. Generally, moist heat is relatively easy to apply and could be utilized in the home as often as required. In subacute and chronic stages of arthritis, moist heat just prior to the exercise program allows enhanced freedom of movement and reduced pain.

Electrical stimulation of muscles is used to relieve pain and muscle spasm and reduce periarticular swelling[67] and has been an effective part of the treatment program to help stretch contractures in patients with spasticity.[45] Another form of electricity, transcutaneous electrical nerve stimulation (TENS) is also valuable in managing the pain of rheumatoid neuropathy and that associated with rheumatoid arthritis.[45,51] TENS is also helpful for managing the pain of osteoarthritis.[67] Patients must be warned, however, not to be lulled into a sense of false confidence when the pain has been significantly reduced. Some patients with good pain relief unduly stress joints, resulting in further damage, and all patients should be warned to be careful not to injure the joints under treatment with TENS.[51] The parameters of TENS are discussed further in Chapter 3.

Massage also relieves pain and stiffness, especially that related to muscle spasm,[67,81] and is effective in promoting relaxation.[85] Massage is given after the application of heat or cold to augment their effects on muscle relaxation and to decrease pain prior to stretching or strengthening exercises.[67] Family members are occasionally instructed in specific techniques for use at home.

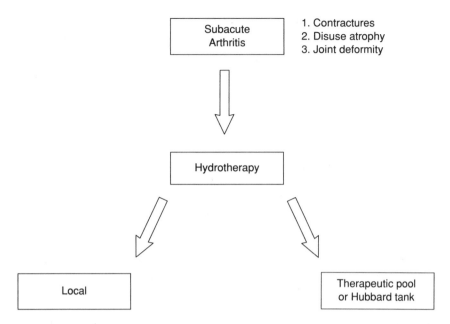

Figure 2-2. Hydrotherapy for arthritis.

Traction relieves pain and decreases flexion contractures of the arthritic knee and hip and relieves pressure on a compressed nerve root.[67] Traction is also effective in the treatment of cervical disc disease and the resultant radiculopathy.[67,90] Static cervical and lumbar traction may be used at home alone or in conjunction with other forms of treatment mentioned above.

Electromyographic biofeedback, combined with home practice of relaxation and various modalities, are frequently helpful in the treatment of chronic pain.[23] Biofeedback for muscular re-education leads to increased strength and reduced pain concomitantly.[47]

Other modalities have been utilized with varying degrees of success, including auriculotherapy, cold laser, and interferential current, which are fairly new additions to the armamentarium of modalities currently available. Their application, partly because of their relatively low availability, is somewhat limited at present. As research and practice prove their value, they will likely be employed more commonly in the management and treatment of arthritis and its sequelae. Additionally, supportive devices to aid in ambulation are frequently used when the joints of the lower extremities are affected. Reliance on a walker, one or two crutches, a four-pronged cane, a straight cane, or other assistive device can provide significant relief from the pain associated with weight bearing. This is accomplished by transferring the stress to less involved joints. Further benefits are gained by obese arthritic patients who lose weight, and attain normal weight for their age and height.[88] Dietary therapy is of questionable help.[31] The significance of this therapy has not been proved through well-controlled studies. The same is true for vitamins and acupuncture.

RADICULITIS

Radiculitis bears much the same relationship to arthritis as bursitis has to tenosynovitis. While patients with nontraumatic, isolated tenosynovitis exist, those with nontraumatic bursitis have had tenosynovitis at some point in the past. While the etiology of all four entities is multifactorial, tenosynovitis facilitates the development of bursitis just as arthritis increases the probability of the later occurrence of radiculitis. Other factors are also important. The intensity, duration, and rate of progression of the arthritis are vital. Moreover, significant central nervous system injury can also produce spinal nerve root dysfunction.[40] Occupational hazards contribute, as do obesity and unexpected trauma (physical and emotional). No one of these items alone will cause radiculitis all the time. Usually, two or three predisposing factors will converge upon the subject and transform him or her into a patient. The force needed to produce localized shearing stress and radiculitis need not be large but is usually strategically applied.[75]

The pathophysiologic barriers that preclude both pain and disability are not so much concerned with etiology as they are with neurologic recovery. Sensory axons have been shown to regenerate in frogs' central nervous systems. Nonetheless, some large diameter axons that regenerated apparently established aberrant connections.[56] Those aberrant axons, formed after complete disruption, would probably not function effectively, limiting the potential for neurologic return. Unreduced, locked posterior vertebral facets also limit spinal nerve root regeneration.[84] Presumably, extra traction is thereby placed upon the nerve root sleeve as it exits the vertebral bony shell. A final barrier exists in the progression of the underlying arthritis. No matter how favorable other factors are, this unrelenting type of inflammation will sooner or later undermine any nerve root recovery. These factors are summarized in Table 2-4.

TABLE 2-4. FACTORS INFLUENCING NEUROLOGIC RECOVERY OF RADICULITIS

Positive	Negative
1. Small disruptive force, no shearing stress	1. Disruptive force is large or so strategically placed as to produce shearing stress
2. Partial nerve root lesions, some nerve fibers intact	2. Complete nerve root rupture, maximal aberrant regeneration
3. Posterior facets unlocked, minimal spinal nerve root sleeve traction	3. Posterior facets locked so that additional traction is placed upon nerve root sleeve
4. Stable or improving arthritis under medical control	4. Progressive, erosive arthritis with surrounding soft tissue inflammation

The clinical picture of radiculitis is often presented as affecting specific dermatomes and deep tendon reflexes. Recording sensory evoked potentials from spinal nerve roots, however, has demonstrated that dermatomal representation varies widely.[9] Other studies of sensory evoked potentials show that stimulation of a single peripheral nerve activates several spinal segments at once.[13] These results would explain shifting myotomal patterns as well. There is no specific nerve root that will innervate a set area of skin or a muscle in every patient. Accordingly, there are no special key cutaneous areas or key muscles to check that in and of themselves will diagnosis radiculitis. Sensory evoked potentials can also wax and wane during an operative procedure. This observation is consistant with the disappearance and reappearance from day to day, and within the same day, of deep tendon reflexes (DTR). The DTR abnormalities are not necessarily permanent and frequently vary.

If dermatomal, myotomal, and deep tendon reflexes in and of themselves cannot help, they can be investigated as part of an overall pattern. Are there sphyncteric disturbances? Has gait been affected? Is there a consistent presentation of sensory loss, flaccid paresis, and reduced DTRs? Every patient with a possible radiculitis deserves a thorough history and physical examination. Overlapping neurologic control of dermatomal and myotomal function can be contrasted and compared. For example, if the biceps DTR is reduced, the C6 dermatome is deficient, and biceps, deltoid, and serratus anterior are all weak and flaccid while the rhomboids are normal, a C6 radiculopathy is probably present. Three principles should be remembered:

1. Relative velocity of blunt objects can be tested across distal cutaneous tissue and is often sensitive to early changes.[38]
2. Muscles should be placed in disadvantageous positions of stretch so that they can be tested for mild weakness.
3. DTRs should be tested on several consecutive days.

Electromyography (EMG) and nerve conduction studies remain valuable confirmatory aids. These are covered in Chapter 12. Nothing has yet replaced the monitoring advantages of EMG. It can be used to estimate neuromuscular physiology across affected nerves or nerve roots.

Somatosensory evoked potentials (SSEP) are frequently cited in the literature.[13,19] They can be used to depict proximal sensory latencies. SSEPs are very sensitive[48] and are frequently abnormal in both radiculopathy and myelopathy but in different ways. SSEP technology, therefore, can be used as one test to help differentiate radiculopathy from myelopathy.

CAT scanning—either alone or in combination with discography—has been used to reduce the number of false negative examinations.[11,62] It also is very sensitive and complements SSEP in that it will depict the anatomic shape of the suspected herniated disk. Both SSEP and CAT scanning share a significant defect: they are sensitive,[18,70] but not necessarily very specific. Consequently, both can render false positive diagnoses. Without specific EMG abnormalities to balance the total picture, too great a reliance on SSEP and CAT scanning might lead to unnecessary surgery. The flow chart for evaluation of CAT scanning is presented in Figure 2-3.

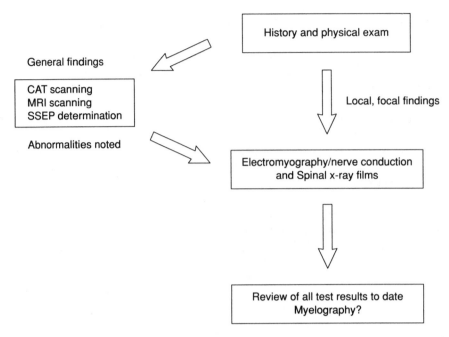

Figure 2-3. Evaluation of CAT scanning.

NEURITIS

Neuritis and neuropathy have been reviewed elsewhere in depth.[4] Accordingly, this section will focus on peripheral neuropathy as a cause of pain and disability. Along with fibrositis, neuritis is one of the most common factors generating pain and disability. Its study is complex because neuritis can be primary or secondary, acute, subacute, or chronic, lancinating or causalgic.

Whether or not the nerve is active has an influence upon its function.[10] Nerves supplying "fast" or type II muscle fibers are especially vulnerable. Indeed, these nerves can show fiber atrophy with disuse. Moreover, vascular metabolic disorders can cause the nerve to become ischemic, and toxins can directly damage the nerve.

Neuropathy can be caused by inflammatory processes affecting the axon, the myelin sheath, or both. Soft tissue inflammation contributing to the neuropathy can be associated with arthritis or collagen disorders. Axonopathies generally lead to lower motor neuron physical and EMG abnormalities, while demyelinative lesions reduce nerve conduction.[22] This type of demyelinative condition block can be primary, secondary to systemic disease, or caused by local and strategic compression. It has been noted in Guillain-Barre syndrome. Secondary neuropathies are heterogeneous and usually very complex disorders. For example, the sensory neuropathy from lung cancer has been associated with a

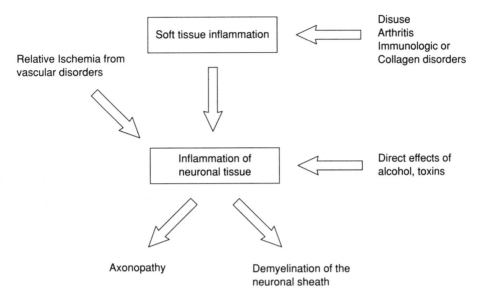

Figure 2-4. Pathogenesis of neuritis.

neuronal antinuclear antibody.[27] The pathogenesis of neuritis is presented in Figure 2-4.

Asbury and Fields have proposed that there are two major types of neuropathic pain—nerve trunk pain and dysesthetic pain—and that each has a separate pathogenesis.[3] Whereas dysesthetic (causalgic) pain is generated by the increasing firing of hyperexcitable, regenerating axons, nerve trunk (lancinating) pain is caused by direct damage or chemical stimulation leading toward increased firing of normal nociceptive fibers. But pain, weakness, numbness, and paralysis are not the only effects of neuritis. Inherited neuropathies, for example, are also associated with recurring fractures and arthropathy.[20] This tendency toward production of fractures could produce its own arthritis and soft tissue inflammation, which could in turn make neuritis more intense.

EMG and nerve conduction findings are covered in Chapter 12. They are very helpful but often quite complicated. Values recorded in studies of ulnar neuritis, for example, partly are dependent upon the length of the nerve segment studied, the angle of flexion of the elbow, and the intensity of electric stimulation.[46] Somatosensory evoked potentials have been used in combination with nerve conduction and EMG studies to investigate brachial plexus injuries.[93]

Diabetes is frequently associated with neuritis. While diabetic sensory peripheral neuropathy is common, so is the combination of diabetes, dysesthesia, only mild nerve conduction abnormalities, depression, and impotence.[1] This particular diabetic neuropathy could be a separate entity. The peripheral neuropathy of diabetes, while often distal and sensory, is usually more intense in the lower extremities and can affect proximal nerve segments.[12] One of the positive aspects of neuritis is the potential for regeneration of the peripheral nerve axon as long as the neuronal body is functioning. In diabetes, however, there is some indication, in animal experiments, that regeneration could be significantly impaired in diabetic nerves.[58]

Alcoholism is also associated with neuritis. It is also distal and can also be mainly demyelinating.[4,32,58] Both alcoholism and diabetes, however, can be associated with some degree of axonopathy. In any case, alcoholic neuritis can resolve, provided alcohol intake is stopped.[32]

A nerve is most likely to be entrapped in certain places.[91] Usually the nerve is confined to a small area, subjected to actual or potential stress or traction and is vulnerable to vascular ischemia phenomenon, and could be pressed by soft tissue against a bony surface. This neuropathy can be single or multiple, primary or secondary (to diabetes, for example), and acute or chronic. Some of the common entrapment neuropathies are presented in Table 2-5.

TABLE 2-5. ENTRAPMENT NEUROPATHIES

Nerve	Where Affected	Characteristics
Median	Across the wrist	Carpal tunnel syndrome is the most common, especially in dominant hand subjected to repeated strain. Thumb-palmar abduction deficits.
	Across the pronator teres muscle	Anterior interosseous and pronator syndromes. Knife wounds and tennis often contribute flexor policis longus deficit.
Radial	In the spiral groove	Saturday night palsy. Prolonged compression contributes. Wrist drop and brachioradialis deficit.
	In the forearm	Posterior interosseous/supinator syndromes. Knife wounds in the forearm can produce both interosseous syndromes and wrist drop.
Ulnar	Across the elbow	Cubital tunnel syndrome/tardy ulnar palsy. Can be immediate or remote trauma with edema, fibrosis, or direct damage. Weak intrinsic hand muscles innervated by ulnar nerve.
Peroneal	Across the head of the fibula	Can be caused by a cast, traction at surgery, sports, trauma. Foot drop.
Tibial	Across ankle and proximal foot areas	Tarsal tunnel syndrome—insidious onset, pain, weakness of intrinsic foot muscles.
Brachial plexus	Upper trunk	Erbs palsy—weakened shoulder abduction, extension, external rotation, and weakened supination and grasp.
	Lower trunk, medial cord	Klumpke's paralysis, thoracic outlet syndrome—weakened grasp, arm, and shoulder pain.

TREATMENT OF RADICULITIS AND NEURITIS

The treatment of radiculitis and neuritis is often similar to or parallel with the treatment of arthritis but is dependent upon the symptomatology accompanying the pathologic condition and the severity of those symptoms. As with arthritis, the subacute or intermediate phases of involvement are most important. As with acute rheumatoid arthritis, bed rest is the most important therapeutic ingredient. But timely judgement has to be made in both areas as to when the acute inflammation is over and mobilization can begin. The same precautions are followed, and clinical decision making follows the stage of the disease. Electrical stimulation is of great help, especially if the nerve is recovering. Once the nerve begins to recover, neuromuscular re-education and therapeutic exercise are appropriate. If nerve root compression is accompanied by pain, traction to help relieve the pressure on the nerve may be indicated. Traction is often accompanied by other modalities, such as moist heat, ultrasound, TENS, and electrical stimulation, to help reduce pain and spasm and allow the distraction to be more effective. TENS is particularly suited to the treatment of neurogenic pain, including peripheral nerve injury, causalgia, intercostal neuritis, radiculopathies, and compression syndromes.[89] If there is acute inflammation or insensitive skin, deep heat is contraindicated. Once past the acute stage, though, ultrasound is recommended in the treatment of both neuritis and radiculitis.[50]

Hydrotherapy is valuable in allowing controlled exercise and remobilization once the acute stage is past. Warm water is soothing and relaxing and has numerous benefits to patients suffering from pain and reduced mobility from most any source. Hydrotherapy, by sedating and releasing weight, is vital to the rehabilitation process.[39] It is, in itself, the most effective therapeutic regimen that can be applied during any subacute condition. As with any modality, however, careful evaluation must precede the institution of hydrotherapy. Therapeutic exercises, gait and transfer training, and activity of daily living exercises must also be part of the plan, but these are less efficient without hydrotherapy. Particularly effective is a therapeutic pool, where "mat" exercises, strengthening, and even gait training can be applied under controlled conditions.

There are contraindications and precautions that must be observed with every device. No single modality can be used with impunity, and the more that are prescribed, the more caution should be exercised to prevent untoward results. The need for certain precautions is particularly important in the treatment of the elderly, as discussed in Chapter 15. In addition, as with other neurologic problems, extreme care must be taken when skin sensation in impaired.

The treatment of many of the entrapment neuropathies is surgical. Treatment of secondary neuropathies is for that of the underlying metabolic disorder.[28] Nonetheless, it is not clear if controlling diabetes alleviates diabetic neuritis—the neuritis might not get worse, but it may not get better. The best treatment may be maintenance of a normal blood sugar level to prevent hypo- or hyperglycemic contributions to the diabetic neuropathy. Skin protection is of utmost importance for all extremities and skin must be regularly examined by the patient. Anticonvulsant tricyclic agents have been advocated to alleviate pain, but results of well-controlled studies have been mixed.

The treatment of pain is discussed above and in Chapter 1, and the treatment of chronic pain is discussed in depth in Chapter 11. However, radiculitis with arthritis is probably the most frequent cause of chronic pain. Complex pain clinics have been formed in response to this difficult problem, and it will probably continue to be a significant cause of disability.

REFERENCES

1. Archer CR, Yeager V: Internal structure of the knee visualized by computed tomography. *J Comput Assist Tomogr* 1978; 2:181.
2. Archer AG, et al: The natural history of acute painful neuropathy in diabetes mellitus. *J Neurol Neurosurg Psychiat* 1983; 46:491.
3. Asbury AK, Fields HL: Pain due to peripheral nerve damage: An hypothesis. *Neurology* 1984; 34:1587.
4. Asbury AK, Gilliatt RW: *Peripheral Nerve Disorders*. Stoneham, Butterworth, 1984.
5. Baker HL, et al: Magnetic resonance imaging in a routine clinical setting. *Mayo Clin Proc* 1985; 60:75.
6. Banwell BF: Exercise and mobility in arthritis. *Nurs Clin North Am* 1984; 19:605.
7. Bennet PH, Burch TA: The epidemiology of rheumatoid arthritis. *Med Clin North Am* 1968; 52:479.
8. Bluestone R: *Practical Rheumatology: Diagnosis and Management*. Reading, Mass. Addison-Wesley, 1908.
9. Buchthal F, Rosenfalck A: Evoked action potentials and conduction velocity in human sensory nerves. *Brain Res* 1966; 3:1.
10. Calder CS, Pollock M: Morphometric effects of use and disuse on peripheral nerves. *Arch Neurol* 1985; 42:868.
11. Cherie Ligniere G, et al: Computerized axial tomography in investigations of the rheumatoid knee. *Clin Exper Rheum* 1985; 3:189.
12. Chopra JS, Hurwitz LJ: A comparative study of peripheral nerve conduction in diabetes and non-diabetic chronic occlusive peripheral vascular disease. *Brain* 1969; 92:83.
13. Cracco RQ: Comparison of single and multiple peripheral nerve stimulation on the human spinal evoked response. *Electroenceph Clin Neurophysiol* 1975; 38:543.
14. Dandy DJ, Jackson RW: The impact of arthroscopy on the management of disorders of the knee. *J Bone Joint Surg (Br)* 1975; 57:346.
15. Dawes PT, et al: Rheumatoid arthritis: Treatment which controls the C-reactive protein and erythrocyte sedimentation rate reduces radiologic progression. *Brit J Rheum* 1986; 25:44.
16. Decker JL, et al: Rheumatoid arthritis: Evolving concepts of pathogenesis and treatment. *Ann Int Med* 1984; 101:810.
17. Dehaven KE, Collins HR: Diagnosis of internal derangements of the knee: The role of arthroscopy. *J Bone Joint Surg (Am)* 1975; 57:802.
18. Del Gado Martins M: A study of the patella using computed tomography. *J Bone Joint Surg (Br)* 1979; 61:443.
19. Dorfman LF, et al: Use of cerebral evoked potentials to evaluate spinal somatosensory function in patients with traumatic and surgical myelopathies. *J Neurosurg* 1980; 52:654.
20. Dyck PJ, et al: Neurogenic arthropathy and recurring fractures with subclinical inherited neuropathy. *Neurology* 1983; 33:357.
21. Ekblom B: Short and long-term physical training in patients with rheumatoid arthritis. *Arth Rheum* 1985; 28:109.
22. Feasby TE, et al: The pathologic basis of conduction block in human neuropathies. *J Neurol Neurosurg Psychiat* 1985; 48:239.
23. Flor H, et al: Efficacy of EMG biofeedback, pseudotherapy, and conventional medical treatment for chronic rheumatic back pain.

Pain 1983; 17:21.

24. Fyfe MC, Chahl LA: The effect of single or repeated applications of "therapeutic" ultrasound on plasma extravasation during silver nitrate induced inflammation of the rat handpaw-ankle joint in vivo. *Ultrasound Med Biol* 1985; 11:273.

25. Gilman AG, et al (editors): *Goodman and Gilman's The Pharmacological Basis of Therapeutics*, 7th ed. New York, Macmillan Pub. Co., 1985.

26. Gloag D: Rehabilitation in rheumatic diseases. *Br Med J* 1985; 290:132.

27. Graus F, Cordon-Cardo C, Posner JB: Neuronal antinuclear antibody in sensory neuronopathy from lung cancer. *Neurology* 1985; 35:538.

28. Hallett M, et al: Treatment of peripheral neuropathies. *J Neurol Neurosurg Psychiat* 1985; 48:1193.

29. Harkcom TM, et al: Therapeutic value of graded aerobic exercise training in rheumatoid arthritis. *Arth Rheum* 1985; 28:32.

30. Hawkes J, et al: Comparison of three physiotherapy regimens for hands with rheumatoid arthritis. *Br Med J* 1985; 291:1016.

31. Hawley DJ: Nontraditional treatments of arthritis. *Nurs Clin N Amer* 1984; 19:663.

32. Hillbom M, Wennberg A: Prognosis of alcoholic peripheral neuropathy. *J Neurol Neurosurg Psychiat* 1984; 47:699.

33. Hollander JL (editor): *The Arthritis Handbook*. West Point, Pa., Merck, Sharp & Dohme, 1974.

34. Hollander JL, Yeostros SJ: The effect of simultaneous variations of humidity and barometric pressure on arthritis. *Bull Ann Meteorol Soc* 1963; 44:489.

35. Jackson RW, Dehaven RE: Arthroscopy of the knee. *Clin Orthop* 1975; 197:87.

36. Jamis R, Hamerman D: Articular cartilage in early arthritis. *Bull Hosp Joint Dis* 1969; 30:136.

37. Johansson M, Sullivan L: Influence of treatment and change of climate in women with rheumatoid arthritis. *Scand J Rheumatol* 1975; Suppl 9:1.

38. Kaplan P: Hemiplegia: Rehabilitation of the

lower extremity. In Kaplan P, Cerullo L (editors): *Stroke Rehabilitation*. Boston, Butterworth, 1986.

39. Kaplan PE: Rheumatoid arthritis and related diseases. In Kaplan PE: *The Practice of Rehabilitation Medicine*. Springfield, Ill., Charles C. Thomas, 1982.

40. Katagama Y, et al: Concussive head injury producing suppression of sensory transmission within the lumbar spinal cord in cats. *J Neurosurg* 1985; 63:97.

41. Katz WA: Modern management of rheumatoid arthritis. *Am J Med*. 1985; 79:24.

42. Kayser V, Gulband G: Further evidence for change in the responsiveness of somatosensory neurons in arthritic rats: A study of the posterior intralaminar region of the thalamus. *Brain Res* 1984; 323:144.

43. Kelly WN, et al (editors): *Textbook of Rheumatology*. Philadelphia, W.B. Saunders Co., 1981.

44. Kelly WN, et al (editors): *Textbook of Rheumatology*, Vol. 1, 2nd ed. Philadelphia, W.B. Saunders Co, 1985.

45. Kelly WN, et al (editors): *Textbook of Rheumatology*, Vol. 2, 2nd ed. Philadelphia, W.B. Saunders Co, 1985.

46. Kincaid JC, et al: The evaluation of suspected ulnar neuropathy at the elbow. *Arch Neurol* 1986; 43:44.

47. King AC, et al: EMG Biofeedback-controlled exercise in chronic arthritic knee pain. *Arch Phys Med Rehabil* 1984; 65:341.

48. Kotani H, et al: Evaluation of cervical cord function in cervical spondylotic myelopathy and/or radiculopathy using both segmental conductive spinal-evoked potentials. *Spine* 1986; 11:185.

49. Kottke FJ: Therapeutic exercise. In Krusen FH, et al (editors): *Handbook of Physical Medicine and Rehabilitation*, 2nd ed. Philadelphia, W.B. Saunders Co, 1971.

50. Krusen FH, et al (editors): *Handbook of Physical Medicine and Rehabilitation*, 2nd ed. Philadelphia, W.B. Saunders Co, 1971.

51. Kumar VN, Redford JB: Transcutaneous nerve stimulation in rheumatoid arthritis. *Arch Phys Med Rehabil* 1982; 63:595.

52. Langohr HD, et al: Muscle wasting in chronic alcoholics: Comparative histochemical and biochemical studies. *J Neurol Neurosurg Psychiat* 1983; 46:248.

53. Leek JC, et al (editors): *Principles of Physical Medicine and Rehabilitation in the Musculoskeletal Diseases*. New York, Grune & Stratton, Inc., 1986.

54. Levine, JD, et al: Intraneuronal substance P contributes to the severity of experimental arthritis. *Science* 1984; 226:547.

55. Licht S (editor): *Therapeutic Exercise*, 2nd ed. Baltimore, Waverly Press, 1965.

56. Liuzzi FJ, Lasek RJ: Regeneration of lumbar dorsal root axons into the spinal cord of adult frogs (Rana pipiens), an HRP study. *J Comp Neurol* 1985; 232:456.

57. Locke M, et al: Ankle and subtalar motion during gait in arthritic patients. *Phys Ther* 1984; 64:504.

58. Longo FM, et al: Delayed nerve regeneration in streptozotocin diabetic rats. *Muscle Nerve* 1986; 9:385.

59. Lorig K, et al: Arthritis self-management: A study for the effectiveness of patient education for the elderly. Gerontologist 1984; 24:455.

60. Malone DG, et al: Immune function in severe, active rheumatoid arthritis: A relationship between peripheral blood mononuclear cell proliferation to soluble antigens and synovial tissue immunohistologic characteristics. *J Clin Invest* 1984; 74:1173.

61. McCarty DJ: *Arthritis and Allied Conditions*, 10th ed. Philadelphia, Lea & Febiger, 1985.

62. McCutcheon ME, Thompson W III: CT scanning of lumbar discography. *Spine* 1986; 11:257.

63. McKnight PT, Schomburg FL: Air pressure splint effects on hand symptoms of patients with rheumatoid arthritis. *Arch Phys Med Rehabil* 1982; 63:560.

64. Millan MJ, et al: Spinal cord dynorphin may modulate nociception via a X-opioid receptor in chronic arthritic rats. *Brain Res* 1985; 340:156.

65. Million R, et al: Long-term study of management of rheumatoid arthritis. *Lancet* 1984; 8381:812.

66. Mizel SB, et al: Stimulation of rheumatoid synovial cell collagenase and prostaglandin production by a partially purified lymphocyte-activating factor (interleukin 1). *Proc Natl Acad Sci USA* 1980; 78:2474.

67. Moskowitz RW, et al (editors): *Osteoarthritis*. Philadelphia, W.B. Saunders Co., 1984.

68. Nicholas JA, et al: Double contrast arthrography of the knee. *J Bone Joint Surg (Am)* 1970; 52:203.

69. Patberg WR, et al: Relation between meterological factors and pain in rheumatoid arthritis in a marine climate. *J Rheumatol* 1985; 12:711.

70. Pavlov M, et al: Computed assisted tomography of the knee. *Invest Radiol* 1978; 13:57.

71. Pavlov M, et al: Computed tomography of the cruciate ligaments. *Radiology* 1979; 132:389.

72. Pettrone FA: Meniscectomy: Arthrotomy versus arthroscopy. *Am J Sports Med* 1982; 10:355.

73. Porter SF, et al: Hand splints. *Am J Nurs* 1983; Feb: 276.

74. Raisy LG, et al: Immunologic factors influencing bone resorption: Role of osteoclasts-activating factor from human lymphocytes and complement-mediated prostaglandin synthesis. In Talmage RV, Owen M, Parsons JA (editors): *Calcium-Regulating Hormones*. Amsterdam, Exerpta Medica, 1975.

75. Raynor RB, Koplik B: Cervical cord trauma. The relationship between clinical syndromes and force of injury. *Spine* 1985; 10:193.

76. Robbins DL, et al: Complement activation by 19SIgM rheumatoid factor: Relationship to disease activity in rheumatoid arthritis. *J Rheum* 1986; 13:33.

77. Rose MB: Effects of weather on rheumatism. *Physiotherapy* 1974; 60:306.

78. Selesnick FH, et al: Internal derangement of the knee: Diagnosis by arthrography, arthroscopy, and arthrotomy. *Clin Orthop* 1985; 198:26.

79. Sibley JT: Weather and arthritis symptoms. *J Rheumatol* 1985; 12:707.

80. Simpson CF: Heat, cold or both? *Am J Nurs* 1983; Feb:270.

81. Simpson CF: Physical and occupational therapy for arthritis. *West J Med* 1985; 142:562.

82. Simpson CF, Dickinson GR: Exercise. *Am J Nurs* 1983; Feb:273.

83. Speigel JS, et al: Rehabilitation for rheumatoid arthritis patients. *Arth Rhuem* 1986; 29:628.

84. Stauffer ES: Neurologic recovery following injuries to the cervical spinal cord and nerve roots. *Spine* 1984; 9:532.

85. Suwalska M: Importance of physical training of rheumatic patients. *Ann Clin Res* 1982; 34:107.

86. Swezey RL: *Arthritis: Rational Therapy and Rehabilitation*. Philadelphia, W. B. Saunders Co., 1978.

87. Tregonning R: Closed partial meniscectomy. *J Bone Joint Surg (Br)* 1983; 65:378.

88. Villaverde MM, MacMillan CW: *Pain*. New York, Van Nostrand Reinhold Co., 1977.

89. Wall PD, Melzack R (editors): *Textbook of Pain*. New York, Churchill Livingstone, 1984.

90. Waylonis GW, et al: Home cervical traction: Evaluation of alternate equipment. *Arch Phys Med Rehabil* 1982; 63:388.

91. Weingarden SI: Rehabilitation of peripheral neuropathies. In Kaplan P, (editor): *The Practice of Physical Medicine*. Springfield, Ill., Charles C. Thomas Pub., 1984.

92. Whipple TL, et al: Arthroscopic meniscectomy: An effective, efficient technique. *Orthop Trans* 1981; 5:336.

93. Yiannikas C, et al: The investigation of traumatic lesions of the brachial plexus by electromyography and short latency somatosensory potentials evoked by stimulation of multiple peripheral nerves. *J Neurol Neurosurg Psychiat* 1984; 46:1014.

94. Zarins B, et al: Knee rehabilitation following arthroscopic meniscectomy. *Clin Orthop* 1985; 198:36.

3

Neck and Upper Back Pain

Paul E. Kaplan and
Ellen D. Tanner

The evaluation of neck and upper back pain has rapidly evolved because of the application of radiodiagnostic, radioisotopic procedures and the results of animal investigations. In particular, the roles of soft tissue anatomy and neurologic control of kinesiologic posturing can be more accurately assessed. As a result, appropriate treatment can be detailed.

PATHOGENESIS

Concepts of motor control have also evolved. Alpha motor neurons have been thought to be "recruited" into an effort requiring accelerating the application of muscular force in an orderly way—slower, smaller units (the muscle supplied by the motor neuron) before larger, faster units. Nevertheless, several factors can modify data about recruitment order:

1. Changes in movement rates
2. Alterations in afferent input
3. Synaptic connections of motor neurons
4. Membrane properties of recruited neurons
5. Singling out of motor units during strong contractions[25]

The discharge frequency of the individual motor neuron determines whether its motor unit will be a fast twitch or a slow twitch. Because of these and other physiologic variables, muscles in the neck and head areas have special aspects of motor control. Histologic and morphometric investigations of these muscles have tended to support differentiation of these head and neck muscles into two distinct groups based on fiber size.[26] The sternomastoid, for example, is similar to the extensor digitorum longus in fiber size. Motor units with high (slow twitch) and low (fast twitch) oxidative levels mix together to form both muscles,[16] but the low (fast twitch) oxidative level motor unit muscle fibers were larger. Vertebral bodies in the cervical spine have concave-convex, inferior-superior surfaces[5] and have an innate lordosis with the arc pivoting around the C4

49

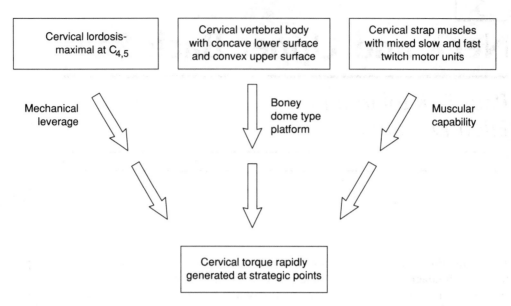

Figure 3-1. Pathogenesis of cervical muscular stress.

level.[33] The lordosis provides cervical strap muscles with mechanical leverage. Cervical strap muscles, therefore, while capable of sustained, postural muscular tone, can also rapidly generate enormous torques directed around strategic areas in the cervical spine. The neck is not just exposed to hyperextension but can actually "wring" itself (Fig. 3-1).

The position of cervical paraspinal muscles influences neurons in more caudal parts of the spinal cord. Rotational torques in the neck of labyrinthectomized cats stimulated L3-6 neurons.[24] Receptors located in cervical paraspinal muscles originate these tonic neck reflexes. Sinusoidal stretching motions of the cervical paraspinal muscles of decerebrate cats have been particularly effective in this regard.[4] The attitudinal posture of the neck, therefore, directly influences neurologic control in the rest of the spinal cord. While these reflexes can be overridden by cerebral hemispheric intervention, most of us adopt a posture without thinking of it in advance. Painful postures in the neck generate direct reactions in the lumbar spine, which compound the patient's impairment. Posture certainly is part of any patient's disability. As locomotion is probably a product of segmental neuronal control with supraspinal reflex modification, lesions in the neck elicit lower extremity reflexive reaction as well as upper extremity function.

DIAGNOSIS

The kinesiologic evaluation must center upon how widespread the lesion is and which structures of neuronal control are affected. Muscles that are end muscles for the nerves supplying them are very strategic:

1. The biceps brachii for the musculocutaneous nerve
2. The deltoid for the axillary nerve
3. The extensor indicis proprius for the radial nerve
4. The adductor pollicis for the ulnar nerve
5. The first and second lumbricals for the median nerve

The only exception to the importance of the end muscle is the anterior interosseous nerve. The pronator teres is not the only forearm pronator, and testing the flexor pollicis longus is much more practical. If these muscles are normal, significant disorders of neurologic control are eliminated. If these muscles are abnormal—weak, flaccid, with reduced deep tendon reflexes—the lesion does not have to be limited to the peripheral nerve.

A more central lesion—plexitis or radiculitis—is still possible. In that event, more than one end muscle will be abnormal. With radiculitis, for example:

1. For a C6 spinal nerve root lesion, the deltoid and biceps muscles will be abnormal.
2. For a C7 spinal nerve root lesion, both the extensor indicis proprius and the flexor pollicis longus will be abnormal.
3. For a C8 spinal nerve root lesion, the extensor indicis proprius and the first lumbrical will be abnormal.
4. For a T1 spinal nerve root lesion, the first lumbrical and the adductor pollicis will be abnormal.

Moreover, these disease processes will be detected earlier using a relatively smaller and more distal muscle. For example, in a C8 spinal nerve root lesion, the extensor indicis proprius is far smaller and is abnormal earlier than the latissimus dorsi. It can be placed in a stretch position and then tested for any weakness. While this maneuver is possible with the latissimus, it is far more difficult. The difference might mean weeks of earlier detection (Fig. 3-2).

Watch out for muscles that have similar names but perform differently. Nomenclature can be deceptive. Two examples illustrate this principle.

1. **Teres major and teres minor.** The teres minor is part of the rotator cuff and therefore participates in external rotation of the humerus. With the deltoid, it is supplied by the axillary nerve. The teres major, however, has actions similar to the latissimus dorsi. It is not part of the rotator cuff but participates in internal rotation of the humerus. Both muscles have similar names and both are ultimately innervated by the C5-6 spinal nerve roots, the upper trunk, and posterior divisions and the posterior trunk of the brachial plexus. Nevertheless, the effects of their contractions are probably equal and opposite.
2. **Extensor carpi radialis longus (ECRL) and extensor carpi radialis brevis (ECRB).** The ECRL usually has a large supply from the C6 spinal nerve root. It is usually active in patients with C6 spinal cord injuries. The ECRL is a radial extensor and is used in tenodesis type orthoses. The ECRB usually has a large supply from the C7 spinal nerve root. It is usually absent in patients with C6 spinal cord injuries and only is powerful in those patients with enough C7 control to use

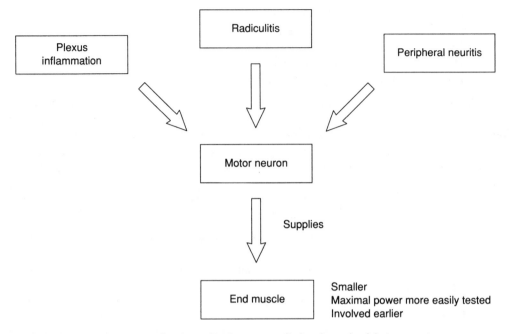

Figure 3-2. The strategic role of muscles that are supplied at the ends of their respective nerves.

their triceps muscles. The ECRB is a power extensor and as such is very active when a fist is clenched. The ECRL and ECRB have similar names and are generally located in the same area of the extensor forearm, and significant kinesiologic differences can be overlooked (Fig. 3-3).

Computed axial tomography (CAT) can be applied to generate 1.5 mm x-ray slices through important areas. Additionally, intravenous contrast material can be introduced to compare vascular tissue with surrounding soft tissue. The data can be projected to sagittal and coronal planes, and the entire procedure lasts approximately 30 minutes.[25] Patients tolerate CAT scanning well, but movement artifacts can obstruct the results. Magnetic resonance imaging (MRI) is supplementing and replacing CAT as a diagnostic procedure partly because it can differentiate as well as CAT and does not require intravenous contrast administration.[26] MRI, however, also is obstructed by movement artifacts and can take a long time to complete.

CAT scanning has become invaluable in the investigation of neck masses.[16] The carotid sheath and its major vascular structures can also be visualized, and CAT scanning has been thought to be of use with obese patients.[26] Between CAT scanning of the neck and the cervical spine, most strategic soft tissue abnormalities in the neck and upper back can be examined noninvasively. Clinicopathologic correlations can then be drawn and differential diagnoses reviewed.

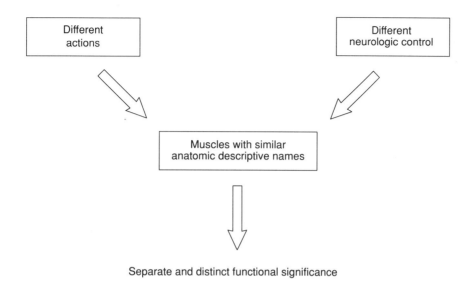

Figure 3-3. Pitfalls of confusing different muscles with similar names.

For the differential diagnoses regarding bursitis, arthritic disorders, and neuritis, please see the first two chapters. Trauma as a source of cervical spine disease is still a specific concern. The frequency of catastrophic cervical injuries connected with college football was so high that in 1975, "spearing tackles" using the helmet facemask were outlawed, and over the next 5 years the offensive tackle was allowed more recourse to arm and hand blocking instead of spear blocking. Subsequently, the incidence of severe head and neck injuries decreased dramatically.[29] A prospective investigation conducted over an 8-year period demonstrated 175 head and neck injuries by 100 performers.[1] Those with pre-existing injuries were twice as likely to be injured. What were these pre-existing injuries? An abnormal physical examination included limitations of ranges of motion and pain or tenderness of axial spinal compression. Radiographic abnormalities included fractures and early osteoarthritic changes. These findings start with high school and college sports. They are enhanced by repetitive industrial motions. Some of the abnormal findings might even be related to wearing seat belts[15] or to specific types of social dancing.[17] The causes of cervical bursitis, osteoarthritis, and neuritis are ambiguous. A careful examination and a full set of x-ray films are merely the first steps. More and more people will also have to have EMG/NCV, CAT with contrast, and MRI studies. A device has been proposed that would allow ambulatory recording of neck rotation over several hours.[18] While the amount of torque generated might be difficult to quantify, one could at least acquire an idea of the qualitative pattern. Combining these results with those of kinesiologic electromyographic (EMG) studies of the cervical paraspinal muscles would allow an evaluation of dynamic movement as well as stable alignment.

Finally, headaches as well as mouth and tongue discomfort can be related to sudden sharp cervical rotary motions.[32] While this neck-tongue syndrome has been noted in patients with osteoarthritis and ankylosing spondylitis, it has also been found in normal young adults. Interconnections between the C2-3 spinal nerve roots and the lingual/hypoglossal nerves are thought to be the basis for this syndrome. These kinds of neural interconnections could easily explain some of the complaints of "whiplash," since, as we have already noted, extreme flexion-extension is commonly associated with cervical rotation.

SOURCES OF CERVICAL PAIN

Pain can emanate from numerous sites in the neck, which makes accurate diagnosis and specific treatment difficult. The type of pain varies as well, from generalized dull ache to sharp focal pain. The tissues that are pain-sensitive include the nerve root, the dura mater, the capsule and synovium of the facet joint, the posterior longitudinal ligament, the anterior longitudinal ligament, the interspinous ligament, and the muscles.[21] These tissues induce pain from being strained, irritated, overused, inflamed, bruised, or infected.

Authors agree that the arthritic changes seen radiologically do not necessarily correlate with physical signs and symptoms.[3,7,11,21] Radiographs are commonly found to demonstrate arthritic changes that are asymptomatic. Arthritis is usually not the sole

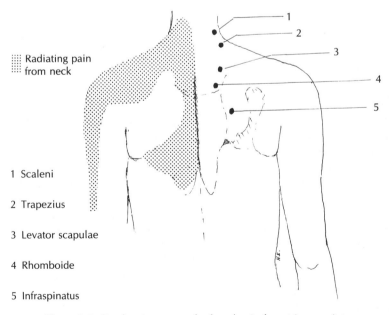

1 Scaleni

2 Trapezius

3 Levator scapulae

4 Rhomboide

5 Infraspinatus

Figure 3-4. Overlapping areas of referred pain from trigger points.

cause of cervical pain, especially when the pain is of acute onset. Arthritic changes take years to evolve, and often cause no pain whatever, despite the presence of associated restriction of movement.

Pain frequently originates from muscle, usually from ischemic changes resulting from tension or overuse, muscle spasm, or tears in the muscle itself. Acute cervical strains are a frequent result of athletic competition.

Cervical strains occur when the muscle-tendon unit is overloaded or stretched, most often involving the sternocleidomastoid, trapezius, rhomboids, erector spinae, scalenes, and levator scapulae.[10] Radiating pain is a familiar concomitant of cervical muscular injury. This pain is frequently felt as headache, especially when the cause of the pain is ischemia, or overuse of the muscles. Alternately, headache can be induced when the periosteal attachment of the muscle is irritated, either through acute injury or sustained traction[3] (Fig. 3-4).

GOALS OF TREATMENT

While the primary reason that treatment is sought is pain, relief of pain is not the singular goal of therapy. Pain is a signal that there is a problem. If the problem is not corrected, pain relief will be temporary at best. Therefore, comprehensive treatment is aimed at promoting healing while restoring normal function. When healing has taken place, normal function should ensue. Function should be maintained within the limits of pain while healing occurs, rather than trying to restore lost function after healing has taken place. A balance should be preserved between enough rest and too much and between pain and mobility. These goals are not easily met, but best results are obtained when there is good rapport and communication between the clinician and the patient.

Education of the patient is no less important. Patients need to know

1. How much pain is too much while exercising
2. How much time to rest
3. The signals of overexertion
4. What to do if those signals appear

No matter how high the quality of care provided, the patient is unsupervised by the clinician most of the time and should be taught responsible self-management. When patients understand the rationale behind what they are asked to do and why it is important, they are more likely to comply with their home programs.

TREATMENT OVERVIEW

Although conservative treatment of many soft tissue cervical injuries is usually tried first, some will not respond to conservative methods. The most severe and potentially disabling cervical injuries require more aggressive intervention. Other injuries that affect the cervical and upper thoracic spine should be treated with at least a trial period of some combination of rest, mechanical and manual modalities, and exercise.

The choice of treatment depends not only on the diagnosis but also on the stage of the problem, acute or chronic. Within the first 48 hours, acute injuries generally respond better to cold than to heat,[30] then later are helped more by heat. The time period defined as "the acute stage" is vague, probably because the severity of the injury tends to determine, to a certain extent, the length of time during which pain is severe, inflammation is great, and motion is most severely restricted by muscle spasm and pain. The appropriateness of treatment also helps determine how long the acute stage lasts, since reduction of pain, spasm, and inflammation by whatever means shortens the time required for healing to take place. Treatment should encourage healing as quickly as possible to prevent the injury from becoming chronic, with long-term pain and disability.

Diazepam has been used for cervical muscle spasm and pain. A well-controlled comparison of diazepam, phenobarbitol, and a placebo, however, did not show significantly positive results.[2] Both drugs are addictive, either physically or psychologically, and produce lethargy and personality alterations. Both remedies are therefore more severe than the disease process itself. They should be used rarely. In fact, in many instances, it is necessary to detoxify patients with neck pain from six or more different and redundant medications. Drugs are usually part of the problem, not the solution, particularly since some patients will collect prescription medications from several physicians. The release of some medicines, which required prescriptions, to the over the counter market has intensified the self-medication problem. Many psychiatric and psychologic centers now offer clinics for alcohol and drug-dependent people, but a massive task of public education awaits regarding the use of pain medication.

Empirically, combinations of different modalities seem to help if they have been customized to individual physical and psychologic needs. The combination selected depends, at least in part, upon the results of the kinesiologic examination. A radiculopathy affects proximal and distal muscles. A more vigorous strengthening effort is required to regain lost function. Hydrotherapy is extremely important. Early in the treatment program, it relieves weight and provides assistive and sedative properties. Later, hydrotherapy provides a medium for gait and for resistive exercises. Muscles can also be overworked, particularly when they are abnormal, and exhaustion must be avoided.

The most commonly agreed upon initial treatment of choice during the acute stage is rest. Rest may be as severely restrictive as absolute bed rest or as mild as "active rest," defined in Chapter 1 as "muscle contraction without load." Rest can be strictly defined, such as reclining in the supine position, or normal activities can be allowed, albeit with certain restrictions. Rest might be at the discretion of the patient or carefully prescribed by the physician or therapist. Finally, rest can include myriad devices, such as collars, pillows, and traction slings, each with its benefits and drawbacks.

Three days of supine, well-positioned, and comfortable bed rest is often sufficient to allow relaxation of muscle spasm and provide significant relief of pain. The collars and static cervical traction devices which are sometimes prescribed for use in bed provide stability and a sense of security for the patient initially, but should be carefully monitored to prevent excessive use. Loss of cervical mobility results from too much of a good thing, making rehabilitation of the patient more difficult when restoration of mobility is begun. A period of 7–10 days[7] of wearing a cervical collar when the patient is upright and active is usually sufficient to restrict painful motion and reduce muscle splinting. Thereafter, the wearing time is gradually reduced, and eventually the collar is worn only when support is most needed to avoid pain. Collars are only used as long as absolutely necessary, or muscle atrophy, fibrosis, and psychologic dependence may result. In some patients, cervical collars cause increased pain and splinting. Sometimes collars are legally considered driving obstructions, depending upon state regulations. Therefore, collars should not be forced upon those patients who, for whatever reason, are unable or unwilling to tolerate them. If these patients drive with their collars on and are involved in an accident, their clinicians could also be liable.

As soon as possible, within the limits of pain, patients are taught gentle, active limbering movements of the neck, shoulders, and arms. Correct cervical posture, avoiding the forward head, is taught at the same time. Progressive rotation to the left and right is done, while avoiding flexion and extension initially.[3] In the early days following injury, these exercises are best accomplished after the application of modalities to reduce pain and spasm, such as ice massage or vapocoolant spray. Heat, if better tolerated, is often effective in reducing pain and muscle spasm, permitting increased range of motion. TENS could also be used to reduce pain and allow freer movement.

MANUAL THERAPY AND MANUAL TRACTION

While there is still controversy about the applicability of manual therapy in the treatment of cervical problems, manual traction and other techniques of manual therapy are frequently of value in the treatment of cervical soft tissue injuries. Especially useful in the acute stage, manual traction provides the advantage of greater control of the position of the head and individualized grading of the quantity and duration of the traction.[3] Manual traction can be used in conjunction with range of motion exercises, providing distraction and stretching of tight muscles and joint structures of the vertebral column. Manual traction is successful in helping decrease pain, muscle spasm, stiffness, and in reducing joint compressive forces.[20] Manual therapy, including manual traction, has a stimulation effect on proprioceptive neural pathways and possibly on complex reflex pathways, which affect pain.[27] Manipulation can provide dramatic relief of pain, spasm, and restriction of motion. Stretching of muscles results in their reflex relaxation, thereby reducing muscle spasm. There is also a reported increase in serum beta-endorphins after manipulation.[8]

Contraindications to the use of manual therapy include bone tumors, severe osteoporosis, hemorrhage, and sepsis.[27] Manipulation and to a certain extent traction can be associated with the onset of stroke symptoms.[14] One of the most serious side effects of manipulation of the cervical spine is vertebral artery occlusion, which can, in the most severe case, cause death.[8] Careful pupillary monitoring prior to manual therapy and use of gentle, precise techniques with a minimal thrust can avoid damage to the vertebral arteries and spinal cord.[27] Knowledge, experience, and judgment add to the safety of the manual therapy techniques. But neither manual therapy nor traction should be applied before the patient has had a thorough physical examination and a full set of cervical spine x-ray films. Home traction applications can also be associated with augmented cervical muscular tension while setting up the apparatus and increased discomfort during treatments.[31] Since the myoelectric activity of the upper trapezius muscles is not altered in

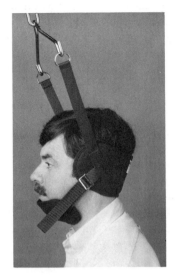
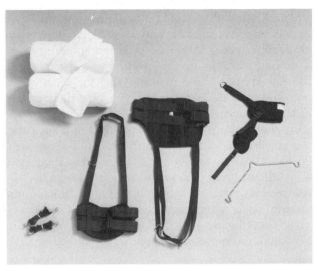

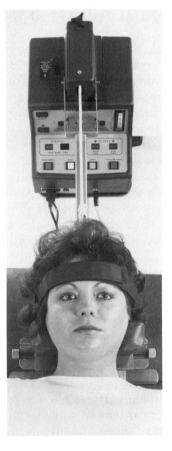

Figure 3-5. Cervical traction devices.

patients with neck pain receiving traction, it might not be a very efficient modality. It would certainly be difficult to create a blind, controlled investigation of either traction or manual therapy.

MOTORIZED CERVICAL TRACTION

Intermittent traction is an important modality in the treatment of cervical pain and injury. Intermittent traction helps improve local circulation, helps prevent the formation of adhesions, and possibly aids in reducing muscle spasm,[9] although there is some question about the induction of muscular relaxation during motorized traction.[12] In the acute stage following cervical injury, however, intermittent cervical traction might cause further injury by fighting the spasm that is splinting acutely inflamed tissues.[3] It is contraindicated in acute sprains for the first 3–5 days and in conditions in which movement is undesirable.[20]

That traction does benefit patients with cervical pain is well established by clinical experience. There is, however, no consensus among clinicians as to technique. Recent literature suggests various positions of the neck for maximum advantage, which might reflect that position significantly affects which joints and tissues are being distracted. Therefore, the position chosen depends on the aim of the treatment. Forces of between 10 and 45 pounds are suggested for separation of the joints of the cervical spine. As little as 10 pounds is required to produce separation at the atlanto-occipital joints,[23] while much greater forces are required to provide distraction in, for example, some spondylitis patients. Many types of halters are on the market, some of which, especially when inappropriately applied, cause undue pressure on the temporomandibular joint with resulting pain and dysfunction. Several newer devices are also available that avoid such pressure. One is a head halter that wraps around the occiput and forehead, avoiding the jaw. The other is an adjustable device that contacts the back of the patient's neck just below the occiput and is stabilized by a forehead strap. Some patients are reported to have as much difficulty tolerating this device as they do tolerating the halter.

Care and judgment are required to appropriately provide traction for those patients for whom it is indicated. No one device, position, or weight can be recommended without reservation. Different devices are available to provide traction, intermittent or static, at home or in the clinic. Cervical traction is recommended, though, alone or in combination with other modalities, by many sources[3,7,9,11,13,19,20-23,34] (Fig. 3-5).

OTHER MODALITIES

Generally one modality alone, no matter how effectively used, is not enough to provide complete pain relief and restore function. Modalities appropriately used in combination, however, relieve symptoms and encourage mobility early in the injury. This early intervention is important because treatment goals include prevention of excess inflammation, swelling, and subsequent fibrosis, which promote reduced mobility.

Modalities are chosen for their effects, not bunched together perfunctorily. Clinical decision-making is based on the stage of the injury, whether the pathologic condition is acute or chronic, the amount of inflammation and swelling present, and the tissue that has been injured, when discernible. When spasm is the most severe symptom, for example, modalities are chosen that are effective in reducing spasm. If pain is restrictive, reducing it will further the goals of treatment by allowing increased mobility, so modalities specific for pain relief are utilized until the pain is controlled. When pain control has been achieved, other treatment goals can be pursued. Further, if one modality proves ineffective after two or three treatments, consideration should be given to changing the treatment program.

There are numerous ways to apply cold. Cold is usually recommended during the early and most acute stage of an injury,[3,9,13,21,22,30,34] which is generally defined as the first 24–48 hours. There are, however, other indications for the application of cold, even when the condition is well past the initial 48 hours after injury. In the case of fibrositis, or "myofascial pain syndrome," for example, cold is applied at any stage in treatment. The effects of cold (discussed in Chapter 4) include reduction of edema and bleeding via vasoconstriction, which reduces hyperemia; reduction of pain; and alleviation of muscle spasm.

Much depends upon the **goal of treatment,** i.e., pain relief (skin anesthesia), reduction of swelling or bleeding, or reduction of spasm. Cold can be applied with a gel-filled pack, ice cubes or crushed ice, ice massage, frozen bandage-type devices, vapocoolant spray, or slush packs of water and alcohol. The most comfortable and well-tolerated method for the patient is probably not the easiest for the clinician. Experience in trying alternatives allows the clinician to determine which of the methods is most often advantageous. Some patients gain greatly from ice massage, and some cannot tolerate it. There are, however, few who cannot tolerate vapocoolant spray, at least in small amounts. If the ice modality is to be applied to the skin and left there for 15–20 minutes (the time required for cooling of the tissues) a moist cloth is placed between the patient's skin and the ice to prevent freezing and damage to the skin (Fig. 3-6).

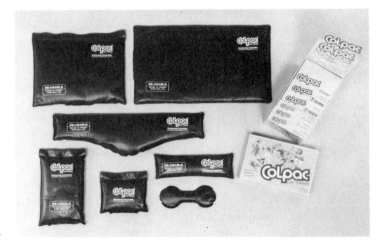

Figure 3-6. Cold packs.

Some form of heat is frequently used, usually after the first 24–48 hours, in the treatment of cervical pain and injuries.[3,7,9,11,20-23,34] The effects of heat (discussed in Chapter 4) include reduction of pain and muscle spasm, increased circulation thereby reducing ischemia, reduction of joint stiffness, and increased tissue distensibility. Heat should be applied with the patient in a comfortable, supported position to promote maximal relaxation. Heat is useful as a precursor to, or during the application of, other techniques, including traction (manual or mechanized) and exercise. As with the application of any other modality, the choice of a heat modality should depend upon the specific goal of treatment and the tissue being treated. For example, if the tissue treated is the deep musculature of the neck, moist heat packs will not reach deeply enough to directly heat those muscles. Ultrasound would in that case be the treatment of choice. The two modalities are often combined, however, to warm both the superficial and the deeper tissues and provide reflex relaxation. Diathermy, while still somewhat controversial, is considered relatively superficial in its penetration but is often effective for heating superficial tissues.

Other electrotherapeutic modalities (discussed in Chapter 14) provide additional benefit over simply applying heat or cold, especially in the area of pain relief.

Electroacupuncture, TENS, and helium-neon laser are valuable adjuncts to the treatment regimen, especially prior to or during exercise. The ability of these devices to reduce pain and break up the pain-spasm cycle, thereby allowing increased mobility, is of great value. Electrical muscle stimulation is often useful for its ability to increase circulation and reduce spasm and ischemia, thereby reducing pain and enhancing mobility. Some patients, however, find electrical stimulation irritating and even exacerbating to their symptoms.

TENS is sometimes difficult to apply to the neck when the location of the pain is more cephalad, as the hairline presents problems with electrode adhesion. Nevertheless, crossed-circuit techniques with alternate electrode placements frequently eliminate the necessity for placing electrodes at the hairline. New products are placed on the market constantly, and this problem could be eliminated by one or more of them.

There are many new ways of delivering TENS, as the pulse width, rate, and amplitude are varied (Table 3-1). New wave forms and techniques of varying them are being introduced constantly by the manufacturers of TENS units, and no definitive research proves one technique more effective than another. Patients often present with unique problems, which may or may not respond to a particular mode of treatment, wave form, or protocol. However, the most important facet of TENS application is electrode place-

TABLE 3-1. TRANSCUTANEOUS ELECTRONIC NERVE STIMULATION (TENS) MODES

	Rate	Pulse Width	Intensity
Conventional	High	Narrow	Low
Acupuncture-like	Low	Wide	High
Brief, intense	High	Wide	High
Pulse-train	Both	Varies	Either

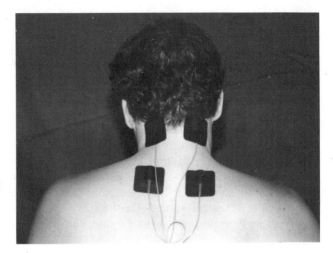

Figure 3-7. One effective TENS electrode placement at the neck. Electrode placement is often effective with left occipital, right trapezius trigger point in one circuit, and right occipital, left trapezius trigger point encompassed in the other circuit. This creates a "crossed circuit" effect.

ment, which is more effective when two circuits are used and the electrodes placed in a criss-cross fashion (Fig. 3-7).

The acutal location of the electrodes depends upon the site of the painful tissue(s). Guidance of electrode placement by acupuncture and trigger points, as well as dermatomes, myotomes, and sclerotomes, increases the effectiveness of the treatment. The more of these "landmarks" incorporated within the circuits, the better the pain relief.

Iontophoresis and phonophoresis can deliver medication to the affected tissue more specifically than can oral ingestion, and with fewer side effects. The medication penetrates from 5–10 cm with phonophoresis, depending on the vibrational frequency.[22]

Biofeedback is of particular benefit for patients with postural or stress-related pain in the neck and upper back, since these sites are most frequently affected by abnormal tension levels. Biofeedback is combined with relaxation exercises, as well as the specific exercises required to stretch or strengthen those muscles determined to be affected.

Massage is valuable for its ability to produce sedation and relaxation and to dissipate accumulated fluids.[3] Because of the increased demand for productivity in many clinics, the time required to provide massage is not as easily dispensed, although the benefits can enhance the treatment significantly. Trigger point and acupuncture point massage, in addition to soft tissue massage, produce further gains in reduction of pain and spasm and improvement of mobility.

CERVICAL DISC

Successful treatment of a herniated nucleus pulposis (HNP) requires a comprehensive regimen, including reduction of the compressive forces through reducing muscle spasm and splinting, refraining from positions that increase intradiscal pressure, and bracing.[23] The first goal, reduction of splinting and spasm, is achieved through the use of carefully prescribed rest, modalities (cold or heat and pain-reducing techniques), medication, and the avoidance of tension-producing situations. Cervical collars and cervical pillows are used to prevent the movements that can cause the protrusion to worsen. Patients are carefully instructed to avoid movements that cause increased pain. Traction is used to reduce the herniation. Traction should be used with caution; while it can be sustained or intermittent, it should be relatively short in duration (10 minutes or less). Saunders emphasizes positioning without flexion of the cervical spine.[23] Afterwards, use of a collar and exercises will help maintain correct positioning. The chin-in, or chin tuck, exercise is usually taught first, and progresses to neck extension exercises as tolerated. Flexion should be avoided.

FACET JOINT

With facet joint impingement, pain radiates down the involved extremity, but true neurologic symptoms such as muscular weakness or paresthesia should not be present. There is usually pain or reduced mobility or both, with no known trauma or minor trauma as the precursor. Sometimes the injury occurs during sleep, and the patient wakes up with a stiff neck.

Dramatic improvement of symptoms of facet joint impingement can result from manual therapy or manual or mechanical traction.[23] Pain and spasm should be reduced prior to any passive elongation of the soft tissue if spasm is moderate or severe. If spasm is mild and the patient is able to relax prior to the manual therapy, moist heat after the treatment is vey helpful in reducing the feeling of stiffness that sometimes follows a successful reduction of facet joint impingement.

When there is a history of trauma, the expectation is that the joint has been damaged. Local treatment is given to help reduce inflammation, pain, and spasm while the injury heals. Rest, the use of a collar, and restriction of movement to within the pain-free range may be necessary. In the acute stage, ice is applied, if tolerated. Gentle heat can be used if ice is uncomfortable. Modalities are added as required to help encourage healing, reduce pain, and spasm, and allow improved mobility. The patient should be encouraged to gradually but continually strive toward increasing the range of motion while not causing increased trauma. Patient education often makes the difference between normal function and hypomobility once healing has taken place.

SOFT TISSUE INJURY

This category includes muscular strain, ligamentous or joint capsule sprain, and contusion. Inflammation is often present, with concomitant pain, spasm, and reduced mobility. Treatment is aimed at reducing inflammation, pain, and spasm; encouraging healing; and restoring normal function. Depending on the severity of the symptoms, rest, protection, and use of modalities are provided until pain begins to ease. Soft tissue injuries often show significant improvement within 1 week.[19] The clinician should pay attention to restoration of movement and function **while healing takes place** and initiate appropriate exercise programs. Careful instruction and supervision is necessary initially, and patient education is helpful in ensuring compliance with a home program. The addition of printed information, a list of exercises, or other simple illustrated instructions can significantly increase compliance.[6]

Painless, gentle active, or active-assisted range of motion is initiated after as little as 24 hours of cryotherapy, applied 15–20 minutes every 1–2 hours, or ice massage for 5–10 minutes every 1–2 hours.[28] Exercises should not, however, cause any exacerbation of symptoms. As healing progresses, isometrics are added to prevent disuse atrophy, especially when a cervical collar is worn. Weaning from the collar starts as early as the third day of use. Patients should be carefully instructed to avoid dependency on the collar.

CERVICAL SPONDYLOSIS

Although radiographic evidence of bony changes in the cervical spine do not necessarily correlate with pathologic conditions,[3,7,11,21] cervical spondylosis is a common cause of neck and arm pain. This pain is often the result of an insult, such as the flexion-extension injury that occurs in auto accidents,[11] or manifests in an older individual who might have had a long history of recurrent cervical pain. Cervical spondylosis is believed to result in part from disc degeneration and the subsequent formation of osteophytes. Thickening of the soft tissues around the vertebral joints also plays a part, and ankylosis results. These changes are usually asymptomatic, but they also can cause narrowing of the spinal canal and impingement on the intervertebral foramen. The nerve root and the vertebral artery thus become compressed.

Treatment consists primarily of cervical traction, whether manual, intermittent motorized, or a home unit, depending upon the specific comfort and needs of the patient. Symptomatic relief of pain and spasm with modalities such as heat, massage, and TENS are often helpful. Careful attention should be paid to the evaluation of cervical mobility. Exercises are prescribed on an individual basis depending on the limitations present. As Saunders points out, when the facet joints are in maximum extension, the axis of rotation is located at the joints themselves, and further extension causes widening of the intervertebral foramen and disc spaces.[23] Therefore, a trial of both flexion and extension exercises is indicated. The exercises that provide relief of symptoms are the ones that should be continued.

Patients also require advice about those activities that aggravate their symptoms, such as lifting, driving, and sometimes sleeping positions. Cervical pillows provide some nocturnal relief, and cervical collars can help initially.

FIBROSITIS (MYOFASCIAL PAIN SYNDROME)

Fibrositis, or the newer term, myofascial pain syndrome, is a common cause of cervical and upper thoracic pain. The neck and upper back area is particularly susceptible because the area is prone to mood overlay and poor positional habits. Pain might be referred, but there are no true neurologic signs, and trigger points are present.[21] Treatment in the cervical and upper thoracic area consists of the spray-and-stretch technique, massage, and modalities, as described in Chapter 1. Relaxation techniques, self-hypnosis, and biofeedback are frequently useful in reducing general body tension.

As with other cervical dysfunctions, the restoration of range of motion with stretching of tight structures and strengthening of weak ones is of extreme importance. As the pain diminishes through treatment, the exercise program is increased gradually, within the limits of pain, with slow, gentle active stretching and isometric or progressive resistive exercises for strengthening.[13]

Education is vital to help reduce patients' anxiety about the condition, to teach them how to avoid stress, and to provide information as to how to deal with the problem in the future. Instructions include changing work habits, taking more frequent breaks, stretching, exercises, and those modalities patients may be able to use on their own.

POSITIONAL PAIN

Cervical pain is frequently caused by sustained contraction of cervical muscles. Pain is probably a result of the accumulation of muscle waste products. The same can be said of tension headaches, which are caused by sustained traction and irritation of the periosteal attachment of the cervical muscles at the base of the skull.[3]

Significant relief is often obtained through most of the modalities at the disposal of the physical therapist.

The goals of treatment are to increase circulation, decrease spasm and pain, and restore full mobility. Poor posture, however, is often the cause of the pain, and careful attention should be paid to correction of postural imbalances. Forward head and round shoulders are common. Exercises are provided to stretch the tight structures and strengthen the weakened muscles. A clavicle splint, "posture" bra or shoulder straps can aid in reminding the patient to keep the shoulders from drooping. Scapular adduction exercises and stretching of the pectoral muscles usually reverse the rounded shoulders problem. Chin-tuck and cervical extension exercises are particularly important in reversing the forward head posture (Fig. 3-8). Patients also need to be educated as to the cause of their pain and how to prevent its recurrence.

EXERCISES

Exercises for cervical and upper thoracic pain are prescribed only after individual evaluation, muscle testing, and observance of indications and contraindications that apply to the individual under treatment. Exercises should not be given arbitrarily as a group but are judiciously chosen for the treatment program. Exercise programs should include a variety of types of movements, incorporating flexibility and relaxation techniques, as well as strengthening, whenever possible.

The usual progression for an exercise program is to start with gentle, relaxing movements with a warm-up period, gradually increasing the demand, and then reversing the sequence with a cool-down period. Strengthening exercises should cause mild exertion but not fatigue. And finally, patients must be carefully instructed in the correct way to do each exercise and how to recognize signs of overexertion. Compliance with exercise programs is significantly increased when instruction is accompanied by printed directions.[6]

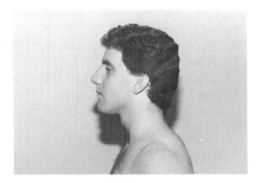 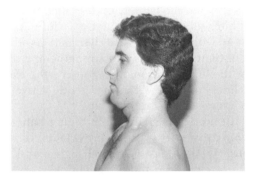

Figure 3-8. Chin Tuck. This exercise can be done upright or supine. The patient is instructed to 'make a double chin' by pulling the chin closer to the neck; no flexion or extension is permitted. The normal cervical lordosis is reduced during this exercise, when correctly performed.

Figure 3-9. Cervical Flexion. The patient is instructed to forward flex the neck while depressing the shoulders, stretching the muscles at the back of the neck. The movement is never forceable, but rather produces gradual elongation of tight structures.

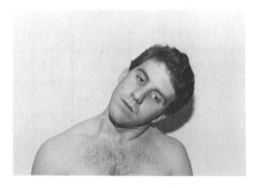 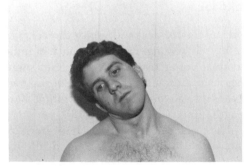

Figure 3-10. Lateral Side-bending. The patient is instructed to avoid rotation during this exercise, gently stretching the lateral cervical structures.

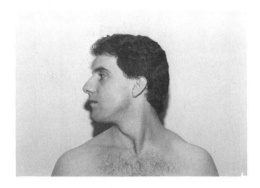 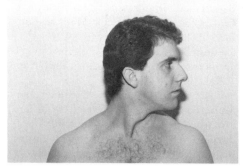

Figure 3-11. Cervical Rotation. Avoiding flexion or extension, the patient rotates the head as far as possible to each side. A gentle stretch is the intended result.

REFERENCES

1. Albright JP, et al: Head and neck injuries in college football: An eight year analysis. *Am J Sports Med* 1985; 13:147.
2. Basmajian JV: Reflex cervical muscle spasm: Treatment by diazepam, barbitol, or placebo. *Arch Phys Med Rehabil* 1983; 64:212.
3. Cailliet R: *Neck and Arm Pain*, 2nd ed. Philadelphia, F. A. Davis Co., 1981.
4. Dutia MB, Hunter MJ: The sagittal vestibulocollic reflex and its interaction with neck proprioceptive afferents in the decerebrate cat. *J Physiol* 1985; 359:17.
5. Fielding JW: Cineroentgenography of the normal cervical spine. *J. Bone Joint Surg* 1957; 39:1280.
6. Glossop ES, et al: Patient compliance in back and neck pain. *Physiotherapy* 1982; 68:225.
7. Gurdjian ES, Thomas LM (editors): *Neckache and Backache*. Springfield, Ill., Charles C. Thomas, 1970.
8. Haldeman S: Spinal manipulative therapy in sports medicine. *Clin Sports Med* 1986; 5:277.
9. Hirsch C, Zotterman Y (editors): *Cervical Pain*. New York, Pergamon Press, 1971.
10. Jackson DW, Lohr FT: Cervical spine injuries. *Clin Sports Med* 1986; 5:373.
11. Jenkins DG: Differential diagnosis and management of neck pain. *Physiotherapy* 1982; 68:252.
12. Jette DU, et al: Effect of intermittent, supine traction on the myoelectric activity of the upper trapezius muscle in subjects with neck pain. *Phys Ther* 1985; 65:1173.
13. Kaplan PE (editor): *The Practice of Physical Medicine*, Springfield, Ill., Charles C. Thomas, 1984.
14. King JC, Thukral S: Stroke associated with neck manipulation: A case study and review of the literature. *Arch Phys Med Rehabil* 1984; 65:660.
15. Lesoin F, et al: Has the safety-belt replaced the hangman's noose? Lancet 1985; 1:1341.
16. Martinez CR, et al: Computed tomography of the neck. *Radiographics* 1983; 3:9.
17. McBride DQ, et al: Break-dancing neck. *New Engl J Med* 1985; 312:186.
18. Murphy C, et al: Continuous recording of neck rotation: Preliminary observations. *Spine* 1984; 9:657.
19. Peterson L, and Renstrom P: *Sports Injuries*. Chicago, Year Book Pub., Inc., 1986.
20. Prentice WE: *Therapeutic Modalities in Sports Medicine*. St. Louis, Times Mirror/Mosby, 1986.

21. Raj PP (editor): *Practical Management of Pain*. Chicago, Year Book Medical Pub., Inc., 1986.
22. Santiesteban AJ: The role of physical agents in the treatment of spine pain. *Clin Orthop* 1983; 179:24.
23. Saunders HD: *Evaluation, Treatment and Prevention of Musculoskeletal Disorders*. Minneapolis, Viking Press, Inc., 1985.
24. Suzuki I, et al: Body position with respect to the head or body position in space is coded by lumbar interneurons. *J Neurophysiol* 1985; 54:123.
25. Swartz JD, et al: High resolution computed tomography of palpable masses of the neck and face. *Radiographics* 1983; 3:645.
26. Swartz JD, et al: High resolution computed tomography, Part 1. Soft tissues of the neck. *Surgery* 1984; 7:73.
27. Swezey RL: The modern thrust of manipulation and traction therapy. *Sem Arth Rheum* 1983; 12:322.
28. Teitz CC, Cook DM: Rehabilitation of neck and low back injuries. *Clin Sports Med* 1985; 4:455.
29. Torg JS: *Athletic Injuries to the Head, Neck and Face*. Philadelphia, Lee & Febiger, 1982.
30. Vinger PF, Hoerner EF (editors): *Sports Injuries*, 2nd ed. Littleton, Colorado PSG Publishing Co., Inc., 1986.
31. Waylonis GW, et al: Home cervical traction: Evaluation of alternative equipment. *Arch Phys Med Rehabil* 1983; 63:388.
32. Webb J, et al: The neck-tongue syndrome: Occurrence with cervical arthritis as well as normals. *J Rheumatol* 1984; 11:4.
33. Weber R: Dysfunction of the neck. In Kaplan PE (editor) *The Practice of Physical Medicine*. Springfield, Ill., Charles C. Thomas, Pub., 1984.
34. Zohn DA, Mennell JM: *Musculoskeletal Pain: Diagnosis and Physical Treatment*. Boston, Little, Brown and Co., 1976.

4

Low Back Pain

Paul E. Kaplan and
Ellen D. Tanner

Low back pain will not increase mortality rates, but it generates as much morbidity as any other single cause of musculoskeletal disability. Many books have been written concentrating upon low back pain evaluation and treatment because of its huge social and economic costs. In the 1970s, in Washington state, low back pain destroyed over one million production days each year.[27] Fourteen billion dollars was spent treating low back pain nationally in one year.[1] Low back pain was, in fact, the most expensive health difficulty for people 30 to 50 years old—people at their peak of earning potential.[31] Moreover, low back pain produces a great deal of severe disability. In California, 24% of compensation cases attributable to low back pain contributed 87% of the total costs.[35] In Washington state, the 4.5% of people with low back pain disability were associated with 36% of the costs—63.5 million dollars.[27] Those 4.5% disabled employees rarely returned to work if they were out for longer than 4 months. Major factors in total direct low back injury costs included temporary disability payments (45% of the total industrial back injury costs), permanent disability awards (43%), and direct medical costs (33%).[36]

The same type of studies performed more recently have similar results. Retrospective investigations in Washington state show that in 1979–1980, back injuries produced 41% of total injury costs with but 19% of the claims.[67] Accidents tended to be more expensive than injuries obtained while handling or lifting weights.[3] Back injuries were also more common during the day shift. These figures are significant. They support the existence of a widespread epidemic of low back failure and dysfunction. Not all of the population is affected. Nonetheless, those affected might otherwise have been productive, energetic, hard-working people at the prime of their careers. If an infection had had these results, a huge outcry would have motivated a drive to cure the problem. That effort has not materialized because there seems to be no one solution that will work in all cases.

Who is at risk? A prospective study of acute back pain patients revealed that those who developed chronic back pain were anxious, had deeper pain sensations, and were not as active as those who did not become chronically affected.[49] Age might play a role. At least one study suggests that older employees have a higher risk of generating high-cost back injuries.[4] That same study also demonstrates that men are at higher risk than women.

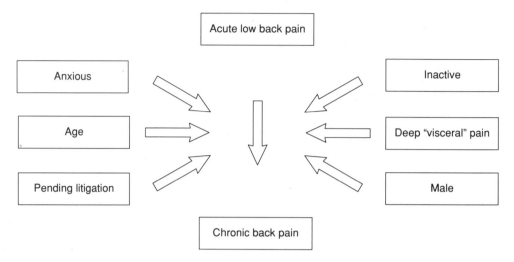

Figure 4-1. Progression from acute to chronic back pain.

Litigation also might make a difference. Patients with pending litigation scored higher on the Minnesota Multiphasic Personality Inventory (MMPI) hypochondriasis and hysteria scales.[70] They also did not improve behaviorally as much as patients without pending litigation. Each of these factors in themselves is probably not decisive, but the combination of several makes chronic back pain much more likely (Fig. 4-1).

PATHOPHYSIOLOGY

What stress drives the progression of back pain? A method of L3 load calculation has been correlated against intravertebral disc pressure data and EMG.[62] During torso or squat lifting, the compressive load on L3 is strong enough to produce vertebral structural failure.[22] Massive crushing forces are routinely generated within the structures of the lumbar spine during activities of daily living. When that structure is modified and weakened by arthritis, it fails more rapidly.

How does the lumbosacral spine handle these large and compressive forces? The bony structure itself helps. The lumbar facets are sagitally oblique,[40] exactly mirroring the orientation of the tough, ligamentous sacroiliac joints. As these facets determine the directions in which the vertebral bodies transmit weight, ligaments spanning the sacroiliac joint distribute the weight over the massive pelvic bony architecture.

There is a second method of stress relief. The lumbar plexus and sacral plexus are not dominated functionally by cords or trunks. Instead, the anterior and posterior divisions are paramount.[29] For nearly every nerve derived from the anterior divisions of the lumbar and sacral plexus, a correlate nerve arises from the posterior division. For example, for the femoral nerve (posterior division), the obturator nerve arises (anterior division). For the common peroneal nerve (anterior division), the tibial nerve arises (posterior division). In mammals, these anterior/posterior divisions translate generally into flexor/extensor muscular control. Neural control is therefore functional, straightforward, and kept at the spinal level. In primates, tibial and femoral nerve torsion has complicated the functional performance. As weight transfer functions of the lower extremities have remained intact, anterior/posterior divisional control is activated, spreading to involve the lower thoracic spinal nerve roots. When a usual lifting effort is mounted, extensor (posterior division), paraspinal muscle control increases. As the force that has to be transferred increases, so does anterior divisional flexor tone. Consequently, the anterior abdominal muscle tone increases, converting the trunk to a rigid cylinder. The apron of a lumbosacral corset with stays mimics this action so that this type of corset can reduce intradiscal pressure by 30%.[34] When a patient is sitting, intradiscal pressure rises, but this progressive divisional supply (and abdominal musculature) is not activated. The increased pressure therefore is only disbursed by the bony architecture, aided by the frame of the chair (Fig. 4-2).

Back pain, therefore, represents a failure of either major mechanism to adequately handle stress. Many factors can intervene. Some have already been mentioned in the discussion about progression from acute to chronic low back pain. Failure is relative to cir-

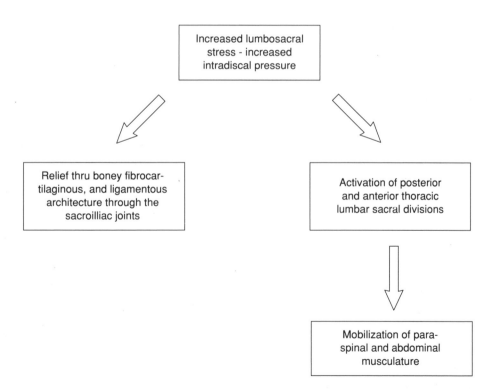

Figure 4-2. Relief of lumbosacral vertebral stress.

cumstances. A football athlete might experience a vicious collision and not feel the pain. A worker, however, could partially lift an item and experience severe pain. Pain cannot be understood out of its milieu. Toward this end, an Activity Discomfort Scale has been devised as a method of predicting response to nonsurgical treatment.[71] Eighteen daily activities are measured, including driving, bending, and walking. The scale provides a profile of pain behavior in patients with low back pain.

To what extent does axial rotation influence weight loading across the L5-S1 vertebral joint? A recent study suggests that axial rotation in normal subjects places extra stress across the L5-S1 vertebral joint.[55] The lordotic lumbar curve was thought to place more stress at the lumbosacral area, but this was not specifically evaluated. Hour to hour modifications of the lumbar lordosis probably occur, thanks to segmental lumbar spinal adjustments. It is not usual for the normal subject to spend much time thinking about posture. Certainly, many normal subjects have marked lumbar lordosis and never have back pain. Coupled with spinal rotation, lateral bending, or axial rotation, lumbar lordosis does not place significant extra stress across the normal lumbosacral area. Studies of spinal motion in patients with chronic back pain confirm that discrepancies between results of straight leg raising and pelvic mobility are often present in patients with deficient motivation.[45] During straight leg raising, the pelvis should be flattened against the plinth, the knee should be extended and the contralateral thigh hyperextended.

DIFFERENTIAL DIAGNOSIS

Lumbosacral plexitis has been noted in patients with metastatic carcinoma,[57] with neurofibroma,[2] and after neonatal injuries.[24] These diagnoses rely upon ultrasound, CT scanning, somatosensory evoked potentials, and EMG. Can the kinesiologic evaluation provide clues as to the presence of a lumbosacral plexitis? Not entirely. A high index of suspicion is a prerequisite. If lumbar and sacral plexes are organized mainly upon anterior/posterior divisions as noted above, radiculitis should affect muscles supplied by both divisions. On the other hand, plexitis can involve only the anterior or the posterior division. Because there is no nerve directly analogous to the long thoracic nerve that can be specifically and easily evaluated by a kinesiologic examination, there is no sure way of functionally differentiating a plexitis, affecting both anterior and posterior divisions, from a radiculitis of the most proximal portions of the sciatic nerve, from an incomplete L5 radiculitis, for example. These conclusions, therefore, must also apply to any functional tests, depending upon performance of these neuronal tracts—nerve conduction, evoked potential evaluation, EMG. In other words, should these involved nerves be activated in any way, the same proximal neural pathways will be affected by a widespread plexitis and an incomplete radiculitis. For example, if EMG demonstrates lower motor neuron abnormalities in the lumbar or sacral paraspinal musculature and abnormalities are noted in appendicular musculature, plexitis of both anterior and posterior divisions additional to the radiculitis cannot be ruled out. To date, the diagnosis of lumbosacral plexitis has been rare. Ultrasound and CT scanning will change this situation by providing much more accurate, clinicopathologic correlations. In any case, a functional examination—and electrodiagnosis—can still differentiate between complete radiculitis and plexitis involving only one of the two divisions.

Even with newer noninvasive tests of anatomic appearance, the evaluation of radiculitis is still dependent upon confirmation by EMG.[38] Nonetheless, EMG is fundamentally another way of documenting a kinesiologic examination. Kinesiologically, the diagnosis of lumbosacral radiculitis is dominated by the kinesiologic correlations between functional control of the upper versus that of the lower extremity.[29] For nearly every nerve or muscles of the lower extremity, there is a correlate in the upper extremity. Consequently, overlapping myotomal investigation establishes spinal nerve root involvement, as it does in the upper extremity. For example, the serratus anterior has a match in the lower extremity—the obturator internus. In the shoulder, the scapula is relatively free floating. The serratus anterior helps direct its position. Evolutionary pressures have necessitated a powerful muscle. The iliac bone—the correlate of the scapula—is fixed strategically distributing weight, as mentioned above. The obturator internus has not been the product of strong evolutionary trends. It is weak and relatively insignificant. As it contracts, its effect is as if the rib cage were mobile and scapula fixed when the serratus anterior contracted. For the clavicle, substitute the pubis; for the ischium, substitute the coracoid process. They look dissimilar because the upper extremity does not have weight-bearing functions. For the femoral nerve, substitute the radial nerve, and for the medial and lateral plantar nerves, substitute the median and ulnar nerves. For the musculocutaneous nerves, substitute the obturator and saphenous nerves. This last relationship generates the same dermatomal-myotomal gap in the foot as in the hand. In the hand, the dermatome of the thenar eminence is usually C6, and the underlying intrinsic muscles are C8 and T1. In the foot, the skin of the great toe is L4, but the underlying intrinsic muscles are S1 and S2. For the vasti, substitute the triceps. For the gluteus maximus, substitute the supraspinatus and infraspinatus. Often, the three gluteal muscles are thought of as one, but their actions are quite different. As the latissimus is active in humeral extension, so the gluteus maximus is active in femoral extension. As the supraspinatus and infraspinatus are rotator cuff muscles of the shoulders, so are the gluteus medius and minimus hip rotator cuff muscles. Here again, the importance of the hip rotator cuff has not become as significant as the shoulder rotator cuff because of differing functions of the hip and shoulder. Mainly, the hip rotator cuff muscles are notable for their abduction and external rotation actions.

Upper and lower extremity correlations are applied to pinpoint the radicular distribution of myotomal abnormalities. Just as C5 is differentiated from C6 or C7 spinal nerve root irritation, so can L4 be separated from L5 or S1 spinal nerve root irritation. For example, in L4 radiculitis, the vasti and adductor muscle groups should be relatively weak and flaccid. But L5 should involve the peroneus longus and brevis and the flexor pollicis longus. S1 radiculitis should affect intrinsic muscles of the foot. Usually, these myotomes can be evaluated separately. "L4 radiculitis" is often a careless conclusion. As far as possible, that famous razor should be used to cut away extraneous diagnosis (Table 4-1).

Unlike the cervical vertebral bodies, there are no pseudojoints in the lumbar area. Moreover, unlike the cervical area, the lumbar vertebral body surfaces are relatively flat and the posterior longitudinal ligament tapers. The chance of having significant anterior-posterior shear forces are greater than in the cervical area. Consequently, the possibilities of disc protrusion and myelopathy are much higher in the lumbar area than the cervical area. While polyradiculitis could be a major source of chronic pain in the cervical area, transverse myelitis is always a possibility in the lumbar area.

TABLE 4-1. UPPER AND LOWER EXTREMITY CORRELATIONS

	Bone		Muscles	Nerves
Upper extremity	Scapula Coracoid process Clavicle	Humerus Radius Ulna	Triceps Palmaris longus Latissimus dorsi	Radial Median Ulnar
Lower extremity	Ilium Ischium Pubis	Femur Fibula Tibia	Vasti Gastrocnemius Gluteus maximus	Femoral Medial plantar Lateral plantar

TREATMENT OVERVIEW

With various studies citing statistics from 50% to 85% of adults who suffer low back pain, there is an increasing need for better quality and more effective treatment for this widespread problem.[8,12] This is especially concerning since there are discouraging reports that potential for return to work after back injury decreases with increased time away from work.[30] Thus, there is considerable pressure to rehabilitate and return back pain patients to work as quickly as possible. In light of the fact that health care workers are among the occupations with the highest incidence of back injury, health care practitioners have a lot at stake.[42]

As high as 90% or more of all low back pain sufferers will recover within 1[47], 2[39], or 3 months[44] with or without treatment. So why bother to treat at all? The first reason is that treatment offers relief from the pain and accompanying muscle spasms that frequently result from low back injury. A number of well-studied modalities can accelerate recovery from injury through various physiological avenues. Treatment has been proven more effective than placebo with both diathermy and mobilization/manipulation, which were both shown to be of higher value than placebo.[15] Even if there were little physiologically demonstrable change, the "laying on of hands" and attention paid to low back pain sufferers probably has significant advantage over the old "take two aspirins and call me in the morning" approach.

Similarly, up to 2 months of suffering, with or without loss of work, is an unacceptably long time to those patients whose symptoms do not remit quickly. Health care practitioners generally cannot stand by idly while their patients suffer needless pain. Even if the treatment proves palliative, it is surely better than no treatment if the patient is made more comfortable while healing occurs. Further, intervention can prevent long-term disability and chronic pain. A combination of therapeutic modalities, manual therapy, and education of the patient is suggested as an effective combination.[51,60]

Finally, how do we know if we are in fact treating that 10%–20% of patients who do not achieve remission of their symptoms spontaneously? Certainly, some of these patients might require surgery or other intervention. Low back pain patients, however, could be prevented from slipping into a chronic pain pattern and from suffering loss of function in the process by appropriate and early therapeutic intervention.[49,56]

A disturbing comment made in a 1966 article is that "greater than 80% of all acute complaints of low back pain have an unknown etiology."[9] One hopes that this is no longer true, in part because of the advances in diagnostic techniques and technologies. There are, however, still a disturbing number of low back pain patients whose pain source is not specifically known and for whom treatment is often hit or miss. Some of these patients eventually become "chronic pain patients." Happily, there are encouraging results from studies of these patients. One study shows that significant improvement in functional capacity can be made, and these gains measured objectively, through use of a Cybex® dynamic trunk strength device.[44] Exciting gains in the battle against long-term disability from back injury may be on the horizon.

For those patients who have known diagnoses and for whom the tissue under treatment is known, techniques for relief of symptoms are sometimes remarkably brief and specific. An example is Travell's treatment of fibrocytic nodules, or "trigger points."[64] One to three treatments of a single site might be all that is required for complete remission. In other cases when it is difficult to determine which tissue has been injured, when diagnostic testing is not definitive, or when multiple tissues have sustained injury, treatment is much more complex.

Some of the tissues that give rise to pain in the back include the periosteum, muscle, nerve, ligament, dural sheath, facet joint capsule, and synovial lining of the facet joint. To further complicate matters, a number of each of these tissues—in some cases a high number—are susceptible to trauma, as in the case of the muscles of the low back area. Although the disc is frequently the offending tissue, its role in the production of symptoms is generally considered to be a result of imposition on one or more of the other innervated tissues.

Making evaluation more complicated is the question of the initiating trauma. It is common for the patient to recall no specific injury or event precipitating the pain, especially in the case of sciatic pain.[5] Since ischemia is one of the mechanisms that induce pain, prolonged static posture is frequently the "event" preceding pain, as in fibrositis (myofascial pain syndrome). Irritation, strain, overuse, inflammation, bruising, and infection are other mechanisms that produce pain, either alone or in combination. It is conceivable that a back injury has two or more tissues injured, and two or more of these causes of the pain occur concomitantly.

TREATMENT GOALS

The patient usually consults a health care practitioner because of pain. Alleviation of this pain is certainly one of the primary goals of treatment but should not be considered exclusively. In order to provide appropriate, comprehensive care, several other important treatment goals are aimed at restoring normal function. Without the restoration of normal function, the chances of recurrence or prolongation of the initial injury are significantly greater.

A common finding is notable reduction in strength and mobility following low back pain. The restoration of strength and mobility to normal can mean the difference be-

TABLE 4-2. COMPREHENSIVE EDUCATION FOR PREVENTING LOW BACK PAIN

Correct posture
    Sitting
    Standing
    Lying
    Walking
    Lifting
Relaxation techniques
Pain and stress management
Appropriate exercises
    Stretching (flexibility)
    Strengthening
    Postural
    Endurance training
    Coordination
Instruction in spinal mechanics (simplified)
    Anatomy
    Biomechanics
Practical instruction in lifting and handling

tween the patient returning to comfortable, functional status, or remaining a chronic pain patient. Also, the amount of impairment remaining after injury seems tied to the ability of the patient to regain premorbid strength and flexibility. While patients are working toward health, they should also be counselled on the application of good body mechanics, learn how to lift and carry safely, and gain an understanding of the normal biomechanics of the body, which helps prevent any recurrence of injury. Back schools have proven effective in achieving these goals in chronic pain patients and probably have an essential role in prevention of back injury. The low return rate of patients after attending a back school (10%) supports these conclusions.[16,23] Back school should be started as soon as healing permits, so that the patient does not decline into chronic low back pain, and the education process is an important part of a comprehensive total back care program. Should no back school be available, the basic concepts taught are readily passed on by health care professionals through careful instruction in as many as possible of the items listed in Table 4-2.

In a study of patient compliance with exercise programs, it was determined that compliance is enhanced when the personalized instruction is accompanied by a booklet or similar educational material.[16] Therefore, whenever possible, a handout consisting of the most essential information should be provided to increase learning. This printed information is referred to by patients following a home program to remind them of the correct techniques.

High-quality communication between the health care practitioner and the patient is also essential. The establishment of a trusting and caring relationship aids in evaluating the results of the treatment program and in making appropriate changes when they become necessary. This relationship of trust also allays fears, reduces stress, and provides the patient with a feeling of security, all of which contribute to more rapid recovery.

ACUTE LOW BACK INJURIES

Rest is the initial treatment of choice after an acute low back injury.[6,8] Often, in the absence of neurologic symptoms, as little as 3 days[5] rest will result in remission. For patients with some disc involvement, 1–2 weeks of bed rest are often effective, even in the presence of "classic" neurologic signs of disc protrusion with nerve root compression.[5] Opinion diverges regarding the strict enforcement of bed rest in the acute stage of low back pain, however. Curtailing and controlling even bathroom privileges has been suggested in patients with acute back pain,[5] while Teitz and Cook[69] recommend standing every 3 hours for patients with suspected HNP to "minimize imbibition of fluids in the nucleus pulposis." Most authors simply describe a period of bed rest without specifying duration or restrictions.[6,8,12,60] The decision on specific restrictions is probably best made on an individual basis, with the diagnosis and severity of the injury as guidelines.

Positioning in bed has an effect upon comfort. The most comfortable position to the patient is "the proper position."[5] It is common for patients to assume the fetal position in bed, since flexion of the low back draws the annulus fibrosis taut and may result in moving the bulge away from the nerve root.[60] Since the flexed position creates higher intradiscal pressure than extension,[60] once patients are able to do so, they should be encouraged to rest in extension. Also to be kept in mind are the well-known serious sequelae to prolonged bed rest, which are to be avoided.

In the presence of an acute trauma to the soft tissues, when there is swelling or muscle spasm, cryotherapy, or ice application, is indicated.[6,61,69] Cold is applied in various ways, including ice massage, ice/slush packs, gel packs, or vapocoolant spray (see Table 4). Once the desired therapeutic effects of cold have been achieved, application of TENS, traction when the strain is not acute, or gentle range of motion exercises[5,69] may follow. The exercises are intended to restore the muscle to its resting length, which is effective in further relieving spasm and pain. The physiological and therapeutic effects of cold are summarized in Table 4-4.

Another early intervention often recommended is exercise, which is done in bed as early as it is comfortable. While controversy rages about the efficacy of flexion versus extension exercises, in the presence of suspected HNP, it is generally agreed that flexion exercises should be avoided.[60,69] Hyperextension exercises have been found to be useful for patients with HNP in the acute stage, but are only done when they do not increase the radicular pain.[60,69]

Controversy is evident in the literature with regard to the use of manual therapy, which encompasses stretching, mobilization, manipulation, and "muscle energy" techniques, each in various styles. There is also disagreement over when manual therapy should be begun, whether in the acute stage or after the injury has progressed to the subacute level. Much depends on the ability and experience of the practitioner and the technique used. Manual therapy can be extremely gentle and not allow the joint to reach the end of its range, or more forceful, and must be used carefully and appropriately. In early visits, passive stretching of the low back has been suggested in the event that reduction in lateral flexibility or a scoliotic posture persists.[5] Many authors, however, believe that manual therapy has a place in the care of low back injuries when pain is present and flexibility is reduced, as long as this modality is applied by a competent practitioner.[5,7,15,17,19,21,51-53,60,68]

LUMBAR DISC

Conservative treatment of intervertebral disc protrusions generally includes bed rest, since the intradiscal pressure is reduced by the horizontal position and patients often report relief of symptoms.[8] Patients are taught the principles of extension and how to avoid aggravation of their symptoms during the early phase of treatment. Positions and activities to be avoided include those that increase intradiscal pressure, including forward flexion in standing or sitting, prolonged sitting, and the Valsalva maneuver.

Modalities and medication are utilized judiciously to reduce muscle spasm, guarding and pain. Either heat or cold may be effective, and the choice should be made by which is more comfortable for the patient. Electrotherapeutic devices are also helpful in reducing pain and spasm, increasing circulation, and introducing medication to the specific area of pain and ischemia. Although considerable controversy exists about the specific effects of traction, it is widely used and believed to be effective in the treatment of intervertebral disc protrusions and other pain associated with the disc.[5,7,20,58,60,72]

Treatment of disc protrusion by manual therapy is claimed to be useful under certain circumstances. Gentle oscillation and nonrotary movements to aid in positional distraction are thought safe, though other techniques used in cases of acute disc problems are not recommended, especially in the presence of severe or worsening neurologic signs.[53] Cyriax[7] described reduction of a small and recent nuclear herniation by sustained pressure, but also admitted that generally, nuclear protrusions do not respond to manual therapy. Meanwhile, Grieve[19] noted severe root pain, involvement of more than one nerve root on one side, or more than two adjacent roots in one lower limb to be absolute contraindications, while prescribing care in the presence of neurologic signs. Without considerable skill and care by the clinician, however, manual therapy has little to offer the patient with nuclear protrusions.

Exercises to strengthen the muscles that maintain normal lumbar extension and stretch those that are tight and pull the spine into flexion are initiated as soon as comfort permits. Once symptoms are reduced or eliminated, more vigorous exercises are introduced to restore the patient to full functional strength and mobility.

FIBROSITIS

Fibrositis, which is commonly referred to as myofascial pain syndrome, is characterized by active and latent "trigger points" that are "self-sustaining hyperirritable foci located in skeletal muscle or its associated fascia."[64] These trigger points, which can occur in other tissues, are frequently associated with referred pain, decreased range of motion, and decreased strength.[64] Once they are located, several specific treatments can "inactivate" these points, permitting restoration of normal function. The factors that perpetuate this syndrome also require attention in order to prevent recurrence. Some of the factors active in the low back area include leg length discrepancy and a small hemipelvis, both of which cause a tilt in the pelvis and resultant muscle strain associated with functional scoliosis.[64] Numerous other systemic factors perpetuate fibrositis, which are described by Simons and Travell,[64,65,66] along with specific muscles and referred pain patterns, which also have been well defined. Trigger points that cause low back pain are found in some of the muscles of the lower extremities.[64,65,66] The trigger points can be palpated clinically and have a characteristic nodular or rope-like feeling with associated tenderness in the affected muscle. Once the trigger point has been identified, several treatments are effective in rendering them inactive, or pain-free. These techniques include spray-and-stretch, injection, ischemic compression, and dry needling, followed by moist heat application and active range of motion exercises.[65,66]

Leg length discrepancy must be corrected by the use of appropriate shoe devices. Other factors that may participate in perpetuating fibrositis include the type A personality, stress, anxiety, and poor sleeping habits.[76] Relaxation techniques, self-hypnosis, and EMG biofeedback combined with specific therapeutic modalities to reduce pain and spasm are useful in dealing with the heightened muscle tone associated with these factors (see Chapter 1).

DEGENERATIVE JOINT DISEASE/OSTEOARTHRITIS

Degenerative joint disease (DJD) or osteoarthritis is considered a "normal" consequence of aging. While it may be true that there are degenerative changes of the joints concomitant with aging, these changes are commonly asymptomatic. There may be reduced joint mobility with osteophyte formation, which is totally nonpainful. As mentioned in Chapter 2, radiographic changes do not correlate with clinical findings. Therefore, when a patient presents with pain of acute onset, it cannot be assumed that the pain is related to the changes noted on x-ray. While degenerative changes may allow injury to occur more easily in the surrounding tissue, the injury itself is comparable to a similar injury in a younger person and is treated accordingly.

An example is a 90-year-old woman who demonstrated moderate to severe DJD of the spine with degenerative disc disease and further radiographic findings of Paget's disease of the proximal femur. When the patient complained of pain in the low back with radiation along the sciatic distribution, the assumption was made that these problems were the cause of the pain. However, the patient had been pain-free until walking significantly farther than normal under poor weather conditions, with acute onset of pain within hours of the incident. Subsequent conservative treatment of the soft tissue involved led to remission of symptoms and restoration of normal function. While it may take longer for an older person to heal, the same soft tissue considerations are applied, albeit with a slightly longer duration of symptoms, and thus treatment may be prolonged.

FACET JOINT PROBLEMS

Facet joint involvement is frequently differentiated from discogenic problems by the positions of increased and decreased pain. If flexion aggravates the symptoms, the likelihood is that the disc is involved. Conversely, if extension causes increased discomfort, the facet joint is the more likely problem.

Several groups have advocated facet joint relief.[40,54,63] Nevertheless, neither the natural history of radiculitis nor myelitis will respond to facet injections of local anesthetics or steroids. Similarly, even facet denervation with phenol will not alter the progression of the disorder. Steroids injected into posterior facet joints might diffuse enough to affect nearby acute irritation of the spinal nerve root, but long-term gains in the face of chronic irritation would be doubtful. Consequently, it should not be surprising that diminishing returns commonly are noted from repeat injections.

In impingement of the facet joint when there is limited involvement of the soft tissue surrounding the joint, manual therapy, in combination with local heat or ice to reduce any soreness, is often effective. When there is a history of trauma, however, the facet joint may be sprained, with accompanying reduction of mobility, joint effusion, and possibly positional faults.[32,60] Facet joint sprain is treated conservatively with rest, modalities for symptom relief, and movement within the pain-free range until healing of the joint has occurred. During the treatment phase, it is essential to impress upon the patient the importance of maintaining normal posture so that faulty posture, weakened muscles, and chronic postural strain are avoided.[60] The health care practitioner should be watchful during the course of treatment for the onset of any postural change so that these changes may be corrected early.

After the acute stage of facet joint sprain has passed, gentle mobilization through manual therapy and range of motion exercises is added to the treatment program. Normal mobility is thus likely to return by the time the joint is healed.[60]

THERAPEUTIC MODALITIES

Numerous modalities are available for treatment of low back pain, all of which, usually in combinations of two or more, are used as the specific patients' needs warrant and symptoms require (Table 4-3).

There is no panacea, no magical modality that will alleviate symptoms in all patients. Individual practitioners will probably find themselves more comfortable in applying some of these devices over others and may acquire special skills in doing so, thus determining them to be of unique value. Individual skills do affect the results, as do other nontangibles, such as patient confidence, trust, and motivation. However, patients and their symptoms are different and are constantly evaluated and re-evaluated during the course of treatment to determine whether the modalities utilized are achieving the intended goals. If not, after a reasonable trial another treatment program is instituted.

The use of modalities is controversial. Some clinicians believe that the fewer modalities used the better. Others seem to believe that the patient responds better when many modalities are used, in an apparent "hit or miss" attempt at finding appropriate treatment devices. Careful assessment, exercise prescription, and judicious use of manual or other techniques is essential. Any device or approach should be considered a part of the armamentarium available to be utilized when it is appropriate to do so, rather than adopting an inflexible routine applied to all patients. In any case, shotgun therapy through the indiscriminate use of therapeutic exercise and therapeutic modalities is deplorable. These practices have lent a bad name to both therapeutic modalities and exercise. Both should

TABLE 4-3. THERAPEUTIC MODALITIES

Cold (cryotherapy)
    Ice/slush/ice and water packs
    Gel packs
    Ice massage
    Ethyl chloride spray

Heat
    Moist heat packs
    Infrared
    Hydrotherapy
    Shortwave diathermy
    Microwave diathermy
    Ultrasound

Electrotherapeutic Modalities
    TENS (transcutaneous electrical nerve
      stimulation)
    Alternating or direct current electrical
      stimulation low or high votage
        Iontophoresis—direct
        Interferential current—alternating

Traction
    Manual
    Motorized
      Static
      Intermittent
    Positional
    Inversion
    Suspension
    Unilateral

Massage
    Friction
    Effleurage
    Petrissage
    Tapotement
    Vibration

Manual therapy
    Stretching
    Mobilization
    Manipulation
    Muscle energy

Exercises
    Passive
    Active-assisted
    Active
    Resistive
      Isotonic
      Isometric
      Isokinetic

Helium-neon laser (cold)

EMG biofeedback

Relaxation techniques

be used as part of a comprehensive, conservative, and individualized approach to each patient with back pain. Effectiveness of treatment is enhanced when the evaluative process, such as that available in a modern medical center, is utilized.

Cryotherapy (Cold)

Cryotherapy may be applied in a number of ways, including ice massage, ice packs, slush packs, gel packs and vapocoolant spray. If a cold pack is to be left in place, a layer of moist toweling is placed between the pack and the patient to prevent freezing of the skin. Once the desired therapeutic effects of cold have been achieved, other modalities may be utilized. Those commonly associated with cold include TENS, traction, or range of motion exercises.[5,69] The physiologic and therapeutic effects of cold are summarized in Table 4-4.

Heat

Therapeutic heat produces both general and local effects, provided by superficial or deep heat. A summary of these effects is provided in Table 4-5. The physiologic response of the body depends upon the size of the area being treated; the thickness of the skin, fat layer, and muscle; the length of treatment time; the intensity of the source of heat; and the ability of the body to dissipate the heat.[11] This is of particular consequence in total body immersion with hydrotherapy, as the body temperature could be raised to a dangerous level.

Locally, the temperature range that is therapeutic is 40–45 degrees C, above which tissue destruction occurs.[61] Clinically, the depth of penetration of the modality and the location of the target tissue are correlated in order to achieve therapeutic goals. Table 4-6 describes the depth of penetration of therapeutic heat modalities.

A very interesting, controlled, partially blind study has shown ultrasound to be an

TABLE 4-4. EFFECTS OF COLD

Physiologic

Lowered cell metabolic rate
Vasoconstriction
Lowered inflammatory response
Lowered nerve conduction velocity
Lowered muscle contractability
Raised pain threshold
Lowered collagen tissue extensibility
Lowered response of muscle spindle to stretch
Increased viscosity of synovial fluid

Therapeutic

Reduced muscle spasm
Reduced spasticity
Reduced local edema
Reduced pain

TABLE 4-5. EFFECTS OF HEAT

Physiologic

Local Effects
> Increased cellular metabolism
> Increased muscle spindle activity
> Increased muscle contractility
> Increased inflammatory response
> Increased extensibility of collagen
> Increased nerve conduction velocity
> Reduction of synovial fluid viscosity
> Vasodilation

General Effects
> Increased pulse rate
> Increased respiratory rate
> Decreased blood pressure
> Perspiration

Therapeutic

Reduction of pain
Reduction of muscle spasm
Increased circulation
Easier stretching of connective tissue
Relaxation

TABLE 4-6. PENETRATION LEVELS OF THERAPEUTIC HEATING MODALITIES

Superficial—Skin and superficial subcutaneous tissues
> Infrared
> Hot packs
> Hydrotherapy

Moderate—Subcutaneous tissues and superficial musculature
> Shortwave diathermy
> Microwave diathermy

Deep—Joints, myofascial interfaces, tendon, tendon sheath, nerve trunks, and fibrous scars within soft tissue
> Ultrasound

effective therapeutic modality.[50] The hypothesis tested is that ultrasound provides an analgesic effect that reduces pain, accelerates resolution of local inflammation, and allows increased range of motion and improved back function. Although more clinician time is required for application than with most other heat modalities, ultrasound heats more deeply and specifically, providing heating of deep structures in a controlled fashion.[13] Microwave units do not absorb clinician time, and some approved units are again available. These units, however, cannot deliver deep heat.

Since both therapeutic heat and cold share some of the same therapeutic effects, it may be confusing to decide which to use in clinical situations. A rule of thumb is that cold is frequently appropriate after acute injury to prevent or decrease edema and hematoma formation and to help provide relief from pain and spasm. Heat is particularly relaxing, and allows stretching of joint contractures to take place more easily, but should not generally be applied in the presence of acute inflammation. In the absence of specific contraindications, the choice often falls to the patient according to individual tolerance and comfort. The precautions and contraindications for both heat and cold are summarized in Table 4-7.

It is not uncommon for more than one heating modality to be applied in a single treatment session. For example, superficial heat, such as a moist heat pack, often provides relaxation and a general sense of well-being; meanwhile, ultrasound heats the affected tissue at the appropriate depth, increasing tissue metabolism, circulation, and pain relief. This, in turn, permits more freedom from muscle spasm and tight connective tissue. The patient then stretches tight structures during exercise or permits passive stretch to take place during manual therapy or traction.

Electrotherapeutic Modalities

Some of the effects of electrical modalities include pain relief, improved circulation, edema and spasm reduction, and the delivery of medication to the affected tissues. These devices are most commonly utilized in combination with other modalities in order to combine the therapeutic effects for an improved result.

TABLE 4-7. PRECAUTIONS AND CONTRAINDICATIONS FOR HEAT AND COLD

Precautions	Contraindications
Heat	
Anesthetic areas	Inadequate vascularity
Sedation or lowered alertness	Hemorrhagic diathesis
Pain upon heating	Metallic implants (diathermy)
Malignancies	Cardiac pacemakers (diathermy)
Pregnancy	Heating the eyes or testicles
Cardiac or respiratory impairment	
Bony prominences	
Total body immersion	
Skin infection	
Open wounds	
Cold	
Ischemic tissues	Cold urticaria
Anesthetic areas	Paroxysmal cold hemoglobinuria
Return to activity too soon after injury	Raynaud's phenomenon
Sedation	Vasculitis
Decreased alertness	Cryoglobulinemia
Skin infection	
Open wounds	

TENS (transcutaneous electrical nerve stimulation) is an effective and safe device of great value in pain control for arthritic and mechanical dysfunction of the low back.[17,41] A recent follow up of over 800 patients with chronic low back pain studied the effects of TENS versus a placebo.[14] Six months after starting TENS, 83% were still being helped. This paper also discussed the difficulties of blind experimental protocol. On the other hand, without significant results from blind studies, autosuggestion could have been a major contributor to relief.[73]

TENS units are readily provided to patients for use at home to help them remain as comfortable as possible while healing occurs. Careful teaching, however, must accompany the provision of the device, for these devices are of little value on a closet shelf. Patients are easily intimidated by the complexity of an electrical "machine" and may prefer not to "figure it out" on their own.

As a secondary result of TENS, while pain is reduced the protective spasm and splinting are also decreased, allowing more normal movement. Concomitantly lowered is the likelihood of the long-term effects of pain and impaired function, including joint contracture and muscle shortening. Additionally, the need for pain medication is often considerably less with TENS.

Electrical stimulation has long been popular in the treatment of low back pain, especially when muscle spasm is a problem. Effects include a pumping action increasing blood supply to the area. This effect improves healing. Pain and spasm are reduced through inhibition.[11] Alternating current is generally more comfortable than low volt galvanic stimulation. Recently high volt galvanic stimulation has become popular because it penetrates more deeply and allows a stronger contraction of the muscle with less pain than the low volt direct current stimulator.[11]

Iontophoresis, originally proposed in the late 1700s,[41] has also been utilized for a long time. By means of either the positive or negative pole, charged ions are repelled from the electrode through the skin and penetrate subcutaneous tissue. Many physiologic principles must be followed in the appropriate application of ion transfer, and numerous parameters significantly affect the outcome of treatment. In competent hands, however, the direction of medication to specific tissues in higher concentrations than would otherwise be possible can provide excellent clinical results. Examples of physiologic responses produced are vasodilation, anesthetic, antiseptic, analgesic, anti-inflammatory, antispasmodic, and sclerolytic effects.[28]

Phonophoresis, a method of driving a drug through a membrane by means of ultrasound, can move entire molecules through tissue by the sound's mechanical properties, reaching depths of 5 cm[61] or more. Preheating allows more ready absorption of the chemical applied.[28] Coupling gel is applied over the chemical, which is applied to the skin over the target tissue.

Interferential current is a relative newcomer to the therapeutic armamentarium. The major advantage claimed over other types of electrical stimulation is that the effects are produced specifically in the tissues where they are desired without skin stimulation, which may be uncomfortable.[28,58] The physiologic effects vary with the intensity of the current, the mode of application (rhythmic or constant), the frequency, and the positioning of the electrodes.[66]

Manual Therapy

Manual therapy techniques have been practiced in some form or another since the time of Hippoccrates.[53] Debate rages over the numerous styles, techniques, philosophies, and effects claimed. It does seem clear, however, that there is at least a short-term benefit, as there is with other modalities, that cannot be disregarded.[5,7,15,17,19,21,51-53,60,68] In the hands of a skilled practitioner and with appropriate evaluation, manual therapy is a valuable addition to our array of therapeutic devices and techniques.

Manual therapy is often effectively combined with one or more other modalities[53] to reduce muscle spasm and guarding and to prevent the stiffness that sometimes occurs the day after manual therapy has been applied. Some health care practitioners apply modalities prior to manual therapy; others reverse the order. Still others attempt to reduce pain and spasm prior to manual therapy and provide comfort and relaxation afterwards. Exercises may or may not be taught at the same time. Nonetheless, no matter the device, technique, or modality applied, without a conscientious attempt at determining the problem, the "solution" is often elusive.

Traction

Traction is believed to distract the vertebral bodies and therefore off-load weight from the intervertebral discs.[7,20,72] Cailliet[5] does not agree that sustained lumbar traction distracts vertebral bodies. As with any form of treatment, ratings of clinical safety and effectiveness should rest upon well-controlled, blind, in vivo studies of living normal subjects and chronic back pain patients. But the majority of studies performed involved essentially in vitro procedures upon cadavers. Only one study demonstrated that intermittent lumbar traction produces distraction.[18] The researchers used the 20 subjects as their own controls, and the study was not at all blind. Still, it does point the way for additional investigations. Additionally, as Swezey[68] points out, "There is a consensus that traction can be a useful adjunct and sometimes an essential component of therapy" for back disorders, despite the fact that the exact mechanism by which traction has its effects is still unknown.

Many techniques are recommended for the application of traction to the low back, including positional, unilateral, three-dimensional, prone, inversion, and suspension, with various parameters of static and intermittent, manual and mechanical, and progressive and regressive styles of application.[18,58,60,] No matter what the parameters of duration, position, and force, however, traction must always be applied according to the responses of the patient,[17,58,60] and constant re-evaluation of signs and symptoms is done before, during, and after each treatment. Careful, specific employment of traction after thorough evaluation is most likely to provide positive results. Utilization of modalities to enhance relaxation and reduce muscle spasm and pain prior to traction frequently provides a better response.

Exercise

Therapeutic exercise centered around modifications of the Williams flexion exercises until recently, with the increasing popularity of extension exercises.[37] Goals frequently noted involve strengthening abdominal musculature and reducing lumbar lordosis.[34] Isometric exercises have also been proposed. There is considerable controversy about the efficacy of flexion versus extension exercises. However, if a nuclear protrusion is suspected it makes sense to avoid flexion exercises, especially in the early treatment phase.[60,69]

Hyperextension exercises have been found to be useful for patients with nuclear protrusion even in the acute stage, but are performed only when no increase in the radicular signs is caused.[60,69] Extension of the lumbar spine causes the posteriorly displaced nuclear protrusion to move anteriorly, back toward the disc.[37] Further, patients with low back pain often have decreased lumbar lordosis.[7] Empirically, therefore, relief from pain and radicular signs is noted with extension in cases of disc protrusion. Extension exercises should be avoided, however, in patients with lumbar facet problems and spinal stenosis, where pain is aggravated by extension and decreased by flexion.[37,60]

There is strong evidence in favor of extension exercises in order to restore or improve extensor strength, unload the disc, and increase the endurance of the extensor muscles.[25] They are not indicated, however, in cases of acute disc prolapse or in the presence of spinal stenosis or other conditions that are worsened by extension.[25,60] Hyperextension exercises may also cause worsening of symptoms in patients with facet joint problems and narrowing of the intervertebral foramen and are shown to cause loading of the disc.[25]

As with any other form of treatment, exercises are recommended on the basis of careful and thorough evaluation of the patient on an individual basis. Attention is paid to flexibility, strength, and endurance. Restoration of these functions is essential if normal activity is to be resumed after back injury. As mentioned earlier, the back school concept, which incorporates these factors into the total rehabilitation of the patient, is an effective program in returning back-injured patients to premorbid functional levels.

The Williams flexion exercises were designed to reduce compression of the nerve root as it passes through the intervertebral foramen by opening the foramen.[26] Currently, there seems to be more attention to protecting the disc through increased intra-abdominal pressure, and the preprogrammed exercise routine commonly utilized was found by Jackson and Brown[25] to be of little value and possibly harmful. The two exercises recommended included the pelvic tilt, done in the subacute stage, and the shoulder-lift sit-up, both of which are thought to strengthen the abdominal muscles appropriately (Figs. 4-6 and 4-8).

Other Useful Treatment Adjuncts

Therapeutic massage has lost favor in the medical field in recent years because of its time-consuming nature, the lack of demonstrable and reproducible benefits, the variability in skill of those applying massage, and perhaps even the rise in the application of massage by other than trained medical professionals.[61] Nevertheless, the "laying on of hands" is empirically of value, and massage, when appropriately given by skilled practitioners, has definite benefits in the treatment of low back pain. Massage causes relaxation and can be applied in such a way as to increase circulation and assist in the removal of metabolic byproducts.

Transverse friction massage, has gained favor because of its unusual ability to mobilize soft tissue. Applied perpendicular to the muscle fibers and gradually increasing in depth of penetration, this technique stretches adhesions that are restrictive and painful. Skill is required, as this technique is potentially very painful.

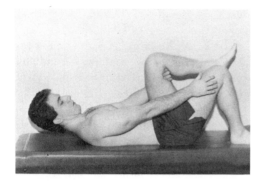

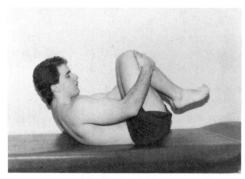

Figure 4-3. Low Back Flexibility. The patient is taught to stretch the lower back by flexing the knees to the chest, one at a time, and pulling the thighs toward the chest, passively stretching the lower back. The knees are lowered to the table one at a time.

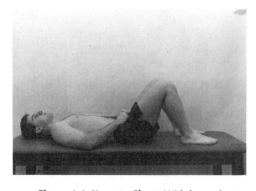

Figure 4-4. Knee to Chest. With knees bent, one leg is brought to the chest, held, and lowered slowly. This exercise is repeated with the other leg, and alternated.

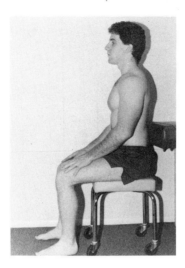 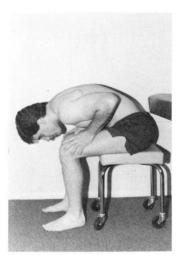

Figure 4-5. Sitting Low Back Stretch. Supporting the upper body weight with the hands on the knees, the patient slowly lowers the chest toward the knees, allowing gravity to provide stretching of the lower back. Upon rising, the patient uses his arms to push up to the neutral position, avoiding undue strain on the muscles of the lower back.

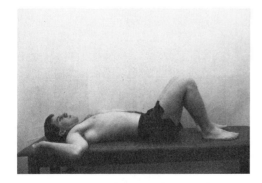 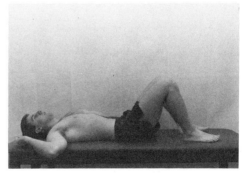

Figure 4-6. Back Flattener. These exercises strengthen the abdominal and buttock muscles, useful in reducing lumbar lordosis.

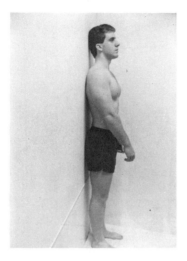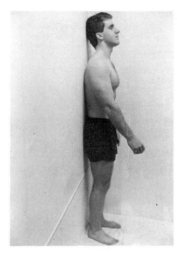

Figure 4-7. Erect Pelvic Tilt. These exercises strengthen the abdominal and buttock muscles, useful in reducing lumbar lordosis.

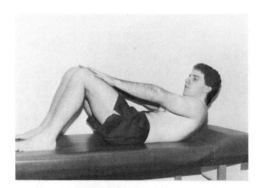

Figure 4-8. Curl-Up. The patient curls slowly, beginning with lifting the head and upper shoulders off the mat, reaching toward the knees, without coming to the sitting position. This exercise avoids stress on the lower back while strengthening the abdominal muscles.

Figure 4-9. Partial Sit-up. The patient is taught to tilt the pelvis, then lift himself slowly toward the sitting position, but not causing more than 40 degrees of trunk flexion. This exercise strengthens the abdominal muscles and provides a gentle stretch for the low back.

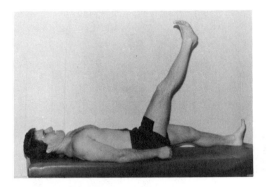

Figure 4-10. Straight Leg Raise for Stretch of Hamstrings. The patient may be taught to flex one knee while raising the other, reducing the strain on the low back. Often a towel, rope, or other device at the foot is used to assist in increasing the stretch on the hamstrings.

Figure 4-11. Sitting Hamstring Stretch. The hamstrings can be stretched in the sitting position, but care is taken to avoid undue strain on the low back by flexing one knee.

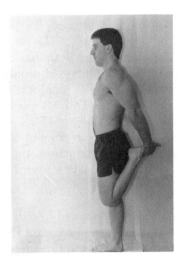

Figure 4-12. Anterior Thigh Stretch. A gentle stretch of the anterior thigh muscles can be accomplished as shown here. Undue lumbar lordosis should be avoided. Stronger stretch is achieved by extension of the hip with flexion of the knee, stretching both the muscles that extend the knee and those that flex the hip.

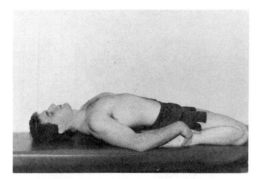

Figure 4-13. Stretching Hip Flexor Muscles. A strong stretch of the hip flexors is provided as shown. Patients who do not tolerate lumbar extension should not attempt this exercise.

Figure 4-14. Heel Cord Stretching. With hands against the wall, one leg is forward and flexed, while the other is behind the body and straight with the heel resting on the floor. Leaning into the wall and flexing the forward leg provides a stretch to the heelcords of the rearmost leg.

A recent study indicates that relaxation or self-hypnosis can be beneficial techniques.[46] One problem that recurs, however, is how to measure the outcome of treatment in patients with chronic back pain. A Functional Rating Scale has been produced that is considered both reliable and responsive.[10] The application of profiles that can be monitored by computers is practical and helpful. For example, profiles could be applied to acupuncture and autoacupressure as well as to hypnosis. This is especially true since these two modalities[33] have also been reported as effective (Fig. 4-15).

The helium-neon laser, or cold laser, has a place in the treatment of low back and other types of pain as well. When combined with electroacupuncture, the pain-relieving results are more rapid and dramatic. As with the other modalities, laser is only one of a number of therapeutic devices and should not be employed alone but as part of a comprehensive treatment program.

Lumbar corsets are a frequently prescribed orthosis designed to reduce the load on the trunk and provide immobilization of the trunk and increased intra-abdominal pressure.[74] This increase in pressure is thought to reduce intradiscal pressure, providing relief for some patients with low back pain of varying etiologies. However, if a lumbar corset is to be used even temporarily,[48] regular isometric lumbar exercises will be required to help maintain muscle strength and tone. The problem with corsets is not that it is difficult to prescribe them, but that it is hard to know when they must be discontinued. A clear and precise contract with the patient is vital before corsets are applied. Seri-

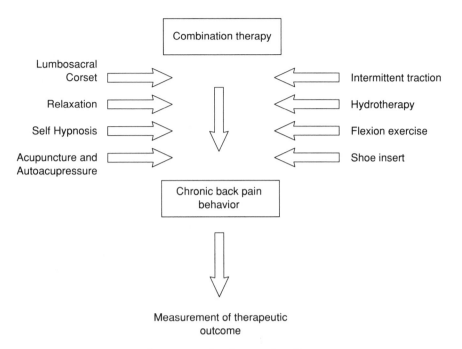

Figure 4-15. The Therapeutic Milieu.

ous weakening of trunk muscles can otherwise result, and physical and psychological dependence result.

Viscoelastic shoe inserts have been found to act as artificial shock absorbers for patients with chronic pain.[75] Their complete contribution to a full service pain clinic needs to be established. Nevertheless, if they reduce the need for wearing a corset or undergoing traction, they will have served their purpose.

The treatment of back pain must be modernized, streamlined, and placed on a scientifically reproducible basis. Accurate, well-blinded, controlled studies of large populations with pain behavior to these differing modalities should be the focus of ongoing research. That is the only way that newer methods and evaluations can be compared.

REFERENCES

1. Akeson WH, Murphy RW: Low back pain. *Clin Orthop* 1977; 129:2.
2. Argyraski A, et al: Solitary neurofibroma of the lumbosacral plexus. *Arch Neurol* 1985; 42:844.
3. Bigos SJ et al: Back injuries in industry: A retrospective study. II. Injury factors. *Spine* 1986; 11:252.
4. Bigos SJ et al: Back injuries in industry: A retrospective study. III. Employee related factors. *Spine* 1986; 11:252.
5. Cailliet R: *Low Back Pain Syndrome,* 3rd ed. Philadelphia, F. A. Davis, 1981.
6. Cantu RC: Low back injuries. In, Vinger PF, Hoerner EF (editors): *Sports Injuries,* 2nd ed. Littleton, PSG Pub. Co., Inc., 1986.
7. Cyriax J: *Textbook of Orthopedic Medicine,*

Vol. 1, 7th ed. London, Balliere Tindall, 1978.

8. Deyo RA: Conservative therapy for low back pain—Distinguishing useful from useless therapy. *JAMA* 1983; 250:1057.

9. Dillane JB, et al: Acute back syndrome—A study from general practice. *Br Med J* 1966; 2:82.

10. Evans JH, Kagan, A II: The development of a functional rating scale to measure the treatment outcome of chronic spinal patients. *Spine* 1986; 11:277.

11. Fairchild VM et al: Physical Therapy. In Raj PP (editor): *Practical Management of Pain*. Chicago, Year Book Medical Pub., Inc., 1986.

12. Fisk JR et al: Back schools. Past, present and future. *Clin Orthop* 1983; 179:18.

13. Forster A, Palastanga N: *Clayton's Electrotherapy: Theory and Practice*. London, Balliere Tindall, 1981. pp 149–154.

14. Fried T, et al: Transcutaneous electrical nerve stimulation: Its role in the control of chronic pain. *Arch Phys Med Rehabil* 1984; 65:228.

15. Gibson T, et al: Controlled comparison of short-wave diathermy treatment with osteopathic treatment in non-specific low back pain. *Lancet* 1985; 8440:1258.

16. Glossop ES, et al: Patient compliance in back and neck pain. *Physiotherapy* 1982; 68:225.

17. Gould JA III, Davies GJ: *Orthopedic and Sports Physical Therapy*. St. Louis, C.V. Mosby Co., 1985.

18. Granakopoulos G, et al: Inversion devices: Their role in producing lumbar distraction. *Arch Phys Med Rehabil* 1985; 66:100.

19. Grieve GP: *Mobilization of the Spine. Notes on Examination, Assessment, Clinical Method*. Handout from mobilization/manipulation course, Atlanta, Ga 1975.

20. Gupta RC, Ramarad SV: Epidurography in reduction of lumbar disc prolapse by traction. *Arch Phys Med Rehabil* 1978; 59:322.

21. Haldeman S: Spinal manipulative therapy in sports medicine. *Clin Sports Med* 1986; 5:277.

22. Hansson TH, et al: The load on the lumbar spine during isometric strength testing. *Spine* 1984; 9:720.

23. Hayne CR: Back schools and total back-care programmes—a review. *Physiotherapy* 1984; 70:14.

24. Hope EE, et al: Neonatal lumbar plexus injury. *J Neurol Neurosurg Psychiat* 1985; 48:844.

25. Jackson CP, Brown MD: Analysis of current approaches and a practical guide to prescription of exercise. *Clin Orthop* 1983; 179:46.

26. Jackson CP, Brown MD: Is there a role for exercise in the treatment of patients with low back pain? *Clin Orthop* 1983; 179:39.

27. Johnson AD: *The Problem Chair, An Approach to Early Identification*. Department of Labor and Industries, State of Washington (mimeographed) 1978.

28. Kahn J: *Low Volt Technique*, 4th ed. New York, Joseph Kahn, 1985.

29. Kaplan PE: Hemiplegia: Rehabilitation of the lower extremity. In Kaplan, P. E. and Cerullo, L. (editors): *Stroke Rehabilitation*. Boston, Butterworth, 1986, p 119.

30. Kelsey JL: An epidemiological study of acute herniated lumbar intervertebral discs. *Rheumatol Rehabil* 1975; 14:144.

31. Kelsey JL, et al: *Musculoskeletal Disorders: Their Frequency of Occurrence and Their Impact on the Population of the United States*. New York, Prodist, 1978.

32. Kessler RM, Hertling D: *Management of Common Musculoskeletal Disorders. Physical Therapy Principles and Methods*. Philadelphia, Harper & Row, 1983, p 507.

33. Kurland HD: *Back Pains: Quick Relief Without Drugs*. New York, Simon & Schuster, 1983.

34. LaBan MM: The lumbosacral pain syndrome. In Kaplan P E (editor): *The Practive of Physical Medicine*. Springfield, Ill., Charles C. Thomas, 1984, p. 107.

35. Leavitt SS, et al: Monitoring the recovery process. Part 1. *Indust Med Surg* 1971; 40:7.

36. Leavitt SS, et al: The process of recovery. Part 1. *Indust Med Surg* 1971; 40:7.

37. Lee CK: The use of exercise and muscle

testing in the rehabilitation of spinal disorders. *Clin Sports Med* 1986; 5:271.

38. Lerman VJ: Radiculopathy and electromyography. *Arch Neurol* 1985; 42:732.

39. Licther RL, et al: Treatment of chronic low-back pain. *Clin Orthop* 1984; 190:115.

40. Lippitt FB: The fact joint and its role in spine pain. Management with facet joint injections. *Spine* 1984; 9:746.

41. Mannheimer JS: TENS: Uses and effectiveness. In Michel TH (editor): *Pain*. New York, Churchill Livingstone, 1985; pp 73–121.

42. Marks LN, Blue BB: Rehabilitating injured backs involves intervention, monitoring. *Occup Health Saf* 1985; 54:24.

43. Mathews J: The effects of spinal traction. *Physiotherapy* 1972; 58:64.

44. Mayer TG, et al: Objective assessment of spine function following industrial injury. *Spine* 1985; 10:482.

45. Mayer TG, et al: Use of noninvasive techniques for qualification of spinal range of motion in normal subjects and chronic low-back dysfunction patients. *Spine* 1984; 9:588.

46. McCauley JD, et al: Hypnosis compared to relaxation in the outpatient management of chronic low back. *Arch Phys Med Rehabil* 1983; 64:548.

47. Moffett JA, et al: A controlled, prospective study to evaluate the effectiveness of a back school in the relief of chronic back pain. *Spine* 1986; 11:120.

48. Morris JM: Low back bracing. *Clin Orthop* 1974; 102:120.

49. Murphy KA, Cornish RD: Prediction of chronicity in acute low back pain. *Arch Phys Med Rehabil* 1984; 65:334.

50. Nwuga VCB: Ultrasound in treatment of back pain resulting from prolapsed intervertebral disc. *Arch Phys Med Rehabil* 1983; 64:88.

51. O'Donoghue CE: Treatment of back pain. *Physiotherapy* 1984; 70:7.

52. Ottenbacher K, Difabio RF: Efficacy of spinal manipulation/mobilization therapy. A meta-analysis. *Spine* 1985; 10:833.

53. Paris SV: Spinal manipulative therapy. *Clin Orthop* 1983; 179:55.

54. Park WM: The place of radiology in the investigation of low back pain. *Clin Rheum Dis* 1980; 61:93.

55. Pearcy MJ, Tibrewal SB: Axial rotation and lateral bending in the normal lumbar spine measured by three-dimensional radiography. *Spine* 1984; 9:582.

56. Pedersen PA: Prognostic indicators in low back pain. *J R Coll Gen Pract* 1981; 31:209.

57. Pettigrew LC, et al: Diagnosis and treatment of lumbosacral plexopathies in patients with cancer. *Arch Neurol* 1984; 41:1282.

58. Prentice WE: *Therapeutic Modalities in Sports Medicine*. St. Louis, Times Mirror/Mosby, 1986.

59. Rogoff JB (editor): *Manipulation, Traction and Massage*. Baltimore, Williams & Wilkins, 1980.

60. Saunders HD: *Evaluation, Treatment and Prevention of Musculoskeletal Disorders*. Minneapolis, Viking Press, Inc., 1985.

61. Sawyer M, Zbieranek CK: The treatment of soft tissue after spinal injury. *Clin Sports Med* 1986; 5:387.

62. Schultz FB, Anderson GB: Analysis of loads on the lumbar spine. *Spine* 1981; 1:76.

63. Selby DK, Paris SV: Anatomy of facet joints and its correlation with low back pain. *Contemp Orthop* 1981; 312:1097.

64. Simons DG, Travell JG: Myofascial origins of low back pain. 1. Principles of diagnosis and treatment. *Postgrad Med* 1983; 73:66.

65. Simons DG, Travell JG: Myofascial origins of low back pain. 2. Torso muscles. *Postgrad Med* 1983; 73:81.

66. Simons DG, Travell JG: Myofascial origins of low back pain. 3. Pelvic and lower extremity muscles. *Postgrad Med* 1983; 73:99.

67. Spengler DM, et al: Back injuries in industry: A retrospective study. I. Overview and cost analysis. *Spine* 1986; 11:241.

68. Swezey, RL: The modern thrust of manipulation and traction therapy. *Semin Arth Rheum* 1983; 12:322.

69. Teitz CC, Cook DM: Rehabilitation of neck and low back injuries. *Clin Sports Med* 1985; 4:455.

70. Trief P, Stein, N: Pending litigation and rehabilitation outcome of chronic back pain. *Arch Phys Med Rehabil* 1985; 66:95.

71. Turner JA, et al: Chronic low back pain: Predicting response to nonsurgical treatment. *Arch Phys Med Rehabil* 1983; 64:560.

72. Twomey LT: Sustained lumbar traction. An experimental study of long spine segments. *Spine* 1985; 10:1416.

73. Vander Ark G, McGrath KF: Transcutaneous electrical stimulation in treatment of postoperative pain. *Am J Surg* 1975; 130:338.

74. Willner S: Effect of a rigid brace on back pain. *Acta Orthop Scand* 1985; 56:40.

75. Wosk J, and Voloshin AS: Low back pain; conservative treatment with artificial shock absorbers. *Arch Phys Med Rehabil* 1985; 66:145.

76. Yunus M, et al: Primary fibromyalgia. *Am Fam Phys* 1982; 25:115.

5
Shoulder Pain

Kip Burkman and
Ellen D. Tanner

In the spectrum of musculoskeletal disorders, low back pain must assume the premier position. Shoulder problems, however, are quite common in physiatry, orthopedics, and sports medicine. The shoulder joint has been designed to provide mobility at the expense of stability. This factor is a key to understanding some of the mechanisms of shoulder pain and injury. This chapter discusses the more commonly encountered of these disorders.

The shoulder should be thought of as a synchronous functioning mechanism rather than just one joint. Calliet[12] provides the example of the shoulder compiled of the following functioning "joints": glenohumeral, sternoclavicular, acromioclavicular, costovertebral, scapulocostal, costosternal and suprahumeral. All of these will act to allow movement at the shoulder in order to place the arm and hand in functional postions.

ANATOMY AND KINESIOLOGY

To appreciate the various mechanisms that produce pain in the shoulder region one must be familiar with the anatomy. The bony components consist of the humerus (covered by hyaline cartilage), ribs, clavicle, scapula, sternum, and vertebrae. Two bursae connect with the glenohumeral joint: the subcoracoid and subscapularis are synovial outpouchings of the joint capsule. The subdeltoid and subacromial bursae are superior to the glenohumeral joint and are important in diagnosis of rotator cuff dysfunction, via arthrography. The glenohumeral joint is covered by a capsule (synovial-lined) that contains redundant tissue known as the glenoid labrum, which allows increased humeral mobility. The superior capsule tension helps prevent downward humeral displacement when the arm hangs down. The superior, middle, and inferior glenohumeral ligaments are arranged anteriorly, and an opening between the upper two (called the foramen of Weitbrecht) may allow anterior dislocations. With advancing age, the shoulder joint capsule becomes less able to handle mechanical stress. Trials of 43 cadaver shoulders versus elbows with traction applied by a Mayes testing machine revealed that the shoulder joint could stretch about twice as far before capsule rupture occurred.[36] These ruptures occurred at the anteroinferior aspect of the capsule. Stability of the joint is secondary to the surrounding muscles.[37] Much of the shoulder support comes from the multiple ligaments: coracoacromial, acromioclavicular, triangular (made up of the trapezoid and conoid ligaments). The transverse humeral ligament helps maintain smooth, gliding motion of the long biceps tendon within the bicipital groove. The coracohumeral ligament is important in consideration of "frozen shoulder." The costoclavicular ligament may produce a fulcrum for shoulder motion at the sternoclavicular joint. Muscles to consider include the rotator cuff (supraspinatus, infraspinatus, teres minor, subscapularis), biceps, triceps, pectoralis major/minor, latissimus dorsi, rhomboids, teres major, levator scapulae, trapezius, subclavius, sternocleidomastoid, and scalenes, all of which affect shoulder movement. The two tendons most commonly involved in shoulder pain belong to the supraspinatus and biceps (which is extrasynovial but intra-articular).

There are six isolated movements at the humerus: flexion, extension, abduction, adduction, internal and external rotations. Six scapular movements also exist: protraction, retraction, depression, elevation, upward rotation (which causes the inferior angle to move laterally/forward and tilts the glenoid fossa upward) and downward rotation (Table 5-1). Circumduction is a complex circular movement of the above. The shoulder acts as a third class lever, since the effort point (i.e., muscle insertion) is between the resistance (i.e., the weight of the arm) and the fulcrum of glenohumeral joint.[4,31] Ivey, et al. tested 31 normal subjects on the isokinetic dynamometer to establish normal values.[32] Of paired antagonist motions the following were the stronger: internal rotation ($p < 0.01$), exten-

---

TABLE 5-1. MUSCLES OF SHOULDER MOTION

1. Humeral flexion: anterior deltoid, clavicular head of pectoralis major, biceps, coracobrachialis
2. Humeral extension: latissimus dorsi, sternal head of pectoralis major, teres major, posterior deltoid
3. Humeral adduction: pectoralis major, subscapularis, coracobrachialis, long head of triceps, latissimus dorsi, teres major, anterior/posterior deltoid
4. Humeral abduction: lateral deltoid, supraspinatus
5. Internal rotation: latissimus dorsi, teres major, subscapularis, anterior deltoid
6. External rotation: teres minor, infraspinatus, posterior deltoid
7. Scapular elevation: upper trapezius, rhomboid major/minor, levator scapulae
8. Scapular depression: latissimus dorsi, serratus anterior, subclavius, pectoralis major/minor, lower trapezius
9. Scapular protraction: serratus anterior, pectoralis major/minor
10. Scapular retraction: rhomboid major/minor, latissimus dorsi, trapezius (as a whole)
11. Scapular upward rotation: upper/lower trapezius, serratus anterior
12. Scapular downward rotation: latissimus dorsi, pectoralis minor, lower pectoralis major, levator scapulae, rhomboid major/minor

sion (p < 0.01), and adduction (p < 0.01); of all the shoulder motions tested, internal rotation was the strongest.[32]

In order to produce such motions a "scapulohumeral rhythm" exists such that for the normal 180 degrees of abduction, approximately 120 degrees occurs at the glenohumeral joint and the remaining 60 degrees takes place at the scapulothoracic articulation.[12] However, this is a nonconstant ratio, since the scapula hardly moves during the first 20 degrees of abduction.[62] In order to enhance the stability of the shoulder, the glenohumeral joint is rotated upward by the serratus anterior and trapezius. The deltoid produces an upward pull upon the dependently positioned humerus, pulling it up toward the coracoacromial arch (containing the supraspinatus tendon, glenohumeral capsule, subacromial bursa, part of the biceps tendon, and the subcoracoid bursa). To prevent this, the rotator cuff muscles depress the head of the humerus, which approximates the glenoid with the humerus. As the humerus abducts, upward glenoid fossa rotation (produced by the upper/lower trapezius and serratus anterior)[31] is needed to allow continuous mechanical contraction efficiency of the deltoid. Downward glenoid motion is produced via the latissimus dorsi, rhomboids, pectoralis major/minor, and levator scapulae.

Shoulder abduction is still possible if the deltoid muscle or axillary nerve is damaged (and confirmed by EMG). This is possible because the infraspinatus externally rotates the humerus, the supraspinatus begins abduction, the biceps and triceps continue abduction up to about 70 degrees, then the pectoralis major also acts to abduct, and finally, the serratus anterior acts upon the scapula to cause arm elevation.[3] Some feel that the biceps in this substituted movement has little power, but one author[3] feels that this is an important "trick" movement, which is important to teach patients as a part of their rehabilitation.

EMG of major shoulder muscles during abduction shows maximum activity (based on action potentials in millivolts) as follows: supraspinatus 120 degrees, infraspinatus 180 degrees, latissimus dorsi 180 degrees, subscapularis 150 degrees, and teres major/minor 90 and 150 degrees.[67] Thus it can be seen that not all the shoulder muscles provide stability during the entire abduction arc.

The glenohumeral joint is unstable because the shape of the humerus does not match the shallow glenoid fossa well, which limits the surface area contact. The glenoid cavity has about one third the articular surface of the humeral head.[67] A classification is based upon the anatomic matching of humeral head and glenoid shapes: type A has a larger glenoid than humeral head radius; type B has virtually the same radii; type C has a larger humeral head radius.[67] Motion studies with cadaver shoulders indicate that during shoulder movement the humerus continually changes its axis.[67]

In order to generate muscle power in the arm with the appropriate shoulder movements, the scapula at times must be stabilized. Other movements include scapular motion. There is a length-tension relationship such that a muscle's largest tension is developed at its resting length.[3,67] Since the scapula rotates during abduction, the deltoid is maintained at a length more like its resting length, thus retaining greater power over a larger abduction arc. The deltoid is multipennate and its fibers are directed angularly, allowing less fiber shortening and thus less loss of tension and greater power.[62] The speed of motion varies inversely with the muscle range and strength.[31]

The glenohumeral joint is innervated by the axillary, suprascapular, subscapular, and musculocutaneous nerves, as well as contributions from the posterior cord of the brachial plexus.[4] Most of the fibers that consistently serve the joint arise from C5-6. These nerves innervate synovium, ligaments, and joint capsule. Of course, in the lateral neck the brachial plexus is exposed and can be easily damaged, causing referred pain and severe muscle dysfunction.

The S-shaped clavicle rotates/elevates to accommodate the decreasing space between the sternoclavicular joint (the fulcrum of motion) and acromion upon shoulder abduction.[62] Part of the shoulder stability is provided by scapular fixation (which is quite diminished if the clavicle is broken, resected, or congenitally absent). The basic job of the clavicle is to act as a strut to keep the shoulders from collapsing toward the chest.

EVALUATION

Pain in the shoulder area can be difficult to evaluate as well as alleviate. Please refer to the differential diagnosis list (Table 5-2). Because the shoulder joint capsule and synovium are richly served with sensory nerve endings, pain in these regions tends to be fairly localized, while pain in surrounding muscles is felt as more diffuse, often aching. Pain radiating through the shoulder and down into the arm and hand is significant for cervical disc/radicular or vascular pathology, rather than shoulder disease.[4] Glenohumeral problems often produce pain in elevating the humerus above the shoulder level, as in flexion and abduction.[4] Poor posture is notorious for producing strain upon the neck and shoulder girdle muscles, and patients will remark that the pain is relieved when they lie down (i.e., when weight-bearing ceases). Discovering what postures and physiologic stresses a person has at work or during sporting activities often clues the clinician into the mechanism of pain production. Of course, shoulder pain after any fall requires questions on the way the fall occurred so that the mechanism of injury can be understood. Careful questioning is often the only way to help in the diagnosis, since patients do not know what information is important in helping the clinician. Timing of when the pain was first noticed after an injury can give an indication of the severity of the process. Occupational and sporting stresses can give insightful data into the pain problem. Knowing which arm is dominant also is helpful. If no injury is recalled, low intensity repetitive trauma to the shoulder mechanism may be happening and should be sought so that work modifications can be arranged. Nonmusculoskeletal etiologies for shoulder pain are not discussed here.

History-taking in painful conditions can be more organized by using the PQRST method.[18] "P" stands for provocative (motion/activity, sports, work, carrying loads, certain arm postions) and palliative factors (rest, heat, cold packs, medications, moving the arm in other ways). "Q" stands for the quality, or character, of the pain (burning, aching, stinging, shock-like, etc.). "R" stands for the region (localized versus generalized) in which the pain is felt and for any associated radiation of pain (down the arm, up the neck, down the back, around the side to the chest). "S" represents pain severity, or intensity (i.e., does the pain limit activities, and if so, is it immediate or is the pain tolerable for long periods?). And finally, "T" stands for the timing of the pain (in relation to the original injury, in relation to any repetitive activity now, duration, relation to weather and time of day, any pain-free periods).

TABLE 5-2. DIFFERENTIAL DIAGNOSIS OF SHOULDER PAIN

1. Neuromusculoskeletal Etiologies

Acromioclavicular joint injuries
Adhesive capsulitis
Ankylosing spondylitis
Biceps tendon (long head) rupture
Brachial plexus injury
Brachial plexitis
Bursitis: subacromial, subdeltoid
Degenerative joint disease
Dislocations/Subluxations: anterior, posterior
Fractures: distal clavicle, acromion, proximal humerus, scapula
Hemiplegic complications
Myofascial syndrome
Osteomyelitis
Osteonecrosis
Overuse syndrome
Referred pain: from cervical radiculopathy, spinal fracture,
      pectoralis minor syndrome, nerve entrapment syndromes
Reflex sympathetic dystrophy syndrome
Rheumatoid arthritis
Rotator cuff tear
Septic arthritis
Steroid arthropathy
Tendinitis: biceps, supraspinatus
Thoracic outlet syndromes
Transverse humoral ligament tear

2. Others

Arterial occlusive disease/Aneurysm
Gout
(Lymphoma)
Phlebitis
Referred pain: from myocardial infarction, irritation of the
      diaphragm, arterial occlusion, gallbladder disease, gastric disease,
      pulmonary infarction, ruptured abdominal viscus
Tumors: superior sulcus tumor, spinal cord, adenitis, bony

General points to cover on the physical examination include inspection (look for muscle imbalance, postural defects, structural defects, bruising, pain behaviors, etc.), palpation (to check for painful areas, dislocations, fractures, trigger points, etc.), and passive/active range of motions and reflexes of the arm. Many problems also require a good neuromusculoskeletal examination and specialized studies, depending upon the situation. These include electrodiagnostic studies (EMG, nerve conduction velocities), radiographs, CT scans, bone scans, and arthrograms. Magnetic resonance scanning (MRI) is noninvasive, but problems of studying the shoulder joint have been encountered, such as poor spatial resolution. This technical problem can be as least partly overcome (to produce sharper images) by the use of special surface coils.[42] It is suggested that any recent injury or painful shoulder should have radiographs taken as part of the initial evaluation. At times, an orthopedic consultation is needed for performance of shoulder arthroscopy for diagnosis and possible repair.

It is important to remember that pain naturally leads to less frequent use of the body part that causes the pain. This can mean progressive weakening of the surrounding musculature over time. This weakness can further predispose to decreasing activity, and a vicious downward spiral is created unless treated. Additionally, weakness of the surrounding muscles can lessen the ability to handle certain physical stresses, thus making the body part more prone to further injury. Certain diseases and medications can also cause muscle weakness, so it is best to determine if the weakness resulted from the pain or if it is an unrelated condition.

EVALUATION OF COMMON PATHOLOGIC CONDITIONS

Acromioclavicular Joint Injuries

Trauma to the superior or lateral aspect of the shoulder is the primary cause of this injury, and the traumatic force is directed inferiorly causing downward rotation of the clavicle and scapula until the clavicle's downward path is stopped by the first rib.[12] No radiographic findings are generally associated unless overt acromioclavicular (AC) separation occurs. Ligamentous stretch, partial tears, or complete tears are possible depending upon the degree of the force that would continue the inferior progression of the scapula. AC ligaments that are stretched or partially torn are called first-degree AC sprains. Only AC joint tenderness on palpation is noted. Localized pain occurs with AC joint such as shoulder shrugging, lying on the affected side, and especially adduction.

If the coracoclavicular ligaments remain stable but there is subluxation of the AC joint, a second-degree injury is said to have taken place. This separation can be appreciated on a stress radiograph as the patient holds a ten pound weight. Conservative measures are usually the treatment for these injuries.

Dislocation can occur if the coracoclavicular ligaments tear and the AC joint capsule ruptures, which allows the acromion to move inferiorly and the clavicle to travel superiorly. Stress radiographs with the patient holding a weight will show the abnormality clearly. Treatment of these cases includes immediate reduction, then open reduction and internal fixation. Repeat pain may be seen in the shoulder after certain operative procedures.

Trauma to the AC joint, with injury to the meniscus, can eventually lead to arthritis.[47] This situation is often seen in the painful arc syndrome, in which the rotator cuff and biceps tendons are involved.[59] Pain is felt if the patient abducts the arm to 90 degrees then moves the arm across the chest during which movement crepitus may occur.[4] "A" clicking at the AC joint with abduction over 90 degrees is suspicious for internal damage. Later development of arthritis can be seen when the arm is abducted more than 90 degrees, and crepitus occurs, caused by grating of the irregular bony surfaces within the narrowed joint space (which can be confirmed radiographically) with an accompanying sharp pain on motion and aching pain at rest. Some of the irregularities of the AC joint may be seen grossly on physical examination. Sports and occupations requiring the arm to be elevated above 90 degrees will exacerbate the symptoms. Athletes who throw or lift heavy objects may complain of pain.[47] Local anesthetic injections into the joint may help in the diagnosis.[47] For patients who do not find relief of the arthritic pain or continue to have decreased function, an acromioclavicular arthroplasty is advised. In this operation, the distal 1/2–3/4 in of the clavicle is resected, but the AC ligaments remain intact to allow stability.[4,47]

Adhesive Capsulitis

Many other names for this process have been given, the most notable of which is "frozen shoulder" (affecting the glenohumeral joint). Rather than a distinct entity, this is a final common pathway of deteriorated glenohumeral status produced by other mechanisms (such as biceps tendon pathology, rotator cuff tears, etc.). The pain upon motion causes the patient to limit range of motion allowing joint capsule contracture and adhesions to form on the synovial surfaces. This in turn produces a further reduction in glenohumeral activity (active and passive) and more pain on use of the arm, perpetuating the pathologic process. Examination of passive range of motion will reveal that motion of the scapulothoracic area and glenohumeral movement are extremely limited. Cailliet feels that the combination of shoulder disuse, low pain threshold, and what he calls a "periarthritic personality" predispose to development of this syndrome, and that the shoulder pain elicits vasospasm in some of these patients.[12]

In a study by Binder, et al. 35 of 38 patients had technetium diphosphonate scans showing increasing uptake (p < 0.0001) on the symptomatic side. However, this test gave no prognostic value; 71% of 42 patients had normal plain x-rays.[5] Of 36 arthrograms on these patients, 15 showed adhesive capsulitis and 11 had rotator cuff injuries.[5]

Pathologically, an inflammatory process sets up in the tissues between the rotator cuff and deltoid, leading to edema, fibrosis, vasospasm, and finally adhesions of subdeltoid bursa layers, within the glenohumeral joint capsule and outside the joint capsule.[12] This area is full of sympathetic nerves and blood vessels. Later, biceps, subscapularis, and shoulder girdle muscles contract to further restrict internal/external rotation initially, then in all directions, with a great reduction in pain.

Radiographs often show humeral head osteoporsis.[64] An arthrogram usually permits 10–20 ml of contrast material to enter the joint, but in adhesive capsulitis, allows only 3–5 ml to enter the joint cavity under pressure. In some cases, enough pressure can be generated to separate the capsule layers and permit improved movement.

One prospective study of 42 patients with adhesive capsulitis followed for a mean of 44 months showed 11 with mild and five with severe loss of objective shoulder range of motion.[6] These people adapted well, with many of them perceiving no functional limitations.

Ankylosing Spondylitis

This is one of the seronegative spondyloarthropathies with a high association with HLA-B27. The joint usually involved is the sacroiliac joint. The spinal column is also usually involved, but about 25% of these patients have peripheral joint problems, and in some cases shoulders can be affected.[29] Pritchett describes a case of a 31-year-old with ankylosing spondylitis and right shoulder pain, which resolved with therapy and medications.[63]

This was considered to be a spondylitic enthesopathy, with lymphocytic and plasma cell infiltration into the ligaments leading to inflammation, later replaced by bone.[2,63]

Biceps Tendon (Long Head) Rupture

The biceps tendon receives frictional stresses with most shoulder motions, so over time may develop attritional changes predisposing it to rupture where trying to handle undue loads.[40,47] Abnormalities of the intertubercular groove and trauma act to initiate many of these cases. A "snap" may be felt followed by visible, palpable upper arm bulge, which is the contracted muscle belly.[47] Onset of localized pain is instantaneous. Weakness (compared with the normal side) can be seen when the patient tries to flex the shoulder against resistance, and pain can be referred to the forearm.[4] Pain often limits shoulder range of motion, which must be remembered in treatment. Palpation of the intertubercular groove reveals an absence of the bicipital tendon. An arthrogram can show leaking of contrast material into the vacant bicipital groove. Acute cases (within 6 weeks of trauma) should be referred to an orthopedist for surgical management, such as a biceps tendon transfer to the coracoid process, to render the biceps more functional and increase its strength.[47] If several months have elapsed from time of the rupture, the tendon will contract enough that mobilization of the tendon is difficult.

Brachial Plexus Injury

Trauma is the most common cause of brachial plexus problems, with motorcycle accidents ranking first. Other causes are: low velocity traction injuries, surgery of the axilla, penetrating injuries, falls, shoulder dislocations, tumor compression, rheumatic conditions, and radiation injury (which takes 9–12 months to develop). Traction usually causes preganglion injury (i.e., proximal to the dorsal root ganglion), while postganglion injuries are more often from direct penetrating violence. Classic patterns include Erb's palsy (or "flail arm"), which is an upper trunk lesion (i.e.,C5-6) and Klumpke's paralysis from a lower trunk lesion (i.e., C8-T1). Two types of neuropathic pain are involved, the more common background burning, tingling, and crushing sensation, and in some patients, intermittent sharp pains radiating from the fingers to the shoulder. Many treatments have been tried but none are consistently helpful. A good history of the mechanism of injury and appropriate physical examination along with electrodiagnostic techniques will delineate the extent of the injury. This is the first step in determining ultimate functioning of the arm for the patient. Mental distraction techniques and transcutaneous nerve stimulator (TENS) units seem to offer better pain relief in some patients.

Kaplan, et al. reported five stroke patients who presented with hemiplegia and flaccid proximal upper extremities that caused traction injuries, as well as pointing out that many of these lesions may take 8–12 months to allow enough reinnervation to strengthen the muscles.[40]

Brachial Plexitis

This entity is also called brachial neuralgia, and neuralgic amyotrophy. The etiology is unclear but may be related to viral infections, trauma, and immunizations.[23,43,61] Clinically, there is intense shoulder, neck, or upper arm pain of sudden onset (sometimes with radiation into the arm), which can be very debilitating. Pain can remain for months or abate within several days. This pain is followed by arm paresis or paralysis (within hours of onset, usually in the upper trunk), loss of sensation patterned by the affected roots, and, later, muscle atrophy. This is bilateral in about 25% of patients. Recovery is likely but may be delayed for up to 2 years. EMG shows positive sharp waves and fibrillations and an interference pattern that is decreased in the affected muscles.[43] Sometimes F-wave studies show deficits in axillary conduction.[43]

Walsh, et al. describe a case of bilateral phrenic nerve involvement during brachial neuritis with the resulting bilateral hemidiaphragm paralysis (confirmed by phrenic nerve conduction studies).[81] This is because of the C5 root shared by the brachial plexus and the phrenic nerve.

Bursitis

Subacromial and subdeltoid bursae may exist as a fused or separate anatomic variant structure.[31] It separates the superior aspect of the supraspinatus muscle from the coracoacromial arch and deltoid muscle.[5,31] Bursitis occurs as a secondary phenomenon in the context of attritional changes of the supraspinatus tendon. As the supraspinatus tendon develops calcium deposits, its intimate contact with the subacromial bursa above allows calcium to break through into the bursa. This sets up an inflammatory response in the bursa producing sudden severe pain (due to stimulation of nearby sympathetic subdeltoid fascia nerve supply), which is exaggerated with shoulder motion.[12] Pain is fairly localized to the anterolateral shoulder area and later can become a dull aching sensation. Since the supraspinatus tendon calcification can erode into the bursa several times in order to empty completely, this pain can be recurrent and over time can cause fibrosis of the bursa against the surrounding structures. The result of this is "frozen shoulder" with a severe loss of joint motion. It should be remembered that other causes of bursitis include overuse syndrome, direct trauma, gout, and rheumatoid arthritis.

Nonsteroidal anti-inflammatory drugs can be used to help with the pain. Barbotage of the calcium crystals/debris (in acute calcific tendonitis) has also been recommended, which can be followed by steroid instillation into the bursa.

Degenerative Joint Disease (Osteoarthritis)

Post-traumatic arthritis is well known. Falls, throwing motions, swimming, many major sporting activities (both contact and noncontact) are examples of trauma involved. Repetition of minor injuries is common in healthy people but since the stresses to the shoulder are often nondramatic, they are not remembered by the patient. Another contributing factor is anterior shoulder joint capsule weakness leading to recurrent subluxations and predisposing to increased degenerative changes.[4] Pain is felt in the glenohumeral joint during motion and there may be associated stiffness. Pain in the shoulder joint itself is a common symptom,[16] and in the later stages, the pain can be felt at rest as well.

Conservative management includes nonsteroidal anti-inflammatory drugs and therapy. If pain becomes intolerable, acromioclavicular arthroplasty may relieve some of the pain and partially restore some of the lost shoulder motion.

Dislocation/Subluxation

This can occur in an anterior or posterior direction. Four major factors of recurrent anterior dislocation are youth, epilepsy, poor muscle coordination and anatomical anomalies.[67] Often, the first episode of shoulder dislocation is traumatic, with the arm placed in a susceptible position with an abnormal stress applied, which levers the head of the humerus away from the glenoid cavity. Examples of this are (1) having the shoulder hit from behind while in the cocked position (as in football) and (2) a fall onto an outstretched, abducted arm with the weight of the body acting as the force. Pain, of course, is instant and localized, but dysesthesias and muscle impairment of the arm are possible if the brachial plexus is compressed by the humeral head.

Anterior dislocations (subcoracoid, subglenoid, subclavicular) occur about nine times more often than posterior dislocations.[12,44] Pain or soreness may last days after the dislocation is reduced. These dislocations may be related to tears in the anterior or inferior joint capsule or capsule laxity.[67] Complications include rotator cuff tear, posterolateral humeral head fracture, humeral greater tuberosity avulsion fractures, and inferior glenoid lip fractures. Clinically the shoulder is painful on attempts at motion, and the normal contour of the shoulder is lost. Radiographs are often helpful; one may need the scapular "Y" view where the head of the humerus overlies the end-on scapula.[22] In people over 50 years old, anterior shoulder dislocations may act differently. In a study of six such patients whose dislocations were reduced, recovery showed that passive shoulder range of motion normalized before active range (although both were normal after 14 weeks), and only one redislocation happened.[83] The authors felt that longitudinal tears of the rotator cuff had occurred but had healed on their own.[83]

Recurrent subluxations are secondary to glenoid labrum tears, which can produce anteroinferior pain and a "catching" sensation.[47] Pain is usually temporary but may radiate down the arm. Recurrent dislocations sometimes are painless but can cause considerable dysfunction and can be stabilized with surgical intervention. In a positive "apprehension test," force is applied to the posterior glenohumeral area while the arm is extended, abducted, and externally rotated, which produces pain and fear of dislocation.[47] A "West Pont" view of the anterior glenoid rim in cases of recurrent dislocation is used to check for rim fractures.[47] These recurrent forms can be assessed by arthrography.

In an EMG study of 11 cases of dislocation (some with humeral head fractures), the primary nerve damaged was the axillary nerve (ten cases); the brachial plexus was damaged in two cases; the musculocutaneous nerve in five cases, and the posterior cord in five cases.[51] The advantage of doing EMGs after dislocation is that it detects true changes in innervation (as opposed to pain leading to false weakness). The musculocutaneous nerve may be damaged by external rotation and traction inferiorly.[51]

Posterior dislocations can be congenital but usually are seen resulting from convulsions and trauma when the humerus is internally rotated with force. Sometimes a patient is unable to give a history of preceding trauma.[4] These patients present acutely with pain plus limitation of external rotation. There can be associated fractures of the humeral head and posterior glenoid rim. Many of these dislocations are missed because the anteroposterior radiograph looks normal. An axillary radiographic view is needed to show the humeral head posterior to the glenoid rim.[44] The axillary nerve is vulnerable to damage from posterior dislocations.[43] Treatment of acute cases involves closed reduction and immobilization via strapping of the affected arm in external rotation for about 4 weeks, followed by mobility exercises. Since some patients are unable to recount the traumatic event that initiated the shoulder pain, it is not uncommon for these patients to present in the emergency room several weeks after the initial dislocation. These cases usually require surgical reduction since the elapsed time has allowed muscles to lock the humeral head against the posterior glenoid after having shortened.

Recurrent posterior dislocations are troublesome because as time goes by, any flexion of the affected humerus (especially if internally rotated) will allow the dislocation to occur. Surgical stabilization is required.

Fractures

This is an obvious case of pain in most cases if the history of trauma (usually from motor vehicle accidents and falls) is elicited and if radiographs are available. If the scapula is suspected, anteroposterior, lateral (parallel to the plane of the scapula), and axillary views are needed. Fragmentation and displacement of the scapula is inhibited by the surrounding muscle coverings. Fractures of the scapular body, glenoid fossa, acromion, scapular spine (all if without significant displacement), and the coracoid process (if with intact ligaments) are treated conservatively. But if these structures are displaced, and there are fractures of the surgical or anatomic necks, then operation is indicated to protect the neurovascular bundle, prevent muscle imbalance, and restore anatomic alignment.[28] Fractures of the distal clavicle are usually easy to diagnose.

Fractures of the proximal humerus present more of a problem because of potential damage of the surrounding neurovascular structures.

Hemiplegic Complications

The hemiplegic limbs begin as flaccid after the acute event. During this phase, the rotator cuff, deltoid, trapezius, levator scapula, and rhomboid muscles are inactive, which predisposes to lateral rotation of the scaplula and glenohumeral subluxation.[12,38,80] This may be a common mechanism of shoulder pain in the hemiplegic.[26] One study indicates over 50% of the hemiplegic patients will have shoulder subluxations.[55] Smith, et al. indicate that 53% of a stroke population studied showed radiographic evidence of glenohumeral subluxation and that those with the worst paralysis had the most chance of subluxation.[74] The painful hemiplegic shoulder is to be distinguished from the reflex sympathetic dystrophy syndrome (because sympathetic changes often are absent). Proper support and positioning is required in the flaccid stage since traction injuries can take place during transfers and attempts at ambulation. If sensation is intact in the affected upper extremity, then the subluxation and traction upon the soft tissues (brachial plexus, biceps tendon, joint capsule, rotator cuff, etc.) causes dull pain in the shoulder area.

Because of the initial shoulder girdle muscle flaccidity after stroke, the scapula may be pulled inferiorly, and consequently the suprascapular nerve may have more traction placed upon it. This nerve might be entrapped in such patients at the suprascapular fossa. Lee and Khunadorn studied 30 male stroke patients by testing nerve latencies to the supraspinatus and infraspinatus muscles.[48] Latencies were virtually identical in all but three patients, suggesting that entrapment of the suprascapular nerve has no effect in producing shoulder pain in stroke patients.[48] Bohannon, et al. studied shoulder pain in relation to the following: spasticity, external rotation, time since stroke onset, weakness, and age.[7] In 36 of 50 consecutive stroke patients, the factor of external shoulder rotation was the most significantly correlated with pain, suggesting a possible relationship with development of adhesive capsulitis.[7]

Rizk, et al. performed arthrography on 30 hemiplegic patients with spasticity.[65] Twenty-three patients had arthrographies indicating adhesive capsulitis, and seven had normal findings, while all had pain and stiffness in the hemiplegic shoulder.[65] All patients in the study had normal EMGs.

When the patient advances to the spastic phase, the scapula can be shown to have a rotation in the forward and down directions.[12] Spasticity may be a significant factor in development of hemiplegic shoulder pain, as pointed out in a study by Van Ouwenaller, et al. in which 85% of 149 spastic patients versus only 18% of eight flaccid patients had pain.[79] As spasticity increases, the arm assumes a position with more internal rotation and adduction thus reducing the overall range of glenohumeral motion, which predisposes to pain. This pain is an inhibitory factor in the patient moving the arm, and adhesive capsulitis can develop, as well as joint capsule contracture, both of which lead to more pain upon attempts to move the arm. In a study by Moskowitz, et al. at 1-year poststroke, approximately half of 518 patients had glenohumeral joint contractures.[56] Besides the pain involved, the patients have less overall functioning.

Other shoulder problems relating to hemiplegia include exacerbation of arthritis, increased incidence of rotator cuff tears, reflex sympathetic dystrophy, heterotopic ossification, and osteoporosis of the humerus (with an increase in the frequency of fractures).[57,80]

Myofascial Syndrome

Multiple synonyms for this condition in the medical literature are confusing. Many clinicians are trained without even knowing that this relatively common condition exists, leading to multiple referrals to psychiatrists and psychologists for these "vague somatic complaints."[73] Myofascial pain causes many of the "aches and pains" associated with overuse syndrome, and its recognition can lead to effective and dramatic relief with proper treatment. Clinicians aware of this condition realize that the patient can undergo all the ramifications of a chronic pain syndrome. When put in this context, one comes to understand that treatment of myofascial pain and its dysfunction offers a great service to patients. Fortunately, myofascial pain is becoming more readily understood in physiatry, rheumatology, dentistry, and orthopedics.

The concept of myofascial pain revolves around what are called "trigger points." These are areas of muscle and fascial coverings that are extra sensitive. The term trigger point is derived from the characteristic of "triggering" or referring pain to a distant body region upon palpation of the sensitive spot. An "active" trigger point refers pain upon palpation, motion of the muscle, and even at rest. Autonomic changes in a referred pattern, potential for a local twitch response, and muscle weakness are seen.[78] In contrast, latent trigger points produce pain upon being compressed; otherwise they are similar. Latent trigger points are a significant factor in producing myofascial dysfunction since they can reactivate if exposed to cold or if the muscle is overstressed, subjected to direct trauma, or excessively stretched.[78] Study has shown that each individual muscle trigger point has a fairly reproducible pattern of referred symptoms (pain, autonomic changes, weakness). This pattern is consistent for most people in a "reference zone," but some may also feel referred pain in the areas surrounding this zone, called the spillover pain zone.[78] Pain in the referred area is mostly a dull ache but may have an associated deep tenderness.[8] Presence of trigger points after the inciting stimulus can follow a protracted course, thus making the diagnostic connection more difficult. The sensitivity of each trigger point varies during different times and stresses, often with associated muscle weakness and stiffness. Indirect stimuli such as psychic stress, visceral pain, and arthritis can send impulses into the spinal cord then out to the referred pain zone and trigger points to heighten their irritability.[77,78]

Physical examination reveals a tight band of muscle tissue, which is palpable and distinct from the surrounding normal muscle. When directly palpated with pressure, the trigger points cause local pain and tenderness at the area of palpation and produce pain (as well as weakness and autonomic changes) in the referred area. Active trigger points account for some of the spontaneous pain in the referred zone. Since the muscle around the trigger point is hypersensitive and tight, any attempts to stretch the muscle cause more pain. This tight muscle band limits the range of motion, and pain can be induced by taking the muscle to the extremes of the range.

Attempts to scientifically delineate trigger points has been disappointing. Laboratory testing shows no conclusive test to indicate presence of trigger points.[73] Results of electrodiagnostic studies are normal when testing resting muscle.[46]

Simons has the following hypothesis to account for development of trigger points after the muscle has undergone trauma.[73] Damage to sarcoplasmic reticulum allows calcium to escape and act as a stimulus to the nearby contractile mechanisms (along with ATP), which then causes increased metabolic activity. The hypersensitivity of sensory nerves is a result of this metabolic change locally. The localized muscle contraction inhibits appropriate blood supply to the muscle, and eventually the muscle contraction is sustained in an energy-deficient state, represented by trigger points. Simons feels that stretching the contractile myofilaments terminates this uncontrolled contraction, restoring the muscle to a normal state.[73]

Multiple treatment methods have been tried with variable success. The attempts here are to inactivate the trigger points, thus decrease the pain and dysfunction they cause. A pressure technique called ischemic compression over the trigger point causes initial ischemia, followed by an increased blood flow to the area.

Use of the spray-and-stretch technique can be very effective, and the procedure described below is that of Travell and Simons.[78] The patient should be kept warm and as relaxed as possible. A vapocoolant spray is directed about 18 in from the skin surface, at a 30 degree angle, parallel with the muscle fibers and directed from normal skin areas over the trigger point and toward the referred zone.[78] As the patient is being sprayed, one end of the muscle is positioned so that it will not move, then the muscle is passively stretched. The purpose of the spray is to distract the patient's awareness so that a more effective stretch (which produces some pain) can be carried out. The spray also stimulates cutaneous afferent nerves to act at a spinal level to overrride pain and tension signals coming from the affected muscle, thus interrupting the pain-spasm cycle. Hot packs are placed over the cooled skin once the stretch is accomplished, to help relax the patient. Patients are then directed to stay relaxed and avoid strenuous activities for the remainder of the day, which may bring on the myofascial pain again.

Another commonly used technique for inactivating trigger points is "needling" or injections. Mechanically breaking up the trigger point is likely a large factor in making this technique effective, since needle contact (without injection of any substances) with the trigger point achieves results similar to injections of local anesthetics, saline, or steroids.[49] The patient is placed in a prone position, the trigger point is located by palpation (with the position marked), the skin is swabbed with an antiseptic, the trigger point is held in place with the noninjecting hand, and the physician probes the area with the needle until the patient feels that the trigger point has been reached (confirmed by producing a referred pain pattern), then the area is infiltrated with a small amount of local anesthetic.[54,78] Before injecting, the plunger should be pulled back to check for any blood aspiration (i.e., inadvertent entry into a blood vessel). The local anesthetic helps decrease the initial pain of the needling. The needle is then withdrawn into the subcutaneous tissue and redirected to surrounding areas to check for and eliminate other trigger points. Immediately after the injection, vapocoolant spray and muscle stretch should be done. Hot packs to the area additionally relieve some of the pain of the injection.

Myofascial pain is common in the shoulder region. A study by Sola, et al. indicates that about half of the 200 people studied had latent trigger points in shoulder-related musculature.[75]

Bonica feels that myofascial syndrome is the most common shoulder musculoskeletal reason for pain.[8] Several muscles are notable for the pattern of referred pain they produce in the shoulder region. The supraspinatus myofascial pain may be similar in presentation to subdeltoid bursitis.[78] Pain in the anterior portion of the shoulder joint can be due to infraspinatus myofascial trigger points.[78] Travell and Simons indicate that subscapularis trigger points often are the initiating cause of idiopathic "frozen" shoulder.[78] The scapulocostal syndrome proposed by Michele, et al. (described in the section on thoracic outlet syndromes) was felt to cause shoulder pain in about one third of patients (which contrasts with the orthopedic view of the glenohumeral joint as the most frequent site of shoulder pain).[4,54] Many other muscles surrounding the shoulder girdle refer pain to the shoulder region, so myofascial pain should remain high on the differential diagnosis list.

Osteomyelitis

This bone infection can result from surface bacterial contamination from direct penetrating trauma or from fractures to the upper humerus, distal clavicle, or scapula. The infection can also occur at the metaphyseal region of the proximal humerus by hematogenous spread. The main form of therapy consists of intravenous antibiotics (depending upon wound or biopsy culture and sensitivity) for an extended period, possibly followed by oral antibiotics. Joint dysfunction with resulting pain can occur in cases of new infection if the patient does not have the antibiotics started within several days. If the clinician plans to aspirate fluid from the glenohumeral joint, cellulitis needs to be ruled out in order not to introduce bacteria (from the cellulitis) into the joint space. Decompression of joint space infections (which spread from bone infection) via needle aspiration (and irrigation of the joint as well in situations with thick purulent material) or even joint suction tubing placement may be needed. In acute cases, surgical intervention should be considered if pain does not improve after 24 hours of intravenous antibiotics or if starting of the antibiotics has been delayed more than 72 hours. Some cases become chronic and can have draining fistulas.

Osteonecrosis

This is also known as avascular or aseptic necrosis is due to loss of blood supply to bone causing degeneration of a segment of that bone. Osteonecrosis has been associated with diabetes, alcoholism, systemic steroid therapy, hemoglobinopathies, trauma, and radiation.[16] Pain is localized, can be fairly severe, and often comes on suddenly. Radiographs may show the "crescent" sign (a sclerotic line of bone indicating microfractures) in the proximal humerus. The use of steroids predisposes to capillary fragility, which could contribute to the ischemic etiology.[16] Technetium pyrophosphate bone scans can aid in the diagnosis by showing decreased uptake.[16]

Overuse Syndrome

This typically occurs in the "weekend athlete" who overuses the shoulder in an activity for which the arm is unprepared. The pain, usually a diffuse aching in the shoulder region, is a combination of muscle strain and acute tendinitis, lasting several days.

A second type of overuse is more chronic and often seen in rehabilitation medicine. This involves nonambulatory patients confined to wheelchairs. These people now use the upper extremities to propel their wheelchairs up to several miles per day on various surfaces and grades, which places mechanical stress on shoulder and other arm joints and soft tissues. These people also use the shoulders as the major weight-bearing joint upon transfers (i.e., sliding board, trapeze, etc.), adding to the wear and tear on these structures. Some of the more common conditions requiring prolonged use of a wheelchair include spinal cord injuries, strokes (in which only one arm may be functional), head injuries, lower extremity fractures, and many others. Even though manual wheelchair skills are standard in rehabilitation education of the patient and use of the arms in sustained physical activity may aid in cardiovascular fitness, the chronic stress upon the shoulder mechanism may be damaging. Some health care professionals use of electric wheelchairs or at least alternating electric and manual chairs, to avoid accelerating osteoarthritis.

Referred Pain

Cervical radiculopathy characteristically produces a shock-like pain in a dermatomal distribution, along with numbness and tingling sensations. Neck motion in extension and rotation often results in the pain radiating below the level of the shoulder. The Valsalva maneuver, sneezing, and coughing increase the pain in many cases. The examiner should look for weakness in muscles supplied by the affected nerve root. Cervial radiographs may show the narrowing of the disc spaces, and oblique films may show the narrowed foramina. EMG studies, MRI, CT scans, and myelography can often provide the diagnosis.

Reflex Sympathetic Dystrophy Syndrome (RSDS)

This entity is common, especially in rehabilitation medicine. It goes by many other names, most commonly "shoulder-hand syndrome," and is seen in about 6%–24% of hemiplegics (onset in 2–6 weeks), 5%–20% of coronary artery disease (onset in 6–8 weeks), 20% of cervical disc disease (with insidious onset), and many peripheral nerve injuries.[21,76] The pathogenesis is unknown, but several theories exist. In the gate theory (Melzak and Wall) specialized inhibitory cells of the dorsal horn substantia gelatinosa[52] are inhibited by small C-fibers (unmyelinated), which "opens the gate" to pain; it is felt that RSDS is carried by the C-fibers. Large A-fibers (myelinated), however, will stimulate the substantia gelatinosa, thus "closing the pain gate." The theory of Doupe, et al. states that trauma causes a connecting artificial synapse (an ephapse) between the sympathetics and the sensory afferents to allow a pain cycle to exist.[20] Chronic painful stimulation could cause an internuncial neuron pool overload leading to intracord bursts of activity, leading in turn to continuous sympathetic and motor discharge, with an eventual reverberating pain cycle.[50]

Onset can be immediately, several hours, or even weeks after the inciting event (thus the etiology may escape discovery). RSDS may last for years. The most common symptom is pain in the affected extremity, which is disabling. It is intense, burning, prolonged, and may be severe enough to cause drug addiction or suicide.[70] There does not seem to be a clear, direct relation to a dermatomal pattern or to a peripheral nerve distribution, possibly because the sympathetics run in a diffuse pattern with the blood vessels. Pain upon squeezing the metacarpal heads together often is an early sign. Edema is next most common, existing mostly on the loose skin of the dorsal surface of the hands, which eventually limits finger flexion and grip. Later, trophic changes occur. Vasodilatation usually occurs early, followed by vasoconstriction of the limb. Joint range of motion becomes limited from pain and edema. The above deficits combine to delay functional recovery. Symptoms can be increased with temperature changes or with increased emotions.

There are three developmental stages. Stage I (acute) presents with pain, edema, vascular changes, and decreased range of motion and can last up to 3 months. Stage II (dystrophy) begins to show the trophic changes (brittle nails, shiny/smooth skin, hair loss) and further loss of joint range. This can last up to 6 months. Stage III (atrophy) shows skin and soft tissue atrophy and bone erosions or osteopenia on radiographs, and the pain may spread proximally.[21] This stage has a poor prognosis. Radiography may show patchy osteoporosis, as well as bony erosions of the metacarpophalangeal, proximal interphalangeal, and distal interphalangeal joints. A triple-phase technetium pyrophosphate bone scan is sensitive, detecting changes earlier than plain films and often showing increased periarticular uptake.[16]

Associated conditions include trauma (fractures, minor trauma, tight-fitting casts), neurologic disorders (hemiplegia, spinal cord injuries, radiculopathies, peripheral nerve lesions, postherpetic neuralgia, herniated disc, brain tumors), postmyocardial infarction, postcoronary artery bypass, arterial occlusion, shoulder and cervical degenerative joint disease, pulmonary tuberculosis, diabetes, and isoniazid and barbiturate use.[21,76] The syndrome, though, can also be idiopathic.[21]

Success in outcome depends upon early diagnosis and treatment. Prognosis is better if treatment is begun within 6 months of onset.

Rheumatoid Arthritis

The shoulder is sometimes affected in patients with rheumatoid arthritis, and usually, as with other joints involved, in a symmetric pattern. The two joints affected by rheumatoid arthritis are the glenohumeral and acromioclavicular joints.[45] Pressure is developed inside the joint capsule from the effusion and results in decreased range of motion (with external rotation and abduction restricted the most). Later, fibrosis of the joint capsule limits motion further. Symptoms include joint pain and limited joint motion with stiffness (especially in the morning).[16] Joint effusion is often difficult to detect because of overlying muscle disguising joint space fullness. Most patients develop joint problems slowly, although acute polyarthritis is well known. The patient seems to have weakness involved as well. This weakness can be secondary to rheumatoid arthritis or to pain inhibiting full muscular activity. A later complication includes subluxation.[16] A case of serratus anterior muscle traumatic avulsion was described by Meythaler, et al.[53] The predisposing factors contributing to rupture in this case included limited shoulder range of motions (thus making serratus anterior rotation of the scapula mechanically stressful), muscle atrophy/ paresis (from the disease and treatment), and a subscapular hematoma.[53] This would, of course, produce pain as well as further dysfunction of the shoulder joint, since the serratus anterior is no longer effective to produce a compensatory scapular rotation upward in cases of glenohumeral restriction.[31]

The case of a patient with 14 years of rheumatoid arthritis who developed a painless swelling, first thought to be an arterial thrombus, has been described by Dejager and Fleming.[19] Vascular studies were negative, but an arthrogram showed joint capsule rupture. The mechanism might be secondary to development of inflammatory cysts and capsule distention by an effusion leading to rupture through the capsule, if shoulder motion produces high enough pressure.

Pathogenesis may relate to either of two mechanisms: the extravascular immune complex hypothesis and the cellular hypersensitivity hypothesis.[66] In the first hypothesis, local antibodies bind with antigens (articular tissue components) to form complexes that trigger the complement cascade, leading to more vascular permeability. This allows local collections of edema and neutrophils that attempt to phagocytose the complexes, but in doing so, release destructive enzymes and free radicals, which themselves perpetuate the inflammatory process. The second hypothesis involves activated T-cells that may mediate the inflammation. This inflammation destroys cartilage, bone, ligaments, and tendons.[66] A fibrovascular granulation tissue called pannus destroys the cartilage, bringing the bony opposing surfaces closer together, and may eventually lead to adhesions, ankylosis, pain on motion, and limited motion—all of which produce severe deformities and progressive loss of functional abilities.

Medical management consists mainly of drug therapy. The "pyramid" method of treatment is considered standard for progression of drug use.[66] The patient (unless medically contraindicated) is started on ASA or nonsteroidal anti-inflammatory drugs. If these fail, step two of the pyramid is use of gold salts or hydorxychloroquine. Next are steroids and penicillamine. The final step is immunosuppressant drugs. The reader is referred to standard rheumatology and pharmacology texts for details of dosages and drug use.

Total or hemiarthroplasty of the shoulder may be needed for refractory pain[4,66] or shoulder instability.[4]

Rotator Cuff Injuries

Along with the aforementioned functions of these four muscles, the rotator cuff also acts to help hold the humerus in approximation to the glenoid cavity. Rotator cuff tears may be caused by a fall on an outstretched arm or direct trauma. However, Neer feels that impingement accounts for 95% of cuff tears.[58] This is often seen in people over 50 years old from repeated trauma (especially in physical endeavors).[14,25] Symptoms progress and continue. Pain is localized to the insertion area of the cuff muscles and occurs mostly at 60–70 degrees of abduction.[4] The patient can have a dull ache several days after the acute rupture.[47] Since the supraspinatus is important in humeral abduction and flexion, these movements are weak, and the patient often hunches that shoulder while trying to produce the movement. External rotation and abduction may produce a "catching" feeling. External rotation is weak because of damage to the infraspinatus and teres minor. Atrophy can occur from inability to use these muscles.

Neer describes three stages of pathogenesis.[58] Stage I deals with trauma or overuse of the shoulder, causing some bleeding and localized swelling. In stage II, bursal fibrosis occurs with repetitive trauma or mechanical irritation, and tendonitis may be associated. Stage III consists of continued mechanical irritation leading to degenerative changes of the cuff (especially the supraspinatus tendon) and possible rupture.

One mechanism felt by some to predispose to tearing is the weight of the arm causing repeated ischemia to the supraspinatus tendon, with the eventual result of weakening this area.[12,25] The idea of a "critical zone" where ischemia develops is challenged by some who feel that the blood supply to the tendon is adequate and that calcium deposition occurs upon live collagen (i.e., not an ischemic area).[4] Since this is a frequently used group of muscles, repetitive microtrauma gradually erodes the tissue. Then a final insulting trauma to the tendon causes it to rupture.

Radiographs may show a superior migration of the head of the humerus, sclerosis, cystic changes, and cuff tendon calcification.[25,47] Anteroposterior and external and internal rotation views are needed to detect associated calcified ligaments.[47] An arthrogram showing contrast material in the subacromial space is diagnostic for a tear of the rotator cuff.[4,12,25,47] A double-contrast arthrogram may reveal even smaller areas of pathology in more detail.[33] If the tear is large, contrast material fills the joint space in an irregular pattern. Even though local anesthetic is often used, an arthrogram can be painful itself, which must be anticipated to avoid misinterpreting it as an extension of pathology. Patients who do not use the muscles because of pain, and thus cannot produce adequate active range of motion are at risk to develop adhesive capsulitis. Therefore, patients with complete tears should be referred to an orthopedist for surgical intervention. This would include anterior acromioplasty, subacromial bursa resection, and coracoacromial ligament division to produce 70%–90% "good" or better surgical results.[25]

Concomitant neuropathy of the suprascapular nerve and rotator cuff tears was shown by Kaplan and Kernahan to confuse the clinical picture of the cuff tears.[39] They showed that in one complete and five partial cuff tears, suprascapular nerve latencies (from Erb's point to the infraspinatus) were prolonged, along with fibrillation potentials, decreased number of motor unit potentials, and positive sharp waves in the infraspinatus muscle on the affected side.[39]

Septic Arthritis

Sudden presentation of an acute arthropathy of a single joint, sometimes with other systemic signs (fever, malaise, diaphoresis, etc.) will lead to this diagnosis, which is a medical emergency. Etiology can be bacterial (including gonorrhea, especially in young adults), tubercular, fungal, syphilitic, and viral. A case of acromioclavicular joint pyarthrosis has been reported by Griffith and Boyadjis.[27] Radiographs help differentiate this from acute calcific tendonitis. Joint aspiration usually shows a white blood cell count over 50,000/ml and over 75%–90% polymorphonucuear leukocytes.[29,45] Pus removal is needed to remove the source of inflammatory debris, which is destructive to the articular cartilage.[68] Saline irrigation should not be too vigorous nor should too large a volume be used, since joint capsule rupture can occur. Joint fluid culture and sensitivity will guide in correct antibiotic therapy.

Steroid Arthropathy

This is caused by prior injections of steroid suspensions directly into the glenohumeral joint. It can cause inflammation of the synovial lining, resulting in pain and decreased shoulder mobility. Later the effects include severe damage to the cartilage.[4]

Tendinitis

Tendon pathology occurs as follows: injury causes hypoxia, which leads to matrix/crosslink disruption plus tendon metaplasia, causing microvascular injury, with consequent tendon weakening and finally rupture. Pain occurs in the shoulder/deltoid area, is common at night, increases with movement, and may be associated with crepitus. Muscle weakness or atrophy may occur secondarily from lack of use of the arm to avoid pain. Symptoms usually are progressive and gradual, and the patient may have pain-free intervals.

Tendinitis is the most common shoulder pain in athletes.[47] In the clinical entity called "impingement syndrome" flexion/abduction in the range of 80–120 degrees compresses the rotator cuff and subacromial bursa between the acromion and the raised humeral head to cause sharp pain.[14,47] Throwing and overhead swimming strokes predispose to this.[14,47] The resulting bursal inflammation causes distension and increases the likelihood of further impingement, as does development of a bony spur on the anteroinferior aspect of the acromion. The experience of nighttime pain may relate to small adhesions formed between cuff and bursa, which lyse with movement.[47] To clinically test for the "impingement sign," the arm is passively or forcefully elevated in flexion, leading to the greater tuberosity impinging on the acromion (while the scapular is held down) and causing pain.[58] To perform the "impingement test" about 10 ml of 1% lidocaine injected under the acromion gives relief.[68] Surgical release of the coracoacromial ligament can be used in refractory cases of impingement in which the cuff is intact in athletes under 20 years old. Older patients with impingement syndrome may need an anterior acromioplasty.[47]

Bicipital tendinitis is fairly common, especially after overuse of the dominant arm in an activity for which the arm is not properly trained. Palpatation of the tendon in the biceps groove usually produces tenderness (which should be compared with the "normal" side). A positive Yergason's test (forearm supination and elbow flexion, both against resistance) aids in the diagnosis.[4,45] Special radiographic views of the bicipital groove may provide clues as to current or future problems related to the tendon. Instillation of a local anesthetic agent (such as 1%–2% lidocaine) with or without steroid into the tendon sheath is often effective management. If steroids are used, one must be careful to inject the suspension into the tendon sheath since injection directly into the tendon (which can be determined because forceful pressure is needed to inject the suspension) can lead to collagen weakening and later rupture of the tendon under stress.[45] Nonsteroidal anti-inflammatory drugs can be used to assist pain relief.

Supraspinatus tendinitis must be understood in the context of the ultimate problem of rotator cuff tears, and the reader is referred to this section on page 113. Initially, hyaline degeneration can be seen in the tendon, and the synovial sheath may begin to fragment; fibrinoid degeneration and fibrosis contribute to loss of tendon strength.[4] Repeated shoulder motion compresses the tendon against the coracoacromial arch structures and causes further fragmentation and calcium deposition by inflammatory cells onto the collagen which can bulge out against the subacromial bursa to cause painful inflammation.[12,59] Calcific tendinitis can cause severe, disabling pain. It is an easy diagnosis to make if the shoulder soft tissue radiographs show the calcium, but if the shoulder is overused or traumatized, the reactive hyperemia around the tendon may cause the calcium deposits to be undetectable.[12] Injection of local anesthetics into the subacromial bursa can be used to help mechanically break up the calcium deposits.[47,59]

Thoracic Outlet Syndromes (TOS)

This actually an aggregate of several mechanisms with the common denominator of compression of nerve or vascular structures in the neck area. If the arterial supply is compressed, the patient may have decreased pulses and pallor in the affected extremity and, later trophic skin and nail changes. Emboli may result from intermittent compression. Nerve symptoms include paresthesias, dysesthesias, a sense of hand swelling, and pain. Weakness may be seen in either type, which if long-standing will progress to atrophy. All patients should be questioned as to the arm position they assume when sleeping. If the arm is held with the humerus against the head, it is reasonable to assume that nighttime or early morning symptoms of pain and paresthesias are due to compression. Counselling on different sleeping positions of the arms may be curative in some cases.

One proposed mechanism is the scapulocostal syndrome.[12,54] As the person progresses through the day, fatigue leads to postural changes. Underdevelopment of strength of shoulder girdle supporting muscles, older age, and paralytic states can predispose to these postural alterations.[4] Among these is a downward migration of the scapula which places traction on the levator scapula muscle, compressing its blood supply and initiating spasm. The medial/superior tip of the scapula rotates laterally. From this pathologic process, a myofascial pain syndrome can develop with trigger points and pain referred in a radiating pattern down the arm and up to the head and neck region.[12] This mechanism puts traction on the prevertebral fascia, which contributes to the pain.[12] Since the scapula rotates such that the glenoid fossa may tip downward, it is conceivable that the rotator cuff may come into increased stress to maintain the head of the humerus against the glenoid cavity. If this occurs, pain from the rotator cuff under traction may confuse the pain problem. Such patients may have rounded, sagging shoulders and describe occupational positional stresses. Trigger point injections or use of the spray-and-stretch technique may benefit the patient.

The most easily understood of these mechanisms is the cervical rib. This is commonly asymptomatic, being an extra finding on chest radiography.[9] It occurs in under 1% of the population.[4,12] Compression involves the subclavian arterial supply to the arm in most cases, but if nerves are also involved, symptoms center around the C8-T1 nerves (i.e., the lower trunk of the brachial plexus), affecting these dermatomes and making it necessary to differentiate this from ulnar neuropathy. Poor posture allows the compression to progress. Reaching, working with the arm above the shoulder level, and lifting exacerbate the pain. A bruit may be heard over the supraclavicular fossa. If conservative measures fail, retrograde arteriography will be needed before performing surgery to resect the cervical rib or scalene division.[12]

In the hyperabduction syndrome, the pectoralis minor muscle compresses the neurovascular structures against the ribs when the shoulder is abducted, elevated, and posterior to the coronal plane. Pressure applied by the examiner in the subcoracoid region over the pectoralis minor (thus manually compressing the neurovascular bundle) may reproduce some of the symptoms.[4] For persistent symptoms, surgical division of this muscle may provide relief.

Another type of thoracic outlet syndrome is the anterior scalene syndrome. The anterior scalene inserts upon the first rib, separating the subclavian artery (which runs posterior) and the subclavian vein. The median scalene muscle attaches to the first rib posterior to the neurovascular bundle. Scalene muscle hypertrophy may increase the chances for compression. The compression occurs between these two muscles and the first rib when (1) the head extends backward and turns toward the affected side, (2) with the arm abducted and externally rotated, and (3) when a deep inspiration is held. This is the basis of the "Adson maneuver," showing the examiner that the radial artery pulse disappears. Symptoms often occur at night, sometimes causing early awakening. Scalene muscle spasms may prolong some of the symptoms. Local anesthetic injections into the scalene muscle may help to relieve pain and reduce spasm. Scalenotomy may be used as a final resort.

In the claviculocostal syndrome, nerves or vessels are compressed between the clavicle and first rib when the shoulders are depressed and retracted. A supraclavicular bruit is sometimes heard. Postural fatigue is a main contributor, but abnormalities such as congenital hemivertebra and cervical scoliosis alter the plane of orientation of the first rib.

The patient is frequently referred for electrodiagnostic studies to help delineate the diagnosis of thoracic outlet syndrome. Stimulation of the brachial plexus above the clavicle in order to calculate nerve conduction across the axilla has several problems; (1) patient pain, (2) the stimulation point can be inferior to the actual site of compression, (3) inaccurately prolonged latencies are seen with submaximal stimulation intensities, and (4) there is the potential for significant nerve length measurement error.[7] Weber and Piero developed a method of predicting F-wave latencies with a multiple regression formula, such that a latency of 2.5 msec above the mean value plus 2 standard deviations is significant to indicate an abnormality.[82] Ulnar F-Wave latency may be prolonged but must be compared with the uninvolved arm.[43] Denervation of the triceps and the muscles of the forearm and hand in the ulnar distribution, and decreased ulnar nerve sensory action potentials may be seen.[43] If the intrinsic muscles of the affected hand show a large number of polyphasic motor unit potentials, this supports the diagnosis of neurologic etiology and is stated to be common in thoracic outlet syndrome.[13] C8 root stimulation with latency measured at the hypothenar eminence can be done. The technique involves testing with an abducted shoulder to 90 degrees and elbow extended. Johnson describes this root-to-distal wrist conduction velocity as 52–56 m/sec, with less than 1 msec difference from the unaffected arm.[34] EMG and NCV's often do not provide the information sought since most commonly, thoracic outlet syndromes are vascular in etiology.

Transverse Humeral Ligament Tear

Rupture of this ligament (which holds the long bicipital tendon in the intertubercular groove) usually is due to trauma. Examination shows medial displacement of the tendon out of the groove with a snap when the shoulder is externally rotated and abducted. If the ligament is torn, operative repair is indicated.

TREATMENT

Because of the complexity of the shoulder mechanism and the proximity of many structures, treatment of the numerous pathologic processes in the shoulder is often challenging. It is difficult to treat one structure in isolation, and consideration is given to the response of the other tissues that are affected by the treatment process. Favorable results are often achieved from treatment of surrounding tissues affected by inflammation or fibrosis. In other situations, appropriate treatment of one tissue stresses another, as in rheumatoid arthritis, in which multiple tissues are involved (Fig. 5-1).

Increasing attention is being paid to prevention of shoulder injuries, especially those resulting from participation in sports. Work-hardening techniques are useful in preventing occupation-related injuries as well. Prevention of shoulder trauma is accomplished through

1. Screening and evaluation of abnormal findings
2. Appropriate stretching of tight structures
3. Strengthening of weak muscles or muscles that are stressed or stretched
4. Correction of postural imbalances

Underlying pathologic findings are treated before the patient is allowed to participate in stressful activities.

Numerous therapeutic strategies, modalities, and techniques are useful in the treatment of the shoulder and are chosen with the affected tissue and desired response in mind. The effects of some of the modalities available are discussed in greater depth in Chapter 4. In treatment of the shoulder, exercises and manual therapy are often the most valuable techniques, requiring skill and experience for best results. As in the treatment of other parts of the body, reduction of pain and inflammation, promotion of healing, and maintenance or restoration of mobility are therapeutic goals. Because however, of the mobility of the shoulder complex required for normal function, special attention must be paid to avoid restricted mobility, which results from limited use after injury. Preventing restriction requires constant vigilance and close patient supervision. Once mobility is lost, restoration of function is difficult. For example, adhesive capsulitis is usually a result of an injury rather than a distinct entity, as previously mentioned, and can be prevented by maintaining the greatest possible range of motion permitted by pain while healing occurs.

Here the necessity of a close working relationship between the patient and the clinician is evident. Good communication and mutual trust enhance patient compliance and so invariably improve the outcome of treatment. Frequent visits initially to teach patients how to help themselves and to oversee the program they are expected to carry out will

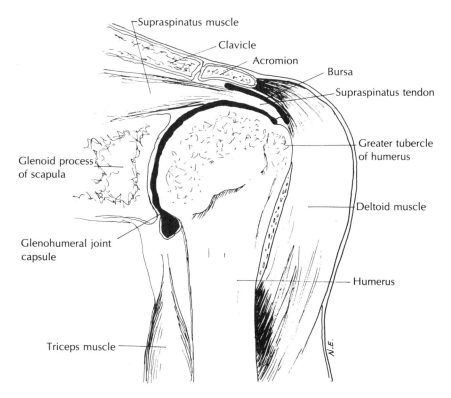

Figure 5–1. Close proximity of multiple tissues precludes isolated treatment of one tissue alone. In the shoulder, as in many other joints, treatment might affect skin, fascia, muscle, tendon, bursa, joint capsule, periosteum, cartilege, and bone.

also improve the outcome. The possibility of incorrect or even dangerous activities being carried out erroneously by the patient will thus be reduced. The clinician can also monitor the progress of the healing (or lack of progress) and make the necessary corrections to the treatment regimen in a timely fashion. This close supervision is most important for the first 2–3 weeks, when patients are in the most pain and most likely to make errors in the home program or to allow themselves reduced pain by not moving the shoulder. Once patients have pain under control, either through medication or modality use, and are able to accomplish the prescribed exercise program correctly, diligence can be relaxed, especially with motivated patients.

Since histologic evidence of fibrosis is detectable almost immediately after an injury and gross evidence of restriction of movement might be demonstrated in as little as 4 days,[45] early intervention after injury is essential in order to prevent complications associated with trauma.

The inflammatory process, which is often accompanied by edema and possibly by reduced circulation, promotes the formation of fibrosis.[45] Patients tend to hold an injured joint in the resting position or in that position in which there is the least pain. As long as there is mobilization of the affected joint at appropriate intervals, healing should be allowed to take place in the resting position. If a sling, or other immobilizing device, is applied, the patient should be cautioned to remove it and mobilize the arm at intervals throughout the day and evening. Patients also should be instructed in the difference between the pain associated with increased trauma and that which accompanies appropriate, therapeutic stretching. They can then better judge how much discomfort to tolerate and how to avoid pain when mobilizing the shoulder, either alone or in therapy.

Treating the shoulder is difficult in cases with radiation of pain from the joint and immediate surrounding structures to the area of the insertion of the deltoid. Patients seek treatment for pain felt at that point, and it can be hard to convince them that the site of the pain is elsewhere. Patient education is essential in establishing a trust relationship. Otherwise, patients may disregard proffered care or fail to follow the prescribed treatment regimen. This is especially true of patients who are candidates for a reflex sympathetic dystrophy, in which disuse is the "predominant instigator"[12] (Fig. 5-2).

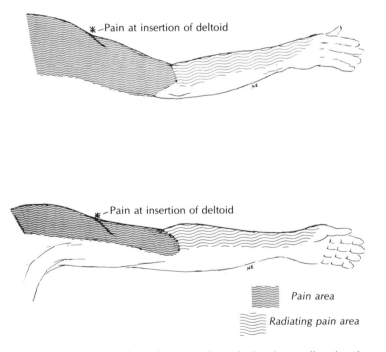

Figure 5–2. The radiation of pain down the arm and into the hand, as well as the often severe pain at the insertion of the deltoid, confuse the clinical picture in evaluating shoulder problems.

Acromioclavicular Joint Injuries

Treatment of the stable, first-degree injury is symptomatic and usually includes a sling for 7–10 days for comfort.[84] Ice is applied during the acute stage to help reduce pain and inflammation. Mobilization is initiated as soon as it can be tolerated by the patient and is graded to avoid stressing the healing joint. Other modalities are added as healing progresses, including heat and TENS to help in pain control. The modalities are effective in reducing the discomfort and stiffness sufficiently to permit increased freedom of movement and subsequent improved range of motion.

In second-degree injuries, immobilization is advocated, with symptomatic care, for 10–14 days until symptoms are reduced.[84] A Kenny-Howard splint is frequently applied to maintain reduction of the joint. This splint requires frequent adjustments[15] and can be difficult to manage.

Pain in the acromioclavicular joint is appreciated locally, rather than referred, and the ligaments are stretched, causing pain, at the ends of the ranges of the shoulder when passively moved.[17] In treatment of a tight joint capsule after healing of the injury has occurred, the acromion is often passively mobilized to permit normal joint play.[85] This movement should be painless, providing anteroposterior glide of the clavicle on the acromion.

The acromioclavicular joint is a synovial joint and may be affected by rheumatoid arthritis or osetoarthritis. Treatment of either entity is usually symptomatic and depends on the stage of the inflammation, as described in Chapter 2.

Adhesive Capsulitis

Adhesive capsulitis is usually the sequela of a shoulder injury that was not mobilized or that required immobilization for prolonged periods of time, often after surgery. Patients suffer no ill effects and find themsleves well adapted after a while. There are, however, many patients who suffer severely on multiple planes as a result of "frozen shoulder." An example is an 82-year-old man who tripped, fracturing his shoulder, and after a period of immobilization, regained only approximately 60 degrees of shoulder flexion, 50 degrees of internal rotation, and 40 degrees of abduction and external rotation. Although he was not significantly impaired in activities of daily living and could dress and feed himself unhindered, he was no longer able to practice his usual exercise routine of 18 holes of golf. As a result, his lifestyle became more sedentary. Over a period of 2–3 years, osteoporosis advanced to the point where he fractured his hip upon twisting his foot and required extensive hospitalization and extended care. The shoulder responded to manual therapy and other conventional therapeutic modalities, and full functional range of motion was regained. Appropriate care after the fracture might have prevented extended and costly hospitalization and reduction of quality of life.

It is dramatically easier to prevent than to "cure" adhesive capsulitis by early, gradual mobilization within the limits of pain. Once significant loss of range of motion and function occurs, under most circumstances functional mobility can be regained through (1) medical management of any inflammation (2) pain control, preferably via TENS so as not to mask pain (3) therapeutic modalities including heat or cold to relax spasm and reduce pain (4) ultrasound to permit greater extensibility of the tight, fibrotic tissues (5) manual therapy to increase joint play and stretch tight structures (6) exercises and stretching techniques for both clinic and home to strengthen weakened muscles and further stretch shortened structures (Figs. 5-4 through 5-9). Electrical stimulation provides assistance in re-educating weakened muscles and alleviating painful muscle spasm and edema. Iontophoresis is sometimes helpful in reducing pain and inflammation. Various other therapeutic modalities are useful as well in relieving pain, reducing inflammation, and increasing range of motion.

The most important aspect of treatment, however, is patient education. There are patients who, after being given a thorough understanding of the problem and how it can be treated, have been able to restore full range of motion without any other intervention. A home pulley system, instruction in how much discomfort is allowable before injury occurs, and motivation on the part of the patient are sufficient in certain situations. Other patients cannot be left to their own resources and are best guided in clinic environs. These patients require much more time and attention in order to achieve the desired outcome of treatment.

Researchers disagree about the most effective treatment protocol for adhesive capsulitis.[11] Reasons for the varying results seen in the studies might include pain threshold, patient education, and motivation. Possible causes for adhesive capsulitis include immobilization, trauma with internal derangement, suprascapular nerve compression neuropathy, autonomic neuropathy, immune abnormalities, and psychogenic disorders.[64] A high incidence of this condition is reported in neurosurgical patients, with significant association between it and hemiparesis.[10] Inflammation and resulting pain can cause the patient to limit movement as in tendinitis and bursitis.[69] Once the cause, if any, of adhesive capsulitis has been discovered and treated, a wide variety of variables complicates treatment of the entity itself, including the skills of the clinician, the patient's physiologic response to the treatment(s) administered, and carry over by the patient.

Brachial Plexus Injury and Plexitis

Once the extent of nerve damage has been assessed, the extent of denervation must be determined. In muscles that have lost innervation, electrical stimulation is required to provide contraction. Electrically assisted exercise must be carried out many times per day to inhibit atrophy and fibrosis. The correct use of a home stimulator can be taught to the patient, who then carries out the program with infrequent clinic visits until reinnervation begins.

In those muscles with partial innervation, electrical stimulation is paired with active contraction until the muscle can move the joint against gravity. Patients are taught active-assistive and active range of motion exercises that, properly done, will prevent joint contractures and muscle shortening. TENS is effective in reducing pain, both during exercise and at rest.

Since reinnervation can take many months to occur, rehabilitation is often prolonged. Until there is sufficient muscular return, supportive measures are continued. Once muscles are under active control, strengthening and endurance training is instituted. Patient education is of great importance during the extended rehabilitation period. Appropriate expectations, especially awareness of the long and sometimes tedious process required before use of the arm is regained, improve patient compliance and increase the odds of a good result of rehabilitation.

Bursitis

Treatment of bursitis is often closely associated with treatment of tendinitis since the two entities are frequently found together, as previously mentioned. Because the two are often clinically indistinguishable the distinction between supraspinatus tendinitis and subdeltoid bursitis is rarely of clinical importance.[72]

Other precipitating factors for bursitis include immobilization, abnormal or unusual movements, and fatigue or weakness of the rotator cuff muscles during activity.[69]

The noncapsular pattern with marked restriction of abduction and little limitation of external rotation is a diagnostic finding.[17] Most contractions are strong and painless with considerable tenderness in the area of the bursa.[41]

In the acute stage, treatment goals include resolution of inflammation and maintenance of range of motion. Ice is the modality of choice, if tolerated, to reduce pain, spasm, and swelling. Superficial heat (e.g., hot packs) may be used instead of ice. Phonophoresis with hydrocortisone or xylocaine or iontophoresis with xylocaine are helpful in reducing pain and inflammation.[35] Codman's pedular exercises, when correctly performed, can prevent joint contracture. These exercises, however, are difficult to teach because of the relaxation required during the movements, which is nearly impossible for many patients to learn. Relaxation during pain and spasm is often an unrealistic goal, and modifications to the exercises may be required. The most important precaution for the patient to be taught is to avoid the painful range during the acute stage. Instructions in written form increase patient compliance and should include a reminder to the patient to work within the pain-free range initially. Support of the arm with a sling can reduce the postural tone in the muscles adjacent to the bursa, which helps reduce pressure on the inflamed area.[41] Frequent mobilization, however, of the noninvolved joints and maintenance of motion at the shoulder is stressed in order to avoid joint contracture.

The acute stage usually lasts only a few days, especially with good management. When the patient no longer experiences pain at rest and shoulder flexion or abduction is possible actively to 90 degrees, ultrasound treatments are started.[41] Ultrasound causes therapeutic elevation of temperature, which produces hyperemia, increased capillary permeability, increased tissue metabolism, elevated enzymatic activity, and increased tissue extensibility[24] and assists in the removal of the inflammatory irritants and debris.[35] If pain continues to restrict movement, gentle manual therapy to restore joint play is initiated. TENS is helpful during exercise if significant restriction in movement continues past the acute stage.

As the condition resolves, the patient is gradually encouraged to increase the intensity of the exercises and to progress toward a strengthening program. Instruction in the exercises is given in the clinic and they are continued by the patient at home. Therapy should be necessary only for a period of 2–3 weeks, unless complications occur.

Osteoarthritis (Degenerative Joint Disease)

Pain, swelling, heat, redness, and crepitation with resultant muscle spasm and restriction of motion are symptoms that may mimic other problems at the shoulder. Pain can be referred as far as the wrist. The capsular pattern is usually present, with the most restriction in external rotation, less in abduction, and less still in internal rotation.[17] Radiographic evidence of bony changes are only occasionally present and is not directly related to the symptomatology.

Treatment of the various stages is discussed in Chapter 2 and should be directed at decreasing inflammation and pain while maintaining or increasing range of motion and strength.

Dislocation, Subluxation, and Fracture

Rather than treat the problem itself, therapeutic intervention is usually directed at the complications or consequences of dislocation, subluxation, or fracture. Treatment is based upon symptomatology and tissues affected. Goals include reduction of pain and spasm, reduction of inflammation, and restoration of strength and range of motion. Modalities are helpful prior to exercise, either active or passive, to help relieve pain, relax spasm, and allow stretching of fibrotic tissue. Manual therapy, when applied with care and skill, is of great value in stretching a tight joint capsule in order to restore joint play and thus range of motion.

Treatment of anterior shoulder dislocation initially consists of a sling, analgesics, and rigid restrictions on the patient's activity level, with emphasis on avoiding external rotation and abduction.[1] Gentle active exercises within the limits of pain can prevent adhesive capsulitis. As pain and healing permit, isometric exercises progressing to isokinetic exercises for internal rotation and adduction are added.

Knowledge about surgical intervention for dislocation or fracture and the goals of the surgery is necessary to set appropriate treatment goals. For example, surgery for recurrent anterior dislocation could be directed at limiting external rotation.[71] In this case, stretching the capsule or working toward increasing external rotation is contraindicated.

Hemiplegic Complications

Careful evaluation to determine what is causing shoulder pain is necessary in hemiplegic patients. The treatment is dependent upon the findings. If adhesive capsulitis is the problem, the shoulder is treated accordingly, with careful instruction to the patient on how to avoid the problem in the future, since the hemiplegia might not improve. Slings are appropriate in patients who have traction injuries, especially while ambulating or sitting with the arm unsupported. Arm troughs with foam padding are helpful to provide a neutral position for rotation of the shoulder and elevation of the hand to prevent dependent edema. Patients must have the ability to notice problems with their arms, as those with neglect tend to ignore the arm and allow further insults to occur. These patients with neglect also tend to have poor compliance with devices due to cognitive deficits. Family members or nursing staff often are responsible for attending to correct positioning of the arm and thus must be taught the appropriate techniques and rationale behind their use.

Overuse

Whatever the cause of overuse, treatment is directed at reducing pain and inflammation and preventing complications, such as adhesive capsulitis. When acute tendinitis and muscle strain are present, as in the case of the weekend athlete, treatment of the specific tendon(s) and muscle(s) involved and general strengthening of the shoulder is the regimen of choice (see section on tendinitis above). Rest of the affected tendon is usually effective, and treatment can result in remission of symptoms in several days.

When rehabilitation patients suffer from chronic overuse, treatment is more difficult, since these patients cannot usually stop the activity that caused the problem. To do so would make the patient immobile and dependent, a choice few would make. In these cases, reduction of the strenuousness of the activities is the best that can be hoped for, and local treatment is geared toward reducing pain and inflammation of the affected tissue(s), while strengthening the musculature around the joint to reduce the stress on the joint. Careful assessment is important in determining the cause of the strain and the structures affected. Treatment can then be specific in both changing the pattern of activity and stress and in targeting the appropriate tissues when intervention is initiated.

Reflex Sympathetic Dystrophy Syndrome (RSDS)

As mentioned earlier, the most successful treatment of this debilitating problem is begun as soon as possible after onset. Better still is prevention of immobility after predisposing factors or commonly associated conditions are recognized. Early in the course of rehabilitation patients can be started on exercise programs that will mobilize the arm and may prevent the syndrome from occurring or progressing. In patients who have one or more of the predisposing or commonly associated conditions, careful regular evaluation is often effective in observing early changes in posture, position, or favoring of the arm. Treatment can then begin early in the course of RSDS. Once trophic changes are established, treatment is much more difficult.

Passive range of motion progressing to active-assistive and then active movement is the most important aspect of treatment. The pain and dysesthesias that accompany RSDS often are so intense that the patient cannot tolerate these exercises, so modalities to reduce pain are employed. TENS, which has little contact with the skin, is one of the most effective means of pain reduction. The electrodes are applied in a crossed-circuit pattern at the proximal extreme of the extremity's pain, usually the shoulder (Fig. 5-3). Although there are many techniques of application and multiple parameters for achieving results, high rate and low pulse width applied with comfortable intensity ("conventional TENS") is most appropriate for treatment of RSDS because of its comfort. Either heat or cold may afford relief. Massage in a centripetal manner, if tolerated, is helpful in aiding venous return, improving circulation, and relaxing the patient in preparation for exercise. Manual therapy is effective and well tolerated in reducing joint contracture and increasing range of motion.

Treatment of this syndrome can be complicated in the patient who has a dependent personality. Cailliet uses the term "periarthritic personality" to describe psychological predisposing factors that may contribute to the development and slow the recovery from RSDS.[12] Since the patient must be made responsible for much of the exercise, both at home and in the clinic, compliance is essential. People who are not active participants in the rehabilitation process have worse outcomes than those who work hard at recovery.

Careful explanation of the cause of the pain and other symptoms, as well as comprehensive teaching of the home program, often improve compliance. Once the patient understands that the pain is real but that it does not warn of impending tissue damage, new methods of coping with the pain can be taught. Family members can be taught to avoid reinforcement of the patient's pain behavior to avoid development of an operant pain problem.[35]

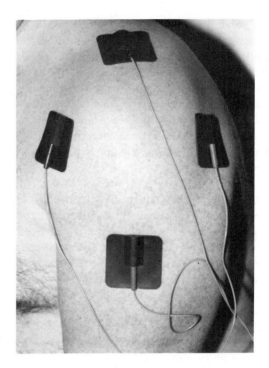

Figure 5–3. The electrodes are placed with one circuit's pads at the anterior and posterior aspects of the pain, and the other cirucit's pads at the superior and inferior locations. (The inferior electrode is often effectively placed at the insertion of the deltoid.) The circuits are thus "crossed," often improving the quality of the pain relief.

Rheumatoid Arthritis

As discussed in Chapter 2, the nonmedical treatment of arthritis is dependent upon the level of inflammation present. The more acutely inflamed the joint, the less stress that should be placed on it and on surrounding structures. During a flare-up, the joint must be protected from undue stress and allowed to rest, while very gentle active and active-assistive movements are carried out once or twice per day to prevent joint contracture. Pain is a limiting factor and should be watched to guide the clinician in assisting the patient with exercises. Cold (cryothermy) application to the joint is often of great value in reducing pain and inflammation, sometimes providing relief of pain for hours. When the joint is not being stressed, TENS may be applied to help reduce pain. The clinician becomes responsible for the decision-making process when inflammation becomes acute, in determining just how much motion is appropriate. Too much movement could cause serious damage. Too little activity can lead to joint contracture and reduced function. Much can depend on the pain tolerance of the patient, because a tolerant patient will allow more motion and require less motivating. There are times when these patients should be reminded that "no pain, no gain" is not applicable and that joint damage is a distinct possibility if stress is not limited.

Rotator Cuff Injuries

Surgical intervention is often necessary for complete or extensive tears of the rotator cuff. Usually, however, the tear is partial and is treated conservatively. Goals of treatment include promotion of healing while allowing pain-free movement to prevent contracture and muscle atrophy.[69] Healing time is usually prolonged, 6–10 weeks, during which the joint is protected from further trauma.[69] Gentle, gradual restoration of function is carefully supervised until full range of motion and strength have been restored. Maintenance of range of motion is essential in order to prevent adhesive capsulitis or other complications. In chronic rotator cuff injuries, strengthening of the rotator cuff muscles is indicated.[69] Once the normal muscular balance has been restored and the patient taught how to avoid insults, the likelihood of further injury is greatly reduced.

Modalities are chosen on the basis of symptomatology. Acute injuries with inflammation and pain are treated with cryothermy (ice packs, ice massage, etc.) and rest. As the symptoms begin to resolve, heat and mobilization exercises are initiated. Pain should be the limiting factor, but if it persists and restricts movement, TENS and gentle manual therapy are effective in reducing pain and increasing range of motion.

Tendinitis

While impingement syndrome is achieving recognition as an important entity, its treatment is the same as for other causative factors that result in tendinitis. The supraspinatus tendon is most frequently affected, and the biceps tendon the second most common site of tendinitis in the shoulder.[72] Subdeltoid bursitis is commonly associated with supraspinatus tendinitis.

Rest[1,30,60,69,72] is widely advocated in the early stage to remove stress from the tendon and avoid the causative insult. Whether it is complete rest (in a sling) or simply avoidance of the causative movements, as in impingement syndrome, depends upon the severity of the pain. If any movement causes pain, complete rest is indicated until some movement is possible in a pain-free range. If pain is only produced with specific movements, activities that produce pain are avoided. Cryothermy (ice packs or ice massage) is applied for the first 24–48 hours and thereafter to reduce inflammation accompanying activity.[1,30,69] Heat increases local blood flow, either superficially, as with hot packs or diathermy, or deeply, as with ultrasound. The increased metabolic exchange produced by the heat and increased perfusion aids in healing and reduces pain and spasm. Ultrasound to the supraspinatus tendon is applied with the arm internally rotated and hyperextended, anterolateral to the tip of the acromion. The overlying tissue and location of the tendon under treatment determines the dosage of the ultrasound. Other useful modalities in the treatment of tendinitis are discussed in Chapter 1.

The most important aspect of rehabilitation from tendinitis is restoration of mobility. Gentle Codman's pendular exercises cause the least trauma, since the muscles are not contracted when the exercises are done correctly. Active motion within the limits of pain are started after the first few days, and painless active-assistive exercises are added to maintain range of motion while healing occurs. Manual therapy becomes necessary if joint contracture ensues.

Once motion is possible without pain, strengthening exercises are initiated for all muscle groups of the shoulder (Figs. 5-7 through 5-9). Manual muscle testing should be performed periodically to be sure that there is no muscular imbalance, which can cause tendinitis and other injuries. Progressive resistive exercises and isokinetic techniques are effective in strengthening the weak muscles. Athletes are taught warm-up exercises prior to heavy activity to prevent stressing the joint and its surrounding structures when they return to normal activity.

Figure 5–4. Stretching the posterior portion of the shoulder joint capsule. With one arm pulling the other at the elbow, the shoulder to be stretched is horizontally adducted across the chest, providing a stretch of the posterior joint capsule.

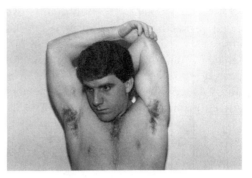

Figure 5–5. Stretching the inferior portion of the shoulder joint capsule. The patient lifts the affected arm over and behind the head, providing stretching with the uninvolved arm, which pulls the elbow up as far as possible.

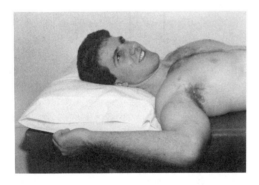

Figure 5–6. Stretching into external rotation. With the elbow in 90 degrees of flexion, the shoulder abducted to 90 degrees, and as much external rotation as possible, the patient reaches the hand back. Gravity provides some force. A weight can be placed in the hand to provide more force. Internal rotation is strengthened in this position, but the hand is raised toward the ceiling while the elbow and shoulder remain in the same position.

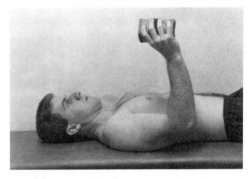

Figure 5–7. Internal rotation and subscapularis strengthening. With the arm at the side and the elbow bent to 90 degrees, the hand is lowered toward the stomach to stretch into internal rotation. When a weight is placed in the hand, the force of gravity is enhanced, providing greater stretching.. When the hand is lifted, rolling the arm into external rotation, the subscapularis is strengthened.

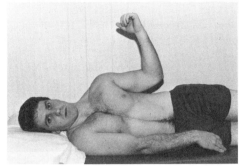

Figure 5–8. Strengthening for the supraspinatus. With the shoulders abducted to 90 degrees, internally rotated, and horizontally flexed to 30 degrees, the supraspinatus is exericsed selectively. Weights are gradually increased as strength improves.

Figure 5–9. Strengthening Teres Minor and Infraspinatus. In the sidelying position with the arm close to the body, the exercise motion is lifting the hand up, moving the arm into external rotation. Weights in the hand are added as indicated. Internal rotation can also be stretched in this position.

REFERENCES

1. Aronen JG: Shoulder rehabilitation. *Clin Sports Med* 1985, 4:477.
2. Ball J: Articular pathology of ankylosing spondylitis. *Clin Orthop* 1983, 173:20.
3. Basmajian JV (editor): *Therapeutic Exercise,* 4th ed. Baltimore, Williams & Wilkins, 1984.
4. Bateman JE: *The Shoulder and Neck,* 2nd ed. Philadelphia, W. B. Saunders, 1978.
5. Binder AI, et al: Frozen shoulder: An arthrographic and radionuclear scan assessment. *Ann Rheum Dis* 1984, 43:365.
6. Binder AI, et al: Frozen shoulder: A long-term prospective study. *Ann Rheum Dis* 1983, 43:361.
7. Bohannon RW, et al. Shoulder pain in hemiplegia: Statistical relationship with five variables. *Arch Phys Med Rehabil* 1986, 67:514.
8. Bonica, JJ: Management of myofascial pain syndromes in general practice. *JAMA* 1957, 164:732.
9. Brown C: Compressive, invasive referred pain to the shoulder. *Clin Orthop* 1983, 173:55.
10. Bruckner FE, Nye CJS: A prospective study of adhesive capsulitis of the shoulder ('frozen shoulder') in a high risk population. *QJ Med* 1981; Series L, No. 198:191.
11. Bulgen DY, et al: Frozen shoulder: Prospective clinical study with an evaluation of three treatment regimens. *Ann Rheum Dis* 1984; 43:353.
12. Cailliet R: *Shoulder Pain,* 2nd ed. Philadelphia, F. A. Davis Co., 1981.
13. Caldwell JW, et al: Nerve conduction studies: Aid in diagnosis of thoracic outlet syndrome. *South Med J* 1971; 64:210.
14. Cone RO, et al: Shoulder impingement syndrome: Radiographic evaluation. *Radiology* 1984; 150:29.
15. Cox JS: The fate of the acromioclavicular joint in athletic injuries. *Am J Sports Med* 1981; 9:50.
16. Curron JF, et al: Rheumatologic aspects of painful conditions affecting the shoulder. *Clin Orthop* 1983; 173:27.
17. Cyriax J: *Textbook of Orthopedic Medicine, Vol. 2, Diagnosis of Soft Tissue Lesions,* 6th ed. Baltimore, Williams & Wilkins Co., 1975.
18. Degowin EL, Degowin RL: *Bedside Diagnostic Examination,* 4th ed. New York, Macmillan Publishing Co., 1981.

19. Dejager JP, Fleming A: Shoulder joint rupture and pseudothrombosis in rheumatoid arthritis. *Ann Rheum Dis* 1984; 43:503.

20. Doupe J, et al: Post-traumatic pain and the causalgic syndrome. *J Neurol Neurosurg Psychiat* 1944; 7:33.

21. Escobar PE: Reflex sympathetic dystrophy. *Orthop Rev* 1986; 15:646.

22. Fodor J, Malott JC: The radiographic evaluation of the dislocated shoulder. *Radiol Technol* 1984; 55:154.

23. Gilroy J, Holliday PL: *Basic Neurology*. New York, Macmillan Publishing Co., 1982.

24. Gorkiewicz R: Ultrasound for subacromial bursitis. *Phys Ther* 1984; 64:46.

25. Goss TP: Rotator cuff injuries. *Orthop Rev* 1986; 15:496.

26. Griffin J, Reddin G: Shoulder pain in patients with hemiplegia. *Phys Ther* 1981; 61:1041.

27. Griffith PH III, Boyadjis TA: Acute pyarthrosis of the acromioclavicular joint: A case report. *Orthopedics* 1984; 7:1727.

28. Hardegger FH, et al: The operative treatment of scapular fractures. *J Bone Joint Surg Am* 1984; 66B:725.

29. *Harrison's Principles of Internal Medicine*, 9th ed. New York, McGraw-Hill Book Co., 1980.

30. Hawkins RJ, Hobeika PE: Impingement syndrome in the athletic shoulder. *Clin Sports Med* 1983; 2:391.

31. Hollinshead WH, Jenkins DB: *Functional Anatomy of the Limbs and Back,* 5th ed. Philadelphia, W. B. Saunders, 1981.

32. Ivey FM, et al: Isokinetic testing of shoulder strength: Normal values. *Arch Phys Med Rehabil* 1985; 66:384.

33. Jobe FW, Jobe CM: Painful athletic injuries of the shoulder. *Clin Orthop* 1983; 173:117.

34. Johnson EW (editor): *Practical Electromyography*. Baltimore, Williams & Wilkins, 1980.

35. Kahn J: *Low Volt Technique—Clinical Electrotherapy*. Syosset, N. Y., J. Kahn, 1985.

36. Kaltsas DS: Comparative study of the properties of the shoulder joint capsule with those of other joint capsules. *Clin Orthop* 1983; 173:20.

37. Kapandji IA: The shoulder. *Clin Rheum Dis* 1982; 8:595.

38. Kaplan PE, Cerullo LJ: *Stroke Rehabilitation* Boston, Butterworth, 1986.

39. Kaplan PE, Kernahan WT: Rotator cuff rupture: Management with suprascapular neuropathy. *Arch Phys Med Rehabil* 1984; 65:273.

40. Kaplan PE, et al: Stroke and brachial plexus injury. *Arch Phys Med Rehabil* 1977; 58:415.

41. Kessler RM, Hertling D: *Management of Common Musculoskeletal Disorders*. Philadelphia, Harper & Row, 1983.

42. Kieft GJ, et al: Normal shoulder: MR imaging. *Radiology* 1986; 159:741.

43. Kimura J: *Electrodiagnosis in Diseases of Nerve and Muscle: Principles and Practice*. Philadelphia, F. A. Davis Co., 1983.

44. Kirchner SG, et al: *Emergency Radiology of the Shoulder Arm and Hand*. Philadelphia, W. B. Saunders, 1981.

45. Kottke FJ et al (editors): *Krusen's Handbook of Physical Medicine and Rehabilitation,* 3rd ed. Philadelphia, W. B. Saunders, 1982.

46. Kraft GH, et al: The fibrositis syndrome. *Arch Phys Med Rehabil* 1968; 49:155.

47. Leach RE, Schepsis AA: Shoulder pain. *Clin Sports Med* 1983; 2:23.

48. Lee KH, Khunadorn F: Painful shoulder in hemiplegic patients: A study of the suprascapular nerve. *Arch Phys Med Rehabil* 1986; 67:818.

49. Lewit K: The needling effect in relief of myofascial pain. *Pain* 1979; 6:83.

50. Livingston WK: *Pain Mechanisms. A Physiologic Interpretation of Causalgia and its Related States*. New York, Macmillan Publishing Co., 1943.

51. Liveson JA: Nerve lesions associated with shoulder dislocation: An electrodiagnostic study of 11 cases. *J Neurol Neurosurg Psychiat* 1984; 47:742.

52. Melzak R, Wall PD: Pain mechanisms: A new theory. *Science* 1965; 150:971.

53. Meythaler JM, et al: Serratus anterior dis-

ruption: A complication of rheumatoid arthritis. *Arch Phys Med Rehabil* 1986; 67:770.

54. Michele AA, et al: Scapulocostal syndrome (Fatigue-postural paradox). *NY State J Med* 1950; 50:1353.

55. Miglietta O, et al: Subluxation of the shoulder in hemiplegic patients. *NY State J Med* 1959; 59:457.

56. Moskowitz E. et al: Long-term follow-up of the poststroke patient. *Arch Phys Med Rehabil* 1972; 53:167.

57. Najenson R, et al: Rotator cuff injury in shoulder joints of hemiplegic patients. *Scand J Rehabil Med* 1971; 3:113.

58. Neer CS: Impingement lesions. *Clin Orthop* 1983; 173:70.

59. Nevaiser RJ: Lesions of the biceps and tendinitis of the shoulder. *Orthop Clin North Am* 1980; 11:343.

60. Nevaiser RJ: Painful conditions affecting the shoulder. *Clin Orthop* 1983; 173:63.

61. Parsonage MJ, Turner JW: Neuralgic amyotrophy: The shoulder girdle syndrome. *Lancet* 1948; 254:973.

62. Post M (editor): *The Shoulder: Surgical and Nonsurgical Management*. Philadelphia, Lea & Febiger, 1978.

63. Prichett JW: Ossification of the coracoclavicular ligaments in ankylosing spondylitis. *J Bone Joint Surg Am* 1983; 65:1017.

64. Rizk TE, et al: Adhesive apsulitis (frozen shoulder): A new approach to its management. *Arch Phys Med Rehabil* 1983; 64:29.

65. Rizk TE, et al: Arthrographic studies in painful hemiplegic shoulders. *Arch Phys Med Rehabil* 1984; 65:254.

66. Rodnan GP, Schumacher HR (editors): *Primer of the Rheumatic Diseases*, 8th ed. Atlanta, Ga. Arthritis Foundation, 1983.

67. Saha AK: *Recurrent Dislocation of the Shoulder,* 2nd ed. Stuttgart, Georg Thieme Verlag, 1981.

68. Sanders TR, Staple TW: Percutaneous catheter drainage of septic shoulder joint. *Radiology* 1983; 147:270.

69. Saunders HD, Woerman AL: *Evaluation, Treatment and Prevention of Musculoskeletal Disorders*. Minneapolis, Viking Press, 1985.

70. Schutzer, SF, Gossling HR: The treatment of reflex sympathetic dystrophy syndrome. *J Bone Joint Surg* 1984; 66A:625.

71. Shands AR, Raney RB: *Handbook of Orthopedic Surgery*, 7th ed. St. Louis, C. V. Mosby Co., 1967.

72. Simkin PA: Tendinitis and bursitis of the shoulder. *Postgrad Med* 1983; 73:177.

73. Simons DG: Myofascial trigger points: A need for understanding. *Arch Phys Med Rehabil* 1981; 62:97.

74. Smith RG, et al: Malalignment of the shoulder after stroke. *Br Med J* 1982; 284:1224.

75. Sola AE, et al: Incidence of hypersensitive areas in posterior shoulder muscles. *Am J Phys Med* 1955; 34:585.

76. Steinbrocker, O: The shoulder-hand syndrome: Present perspective. *Arch Phys Med Rehabil* 1968; 18:38.

77. Travell JG, Rinzler SH: The myofascial genesis of pain. *Postgrad Med* 1952; 2:425.

78. Travell JG, Simons DG: *Myofascial Pain and Dysfunction: The Trigger Point Manual*. Baltimore, Williams & Wilkins, 1983.

79. Van Ouwenaller C, et al: Painful shoulder in hemiplegia. *Arch Phys Med Rehabil* 1986; 67:23.

80. Varghese G: Evaluation and management: Shoulder complications in hemiplegia. *J Kansas Med Soc* 1981; 82:451.

81. Walsh NE, et al: Brachial neuritis involving the bilateral phrenic nerves. *Arch Phys Med Rehabil* 1987; 68:46.

82. Weber RJ, Piero DL: F Wave evaluation of thoracic outlet syndrome: A multiple regression derived F wave latency predicting technique. *Arch Phys Med Rehabil* 1978; 59:464.

83. Wenner SM: Anterior dislocation of the shoulder in patients over fifty years of age. *Orthopedics* 1985; 8:1155.

84. Wickiewicz TL: Acromioclavicular and sternoclavicular joint injuries. *Clin Sports Med* 1983; 2:429.

85. Zohn DA, Mennell JM: *Musculoskeletal Pain: Diagnosis and Physical Treatment*. Boston, Little, Brown & Co., 1976.

6

Elbow Pain

Robert R. Conway and
Ellen D. Tanner

The elbow functions with the shoulder and wrist to position the hand in space. In addition to acute trauma and arthritis, elbow function can be affected by overuse and entrapment syndromes.

ANATOMY

The elbow joint consists of two types of articulations. The ulnohumeral joint is a ginglymus, or hinge, joint, allowing 150 degrees of flexion-extension. The radiohumeral and proximal radioulnar joints function as a trochoid or pivot joint, allowing 85 degrees of supination and 75 degrees of pronation. These are synovial joints with hyaline cartilage lining the articular surfaces. Medial stability is provided by the medial collateral ligament complex, which consists of the radial collateral, ulnar collateral, and annular ligaments. Several consistent, and some variably occurring bursae are associated with the elbow joint. The most commonly involved clinically is the olecranon bursa, located posteriorly between the olecranon process and the subcutaneous tissue.

The flexors of the elbow include the brachialis, biceps brachiae, and brachioradialis muscles, innervated by the radial nerve. The extensors are the triceps brachiae and anconeous muscles. Supination of the forearm is accomplished by the biceps brachiae and supinator muscles. Pronation is accomplished by the pronator quadratus and the pronator teres muscles.

EPICONDYLITIS

Etiology

Age. Epicondylitis is seen most frequently between the ages of 30 and 50 years with a median age of 43 years.[15]

Overuse. An insidious onset may be seen in individuals whose jobs or hobbies require prolonged, repetitive wrist and forearm movements, particularly against resistance. Lateral epicondylitis is seen when extension or supination is involved, whereas medial epicondylitis is seen when flexion and pronation are involved.

Trauma. An acute onset may be seen following an acute strain or direct trauma.

Rheumatologic Disorders. Reiter's disease and fibromyalgia may cause epicondylitis.

Pathology
There is probably more than one pathologic entity that can present with epicondylitis. Epicondylitis is thought by some to be an enthesopathy involving inflammation of the insertion of tendon of the extensor carpi radialis longus or brevis into the periosteum. Others report a degenerative lesion of the common extensor or flexor tendons themselves.[3]

Diagnosis
The diagnosis of epicondylitis is made on clinical examination. Range of motion and joint examination are normal. There is no evidence of neurologic deficit. In lateral epicondylitis there is tenderness to palpation localized to the lateral epicondyle and often pain on dorsiflexion of the wrist. In medial epicondylitis, the localized tenderness is over the medial epicondyle. Relief of symptoms with local injection of lidocaine helps confirm the diagnosis.

Prognosis
Some investigators, such as Kivi,[15] have found a good prognosis, with 82% of patients being asymptomatic after 6 months, 90% asymptomatic after 1 year, and 98% were improved. Only 2% received surgery.

Binder and Hazelman,[3] however, showed only 33% asymptomatic at 3–4 months and found recurrences to be common. Only 28% were asymptomatic at 1 year. Manual workers were the most susceptible to persistent or recurrent discomfort.

Treatment
Multiple treatments have been advocated, including immobilization,[8] resistive exercise,[11] shortwave diathermy,[29] ultrasound,[1,29] steroid injections,[6,22,29] and nonsteroidal anti-inflammatory drugs.[14] Unfortunately, there are few well-controlled studies evaluating these treatments.

OLECRANON BURSITIS

Etiology
Olecranon bursitis is usually traumatic in origin but can also be associated with infection, gout, pseudogout, or rheumatoid arthritis.[2] Multiple infectious organisms have been isolated, the most common being *Staphylococcus aureus*.[9]

Pathology

The bursa is a sac interposed between gliding surfaces and bony prominences for facilitating mobility of the soft tissues in areas of function. In acute bursitis, the bursa fills with watery, mucoid fluid. The fluid later becomes darker and filled with calcific densities. With chronic inflammation the bursa wall becomes thickened.[24]

In septic bursitis, leukocyte count of the bursal aspirate is increased, and an organism might be identified on Gram stain. Fluid should be examined for crystals if gout or pseudogout is suspected.[10]

Clinical Presentation

The usual presentation is painless, with a slow, gradual onset and swelling noted over the area of the olecranon. A more sudden onset is sometimes seen with septic bursitis.

Diagnosis

Differentiation between traumatic, septic, and inflammatory bursitis is based on clinical evaluation and analysis of fluid from the bursal aspirate.

Treatment

Traumatic Bursitis. This is best treated with nonsteroidal anti-inflammatory drugs, compression dressing, and avoidance of further trauma to the area. While intrabursal injection of steroid could hasten recovery, complications including infection, skin atrophy, and chronic local pain occur in a significant number of patients. Since spontaneous resolution can be expected, the use of steroid injection should be questioned.[32]

Septic Bursitis. This is treated with appropriate antibiotics based on the culture and sensitivity studies of the bursal aspirate. Frequent aspiration of the bursa might also be required.[16]

ARTHRITIS

The elbow joint is most frequently involved in rheumatoid arthritis, gout, or pseudogout. Nonetheless, Reiter's syndrome, septic arthritis, psoriatic arthritis, and osteoarthritis can also involve the elbow.[25] Synovitis is best detected by the presence of warmth and tenderness with limitation of full extension. With advanced synovitis, swelling is often detectable by palpation of the recesses on either side of the olecranon process. This sign should be distinguished from the swelling seen over the olecranon process in olecranon bursitis. Radiographs and laboratory studies are helpful in making the diagnosis. Joint aspiration with analysis and cultures may be indicated in an inflamed joint. For medical treatment, please refer to other texts.[25]

NERVE ENTRAPMENT SYNDROMES

Ulnar Nerve Entrapment

The ulnar nerve branches off the medial cord of the brachial plexus and travels down the medial arm. This nerve crosses the elbow in the cubital tunnel, a fibro-osseous tunnel, bordered laterally by the elbow joint with its transverse ligament and medially by the aponeurosis between the two heads of the flexor carpi ulnaris. The most common cause for an idiopathic ulnar neuropathy is entrapment of the nerve at the cubital tunnel.[30] Other etiologies of ulnar nerve entrapment at the elbow include ganglion, hypertrophic arthritis, and recurrent subluxation or dislocation of the ulnar nerve.

Clinical Presentation. Patients commonly present with complaints of elbow or forearm pain with paresthesias or numbness of the fourth and fifth digits. Additionally, they frequently complain of hand weakness.

The clinician can often elicit a history of prolonged bed rest, extensive driving with use of arm rests, habitual elbow leaning, or sleeping with the elbow in extreme flexion. All of these activities, in addition to the etiologic factors mentioned earlier, aggravate ulnar nerve entrapment at the elbow. On physical examination, findings include weakness of ulnar hand intrinsics and the flexor carpi ulnaris muscle. Sensation is decreased in an ulnar distribution. A Tinel's sign can often be elicited at the elbow. The elbow should be palpated for evidence of synovitis, bony irregularity, or ulnar nerve dislocation.

Diagnosis. Differential diagnosis includes C8 radiculopathy, lower trunk medial cord injury, or ulnar entrapment in Guyon's canal at the wrist. Diagnosis is made based on localization to an appropriate level and is confirmed by EMG and nerve conduction studies. As in all entrapments, one must consider the possibility of an underlying peripheral neuropathy. Radiographs of the elbow are usually indicated to rule out bony pathology.

Treatment. Conservative management should be tried initially. This treatment consists of education of the patient to avoid aggravating postures or activities, such as leaning on the elbow. Soft elbow pads are commonly helpful.

Surgical management in the past has included translocation of the ulnar nerve anteriorly or excision of the medial epicondyle. Currently, simple excision of the fibrous arch between the two heads of the flexor carpi ulnaris is the procedure of choice in most cases. Ulnar nerve translocation is reserved for recurrent subluxations or dislocations of the ulnar nerve, advanced osteoarthritis, and cubitus valgus deformity.[17,23,30]

Posterior Interosseous Syndrome

The radial nerve branches off the posterior cord of the brachial plexus and travels along the posterior aspect of the humerus in the spiral groove, prior to coursing anterolaterally and piercing the intermuscular septum proximal to the elbow. This nerve crosses the elbow joint anterolaterally, passing under the brachioradialis and extensor carpi radialis longus and brevis. At the supinator, it splits into the superficial radial nerve, a pure sensory branch that travels down the ventrolateral forearm, and the posterior interosseous nerve, a pure motor branch. The posterior interosseous nerve dives dorsally, piercing the supinator muscle and the interosseous septum near the arcade of Frohse. It supplies the extensor digitorum communis, extensor carpi ulnaris, extensor indicis, abductor pollicis longus, and extensor pollicis longus and brevis. Entrapment of the nerve could occur as it pierces the supinator at the arcade of Frohse.[30]

Etiology. Injury to the nerve may be caused by local trauma, by fracture dislocation of the radial head, or occasionally in rheumatoid arthritis by a compressive synovitis.[30]

Clinical Presentation. Typically, presentation is with pain and tenderness in the area of the supinator. There is weakness of the finger extensors and the ulnar wrist extensors, without sensory deficit. Differential diagnosis includes lateral epicondylitis and rupture of the extensor tendons. Electrodiagnostic studies are helpful in making the diagnosis.

Treatment. Conservative management consisting of local steroid injections, anti-inflammatory medication, and splinting has been successful in treating some cases.[12] In those not responding to conservative management, surgical decompression is indicated and effective.[7,26,33]

Median Nerve Entrapment

The median nerve is formed by branches of the medial and lateral cords. This nerve crosses the anteromedial aspect of the elbow adjacent to the brachial artery. Just distal to the elbow, it pierces the heads of the pronator teres and gives off a motor branch, the anterior interosseous nerve, which innervates the flexor pollicis longus, pronator quadratus, and the flexor digitorum profundus to the second and third digits.

Ligament of Struthers

Entrapment of the median nerve and brachial artery can occasionally occur as they cross the medial epicondyle by the ligament of Struthers, proximal to the innervation of the pronator teres. The patient presents with sensory loss and weakness in a median distribution. Full extension of the elbow causes compression of the radial pulse.[14] Electrodiagnostic studies are valuable in differentiating this from a more distal medial entrapment. Treatment involves surgical release.

Pronator Syndrome

The medial nerve is often injured or compressed as it pierces the two heads of the pronator teres.[30]

Etiology. Injury usually occurs by direct trauma or fracture. Compression may be associated with hypertrophy of the pronator teres or by an anomalous fibrous band between the pronator teres and flexor digitorum sublimis,[30] particularly in individuals with jobs or activities requiring repetitive supination-pronation.

Clinical Presentation. Patients commonly present with forearm pain and a median sensory and motor deficit. Pronator teres is usually not involved, since it receives its innervation proximal to the site of entrapment. This fact is useful in differentiating this syndrome from a more proximal entrapment. Electrodiagnostic studies are helpful in making the diagnosis. Surgical decompression is usually required for definitive treatment.

Anterior Interosseous Syndrome

Entrapment of the anterior interosseous nerve can occur as it branches off the median nerve just distal to the elbow.

Etiology. Multiple causes of anterior nerve compression or injury have been reported. They include elbow dislocation, bicipital bursitis, direct trauma, neuralgic amyotrophy, and compression by tendons of the forearm muscles during repetitive supination-pronation.[30]

Clinical Presentation. Patients commonly present with vague forearm or elbow pain without sensory loss. On examination, one commonly observes weakness of the flexors of the distal interphalangeal joints (DIPs) of the first three digits and pronator weakness with the elbow flexed. When asked to make an "OK" sign, they will form a triangle instead of a circle. Sensation is intact. Electrodiagnostic studies are useful in diagnosis.

Treatment. Treatment is dependent upon etiology but often requires surgical decompression.[27,30]

CONSERVATIVE TREATMENT OVERVIEW

Because the elbow is a complex joint with three articulations, during evaluation careful attention is given to restrictions in range and any limitations in function. Cyriax[5] describes a specific examination routine for the elbow that can help in determining the structure involved. Elbow flexion and extension, supination and pronation are accomplished passively. The same movements are then resisted. The wrist is then resisted in flexion and extension. The offending structure can often be determined by whether or not the movements are limited and painful. If resisted movements are painful, and passive movements are not, the contractile unit is usually at fault. If the opposite is the case, the joint and its surrounding structures are suspected.

Passive assessment of accessory movements in a nonacute elbow provides information about the location of the restrictive tissue, or fibrosis. The same movements when applied for treatment of a hypomobile joint aids in restoring lost range of motion (manual therapy). Abduction and adduction of the ulna on the humerus and the medial and lateral glide and distraction of the ulna occur at the humeroulnar joint; distraction and compression of the radius, and dorsal and ventral glide of the radius on the humerus occur at the humeroradial joint; ventromedial and dorsolateral glide of the radius at the proximal radioulnar joint are the accessory movements around the elbow.[4]

As in any musculoskeletal problem, treatment should be specific, directed at the affected tissue(s) rather than "shotgun" style. Therefore, careful evaluation and identification of the specific problem are required before treatment is initiated. Although there is controversy about the terminology and etiology of elbow pain, when the structure involved is known and the level of acuteness is determined, there is little divergence of opinion about the treatment.

Since the same principles apply as in treatment of other musculoskeletal problems, the goals of treatment are also quite similar. Generally, reduction of pain and inflammation, encouragement of healing, and restoration of function are desired outcomes of treatment. When motion has been lost, restoring range of motion, strength, endurance, flexibility and coordination may be necessary before function is returned. A balance is necessary, however, since strengthening of one muscle group leaves the joint less protected in other directions, which might allow injury to the joint. Strengthening of one muscle group is therefore not sufficient, and rehabilitation should include strengthening of all weak or dysfunctional muscles identified in the entire arm, since the elbow does not exist in isolation.

As stated by Nirschl and Morrey,[20] rehabilitation of an injury starts at the time of the injury, since minimizing the inflammatory phase of healing results in more complete and rapid rehabilitation. The treatment regimen should include (1) early exercise of noninvolved adjacent structures, (2) protection of the injured area from further insult, (3) reduction of inflammation and pain, and (4) restoration of normal function.

Prolonged immobilization often leaves the patient with a joint that lacks full range of motion. Restoration of motion of the elbow can be costly in both time and effort and ultimately might not succeed. It is best, therefore, to limit immobilization as much as possible and begin remobilization as soon as it is safe to do so. The sooner motion is begun, the less the likelihood of restricted function.

Epicondylitis

"Tennis elbow" has been used as a misleading catch-all name for pain at the elbow. The term is frequently applied to lateral humeral epicondylitis[21,28] and to tendinitis.[19] The tendinitis is often of the extensor tendons, usually of the extensor carpi radialis brevis near its insertion at the lateral epicondyle.[5,13] Other tendons, including the extensor communis, the extensor carpi radialis longus, the extensor carpi ulnaris on the lateral epicondyle, and the common flexor tendon on the medial epicondyle may be affected by the tendinitis called "tennis elbow."[13,19] As Zohn and Mennell point out, however, there are a number of other pathologic conditions that occur around the elbow involving trigger points, tendons, bursae, joint meniscus, periosteum, synovium, and bone.[34] Cyriax[5] collected a list of 26 different lesions to which the condition had been attributed. Further, initial insult to one tissue can be complicated, as in the case of chronic tendinitis, with adhesions. Therefore, it is important to first define the tissue or tissues affected when evaluating or treating a "tennis elbow" (Fig. 6-1).

Tendinitis

Specific treatment of tendinitis depends upon whether it is in the acute or chronic phase, as discussed in Chapter 1. The acute phase is treated by ice several times per day and by avoiding movements that cause pain, i.e., resting the hand and wrist and exercising actively but gently through the range of motion several times per day. Once the acute phase has passed, usually within several days, ultrasound and other modalities such as electrical stimulation, iontophoresis, or phonophoresis are initiated to help increase local

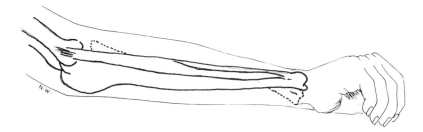

Figure 6-1. Lateral epicondylitis occurs frequently at the lateral humeral epicondyle, where several tendons, including the flexor carpi radialis, flexor carpi ulnaris, extensor carpi radialis longus and brevis, and extensor carpi ulnaris insert.

circulation and promote healing. Avoidance of activities that reproduce the pain is advised until there is little or no pain on resisted isometric wrist extension and little or no pain upon passive stretch of the tendon. Gradual resumption of normal activity is then begun, while preventing further injury to the tendon. The use of an elbow band or cuff helps attenuate the stress on the tendon, protecting it from strain. Friction massage across the longitudinal fibers causes local hyperemia and may assist in tissue maturation while preventing restrictive fibrosis. Friction massage is initially very gentle and should not cause significant pain. Tenderness usually abates, permitting deeper and deeper pressure, until the target tissue is affected. Care is taken to treat over the site of the lesion rather than the site of pain, which might not coincide.[13]

Once symptoms are under control and activities resumed, it is important to restore any extensibility that may have been lost by gentle, gradual stretch of the muscle and tendon. Strengthening exercises are also initiated at this point. Increased discomfort should be the limiting factor.

Bursitis

With the exception of the olecranon bursa, bursitis is uncommon at the elbow.[18] There have been 11 or more bursae described around the elbow, some of which are superficial and some deep, between muscle and muscle or muscle and bone, making clinical diagnosis difficult[18] (Fig. 6-2).

Although often associated with injury, either acute or small repeated insults, olecranon bursitis frequently presents with a distended bursa that causes little discomfort. Acute olecranon bursitis might not be painful, and protection from injury is all that is required. Morrey recommends a resting splint and compression in the early stages.[18] If pain and inflammation are present, they are usually treated symptomatically. Ice initially, in the first 24–48 hours, is valuable in reducing pain and inflammation. Heat is applied thereafter to increase circulation and promote healing. Range of motion exercises are instituted as soon as pain permits, as flexion is often limited to 90 degrees. This limitation of motion is probably due to increased pressure on the bursa with increasing flexion of the elbow. Care is required in order to prevent the formation of dense, inflexible fibrous tissue (scar), which can restrict normal movement.

In the case of chronic olecranon bursitis the bursa is sometimes so enlarged that it interferes with activities of daily living or occupation and occasionally requires surgical

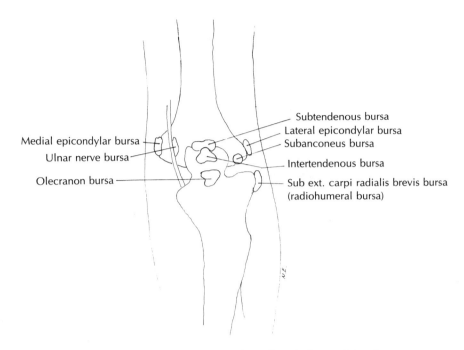

Subtendenous bursa
Lateral epicondylar bursa
Subanconeus bursa
Intertendenous bursa
Sub ext. carpi radialis brevis bursa
(radiohumeral bursa)

Medial epicondylar bursa
Ulnar nerve bursa
Olecranon bursa

Figure 6-2. Some of the bursae that have been described around the elbow.

excision.[21] Calcium deposits further complicate chronic olecranon bursitis, making it painful to rest weight on the elbow. Complications after surgery include joint contracture of the elbow and of the shoulder due to prolonged or inappropriate immobilization.

REFERENCES

1. Aldes JH: Ultrasonic radiation in treatment of epicondylitis. *Gen Pract* 1956; 13:89.

2. Benson BG, et al: Hemochromatotic arthropathy mimicking rheumatoid arthritis. *Arth Rheum* 1978; 21:844.

3. Binder AI, Hazelman BL: Lateral, humeral epicondylitis—a study of the natural history and the effects of conservative treatment. *Br J Rheum* 1983; 22:73.

4. Bowling RW, Rockar P: The elbow complex. In Gould JA, Davies GJ (editors): *Orthopedic and Sports Physical Therapy*. St. Louis, C. V. Mosby Co., 1985.

5. Cyriax J: *Textbook of Orthopedic Medicine*, Vol. 1. Baltimore, Williams & Wilkins Co., 1975.

6. Day BH, et al: Corticosteroid injections in the treatment of tennis elbow. *Practitioner* 1978; 220:459.

7. Dewey P: Posterior interosseous nerve and resistant tennis elbow. *J Bone Joint Surg* 1973; 55b:435.

8. Hansson KG, Horowich ID: Epicondylitis, humeri. *JAMA* 1930; 94:1557.

9. Ho G, et al: Septic bursitis in the prepatellar and olecranon bursae. *Ann Intern Med* 1978; 89:21.

10. Jaffe L, Fetto JF: Olecranon bursitis. *Contemp Orthop* 1984; 8:51.

11. Johnson EW: Treatment of tennis elbow. Personal communication.

12. Kaplan PE: Posterior interosseous neu-

ropathies: Natural history. *Arch Phys Med Rehabil* 1984; 65:399.

13. Kessler R: The elbow. In Kessler RM, Hertling D (editors): Management of Common Musculoskeletal Disorders. Philadelphia, Harper & Row, 1983.

14. Kimura J: *Electrodiagnosis in Diseases of Nerve and Muscle: Principles and Practice.* Philadelphia, F. A. Davis, 1983.

15. Kivi T: The etiology and conservative treatment of humeral epicondylitis. *Scand J Rehab Med* 1982; 15:37.

16. Knight JM, et al: Treatment of septic olecranon and prepatellar bursitis with percutaneous placement of a suction-irrigation system. A report of 12 cases. *Clin Orthop* 1986; 206:90.

17. Lugnegard H, et al: Ulnar neuropathy at the elbow treated with decompression. A clinical and electrophysiological investigation. *Scand J Plast Reconstr Surg* 1982; 16:195.

18. Morrey BF: Bursitis. In Morrey BF (editor): *The Elbow and its Disorders.* Philadelphia, W. B. Saunders Co., 1985.

19. Nirschl RP: Muscle and tendon trauma: Tennis elbow. In Morrey BF (editor): *The Elbow and its Disorders.* Philadelphia, W. B. Saunders Co., 1985.

20. Nirschl RP, Morrey BF: Rehabilitation. In Morrey BF (editor): *The Elbow and its Disorders.* Philadelphia, W. B. Saunders Co., 1985.

21. O'Donoghue DH: *Treatment of Injuries to Athletes.* Philadelphia, W. B. Saunders Co., 1970.

22. Quin CE, Binks FA: Tennis elbow (epi-condylalgia externia): Treatment with hydrocortisone. *Lancet* 1954; 2:221.

23. Reddy MP: Ulnar nerve entrapment syndrome at the elbow. *Orthop Rev* 1983; 12:69.

24. Robbins SL: *Pathologic Basis of Disease.* Philadelphia, W. B. Saunders, 1974.

25. Rodnan GP, et al: *Primer on the Rheumatic Diseases,* 8th ed. Atlanta, Ga., Arthritis Foundation, 1983.

26. Roles NC, Maudsley RH: Radial tunnel syndrome: Resistant tennis elbow as a nerve entrapment. *J Bone Joint Surg* 1972; 54b:499.

27. Saeed MA, Gatenns PF: Anterior interosseous nerve syndrome: Unusual etiologies. *Arch Phys Med Rehabil* 1983; 64:182.

28. Saunders HD: *Evaluation, Treatment and Prevention of Musculoskeletal Disorders.* Minneapolis, Viking Press, Inc., 1985.

29. Sinclair A: Tennis elbow and industry. *Br J Int Med* 1965; 22:144.

30. Spinner M: *Injuries to the Major Branches of Peripheral Nerves of the Forearm,* 2nd ed. Philadelphia, W. B. Saunders, 1978.

31. Steiner C: Tennis elbow. *Am Osteop Assn* 1976; 75:575.

32. Weinstein PS, et al: Long-term follow-up of corticosteroid injection for traumatic olecranon bursitis. *Ann Rheum Dis* 1984; 43:44.

33. Werner CO: Lateral elbow pain and posterior interosseous nerve entrapment. *Acta Orthop Scand* 1979; Suppl 174:1.

34. Zohn DA, Mennell JM: *Musculoskeletal Pain: Diagnosis and Physical Treatment.* Boston, Little, Brown & Co., 1976.

7

Wrist and Hand Pain

*Paul E. Kaplan and
Mark Walsh*

No other part of the body regularly performs such complex, complicated activities as the hand and wrist. The degree of fine motor coordination utilized is so high that even a relatively small amount of wrist or hand discomfort generates great feelings of impairment and loss. Quadriplegic spinally injured patients with central cord syndrome cannot ambulate and yet are intensely frustrated because of poor hand function. For better or worse, our feelings are rapidly focused on even small amounts of hand disability. The feeling of loss is not even cosmetic. Patients with erosive rheumatoid arthritis are usually more disturbed by the pain and weakness than they are by how their hands look. Children with hand amputations will use hooks if they are functional. To be sure, looks do count—children with congenital hand defects are often profoundly concerned with body image issues—but function and comfort count even more.

CLINICAL PATHOPHYSIOLOGY

For the hand to be effective, the wrist must precisely place it. When prosthetic hand-wrist units are employed, the average total of wrist flexion and extension is 55–64 degrees.[8,13,24,25] The minimum efficient range is probably 10 degrees of flexion and 35 degrees of extension.[10] When the wrist is fused, 10 degrees of extension is optimal. Functionally, the distal radius revolves about the pivotal distal ulna.[15,42,44,47] When the forearm is axially loaded, the radial-lateral carpus transitional articulation will carry 80% of the weight.[45] Furthermore, the stability of the wrist depends upon an intact triangular fibrocartilaginous cover of the ulnar head and of the radioulnar joint. The distal radioulnar joint is designed for both flexibility and spatial weight transfer. The bony and cartilaginous structure are essential, and usually muscles do not appreciably add to these functions. They aid only by directing the force vectors produced through skeletal actions.

The range of radiographic investigative procedures ensures that most lesions altering the bony structure of the hand and wrist are diagnosed. Should routine views be normal, fluoroscopy could be useful.[26] The next line of study would involve scanning or arthrography. CT scanning or MRI are also available and might accurately pinpoint the offending anatomic abnormality.[14,43] The rate-limiting stage lies not in the availability of evaluations delineating anatomic lesions but in organizing the information to explain the etiology of the functional deficit (Fig. 7-1).

In examining the painful hand or wrist, depending upon the patient, locating exactly where the pain is can be difficult. In such cases, the clinician usually evokes Tinel's sign either alone or accompanied by other attempts to mechanically stimulate or stress the involved nerve, muscle, or ligament. Many otherwise accomplished observers are misled by these clumsy and antiquated maneuvers. For example, neuromas are often extremely painful, but neural activity recorded from a baboon's neuroma is disappointingly low.[40] Only 18% of C-fibers and A1-fibers were mechanically sensitive when the neuroma was 1 or 2 months old. When the neuroma was 7 months old, the percentage fell to 4%. Tinel's sign is not accurate for the same reason that painful site location is misleading—both are influenced by the emotional reaction to the discomfort and loss of function. Moreover, dermatomal areas vary with overlapping nerve supply and shifting spinal nerve root contributions.[37] A detailed sensory examination revealing an area of deficit does not in itself necessarily point the way to a specific neuritis (Fig. 7-2).

The clinical pathophysiology, therefore, centers on the vast amounts of torque gen-

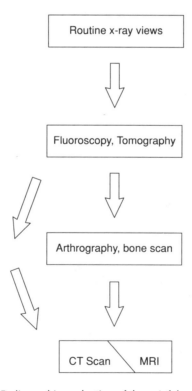

Figure 7-1. Radiographic evaluation of the painful wrist or hand.

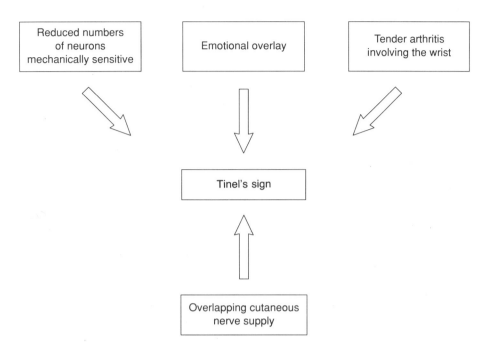

Figure 7-2. Factors altering the accurate response to a Tinel's sign.

erated at the distal forearm and wrist in order to accurately place the hand in space. The ligaments about the carpal bones help transmit rotational force and diffuse these stresses. The transverse carpal ligament is especially helpful in that it restrains long flexor tendons and acts as a pulley. Those tendons thus acquire an added mechanical advantage.[24] Additionally, the intrinsic carpal ligaments align the terminal carpal bones more closely together so that in effect they are dynamically one body.

Another factor is that nerves supplying the wrist and hand branch as these nerves proceed distally.[29,30,31,54] As a result, the number of fascicles increases so that interfascicular and perifascicular connective tissue occupies about 50% of a single section of the distal ulnar nerve.[7] This situation helps explain inconsistent clinical sequelae after neurosurgery. These smaller branches are more vulnerable to trauma than larger trunks and heal slowly.

To make matters more uneven, some people, because of congenital variations in nerve supply, are more at risk than others. For example, the ulnar nerve in the hand can divide to two or three branches or even have an anomalous sensory branch.[8] Those congenital variations that result in more but smaller distal branches, or in branches strategically misplaced, are at greater risk for developing a neuropathy if chronic trauma or vasculitis is applied. Even if the anomaly is moderately protective, subsequent neuropathy can occur. The motor division of the median nerve can join the ulnar nerve in the forearm, wrist, or hand.[36] Nonetheless, these patients can still develop carpal tunnel syndromes of the sensory division of the median nerve left behind (Fig. 7-3).

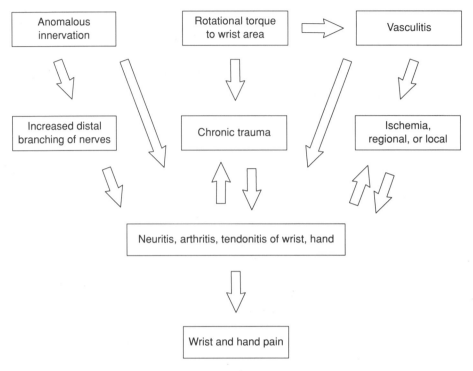

Figure 7-3. Pathophysiology of wrist and hand pain.

DIFFERENTIAL DIAGNOSIS

Muscles that are both the last supplied by their innervative nerves and also reasonably available for testing acquire strategic importance. Most of these muscles are found in the wrist and hand area. An example is the extensor indicis. The last supplied muscle of the radial nerve, it is the first affected by any lower motor neuron process. The entire radial nerve system can be screened by checking this one muscle. Since it also has an extremely strong and almost exclusive spinal nerve root contribution—C8—it is also helpful if C8 or T1 spinal nerve root abnormalities are suspected. While radiculitis might affect only the posterior primary ramus, most patients with wrist and hand pain from radiculitis probably have extensor indicis involvement if C8 radiculitis is the cause of the discomfort. Table 7-1 gives other examples of this principle.

All of these muscles are small, distal, type II muscles. All are easily overcome and therefore can be tested for signs of early weakness. All should be tested individually and at disadvantageous positions. For example, the lumbricals should be stretched by extending the metacarpophalangeal joint in question then testing interphalangeal extension. Positioning each muscle and testing it individually takes some time if done correctly. Actions of many muscles, as in grip, should be evaluated only after each individual muscle

TABLE 7-1. STRATEGIC HAND AND WRIST MUSCLES

Muscle	Nerve	Spinal Nerve Root
Extensor indicis	Radial	C8
First and second lumbricals	Median	C8, T1
Adductor pollicis First dorsal interosseous	Ulnar	T1

has been examined. The hand and wrist kinesiologic evaluation should take a long time; brief examinations are incorrect examinations (Fig. 7-4).

Two further examples illustrate this process. The pronator quadratus is not the type of strategic muscle discussed above. Because several other muscles also participate in pronation of the forearm, it cannot be individually identified for examination. The flexor pollicis longus, however, has no such disadvantage. In fact, the most distal muscle of the anterior interosseous nerve that is also strategic is the flexor pollicis longus. As this nerve is often injured while defending against knife thrusts, its injury is common to any inner city emergency room. Loss of thumb function that ensues is significant because flexion of the distal thumb phalanx is affected. The OK sign cannot be made with the thumb and index finger. As the posterior interosseous nerve can also be interrupted in this way,

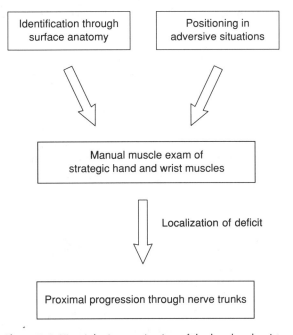

Figure 7-4. Kinesiologic examination of the hand and wrist.

loss of thumb function is often accompanied by a loss of finger and thumb extension (Fig. 7-5).

The second example concerns abduction of the thumb. Two types of abduction are possible, radial and palmar. Each has its own character. Radial abduction is a joint function of the radial and median nerves. As the radially innervated muscle—the abductor pollicis longus—is also a flexor of the wrist (the only "extensor" muscle that is also a flexor), it can be separated by testing radial thumb abduction with the wrist flexed and then extended. Radial abduction has a correlative movement in the feet. Its functional significance is not vital.

Palmar abduction is entirely different. It is the only median nerve mediated action that is unique. Thumb opposition can be performed using an intact carpometacarpal joint and the adductor pollicis, an ulnar innervated muscle. Thumb palmar abduction underlies both precision and power grips and is the engine behind the three-point fingertip

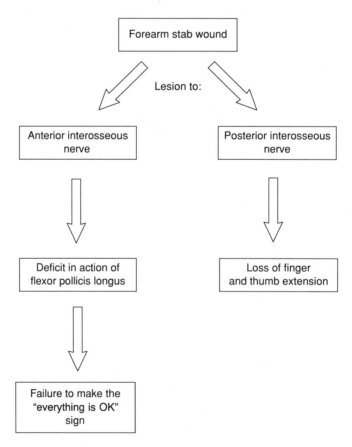

Figure 7-5. Interosseous nerve syndrome.

grasp that gives the hand its special function. There is no correlate function in human feet, although this function is present in the feet of certain primates. It probably helped distinguish and aid humanoid evolution compared to that of other mammals. Grasping motions are often noted in fetal stages.[28] The best way to evaluate this activity is resistive. The thenar muscles are the short, type II variety and are easily overcome if the muscle is first placed in some stretch. Resistive isometric contraction is the best position from which to assess atrophy. The muscle's profile is scalloped if atrophied. Palmar abduction can be compared to flexion of the fifth finger—it should be stronger. Dermatomal sensory deficits can help confirm a diagnosis but can otherwise mislead. Dermatomes of the hand and wrist are not chiseled in stone, but are notoriously variable. Palmar abduction of the thumb is both reliably tested and vitally significant for forming rehabilitation potential estimates (Table 7-2).

Arthritides provide the second source of wrist and hand discomfort. It is not enough to document the qualitative pattern of inflammatory distribution. Quantitative measurements of the involved joint(s) must also be available on a serial basis. A jeweler's ring sizer or a simple loop measurement device will do for phalangeal joints. Strength of hand functions can likewise be quantitatively determined serially. The Jamar dynamometer reliably measures grip strength, and the B and L pinch gauge is consistent for pinch strength.[38] If heating modalities are applied, these size and function measurements should be repeated to document the patient's response to this treatment. In this way a modality tolerance test could screen reactions to therapy before a long course of treatment has been ordered.

In conjunction with arthritic diseases of the hand and wrist, long flexor and extensor tendons must be individually evaluated for secondary inflammation. Fortunately, the range of these tendons is relatively short. This fact helps locate and then isolate these tendons. For example, if the fourth flexor digitorum superficialis tendon is tested, fully extending the third and fifth fingers completely neutralizes the fourth flexor digitorum profundus (Table 7-3).

TABLE 7-2. ABDUCTION OF THE THUMB

Type	Characteristics
Radial	Supplied by radial and median nerves
	Power associated with relative flexion of the wrist
	Has correlate action with that of the great toe
Palmar	Supplied only by the median nerve
	Vital to hand function
	Powers precision grasp
	Affected by fine motor coordination deficits
	Dysfunctional in the parietal hand seen in stroke or head trauma

TABLE 7-3. EVALUATION OF ARTHRITIC HAND AND WRIST

Test	
Qualitative	Which joints are inflamed?
	Which tendons are inflamed?
	What deformities are present?
	Swan-neck? Boutonniere?
	What is the manual muscle test?
	What is the radiographic appearance?
Quantitative	What is the joint circumference?
	What is the joint range of motion?
	What is the grip strength?
	What is the pinch strength?

EVALUATION

After the diagnosis has been made, and before proceeding with treatment, specific localization of the tissues involved through a logical course of evaluation assists the rehabilitation team in establishing a baseline by which to measure the effectiveness of the treatment program and determine the functional level of the patient. A careful history identifies causative factors that contribute to the patient's hand or wrist dysfunction. Problems that appear to have an insidious onset may be due to cumulative or repetitive trauma in the workplace or recreational arena. The patient's age, sex, specific occupational tasks, and recreational activities might also be contributory factors to the patient's dysfunction. It is also important to ascertain the length of time the dysfunction has been present. In general, there is a proportional relationship between the duration of the dysfunction and the time needed to alleviate the dysfunction. The longer the problem has been present, the more recalcitrant it usually is to treatment.

Functional evaluation includes inspection of the hand and wrist, with special notation of inflammation, edema, atrophy, soft tissue atrophic changes, or deformities. Edema can be objectively measured through circumferential measurements of the wrist, palm, and each digit, taking care to standardize measurements by repeatedly performing them at the same location. Hand volumes can also be taken using the hand volumeter, although this is only accurate to within 10 mm[57] (Fig. 7-6). Therefore, edema involving isolated digits often does not demonstrate any significant change with hand volumetry, so individual circumferences should be utilized. Active and passive range of motion of the entire upper extremity should be carefully examined. Specific tests for intrinsic, extrinsic, and periarticular tightness, an Allen test for circulation, and any other specific clinical tests for particular pathologic conditions should be performed. Resistive motion testing assists in isolating the involved tissue into contractile or noncontractile components of dysfunction and screening individual muscles' strength. Functional strength is usually assessed with the Jaymar dynamometer for gross grasp and a pinch gauge for lateral, palmar, or tip-to-tip prehensile strengths[38] (Fig. 7-7). Care should be taken to perform these tests in the same position repeatedly for standardization.[53]

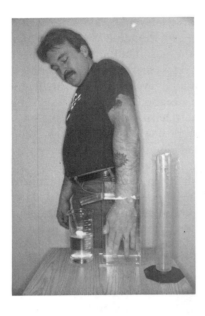

Figure 7-6. Hand volumetry based on water displacement. Note standardized position of thumb facing spout and dowel at middle-ring finger web space.

A

B

C

Figure 7-7. Gross grasp and prehensile strength measurement. Standardization with the humerus at the side, elbow at a 90 degree angle, and forearm and instrument unsupported. For A) gross grasp, and B) lateral prehension, the instrument is placed on the middle phalanx. For C) palmar prehension, the pad of each finger is placed on the instrument.

Perhaps the most important function our hands perform is sensibility. Careful sensory evaluation is necessary and assists the clinician in determining whether neurologic involvement is mediated through the peripheral or central nervous system. This distinction can be made by a demonstrated loss or alteration in sensation in a dermatomal or peripheral cutaneous nerve distribution. Evaluation should include testing for threshold stimulus using monofilaments[58] or vibration of 256 cycles per second.[17] These evaluation tools have been found to be the most sensitive in diagnosing nerve entrapment and compression and in monitoring sensory recovery after laceration. Innervation density tests of moving two-point discrimination and static two-point discrimination have been standardized and also assist in determining the level of sensory compromise or return.[18]

In addition to traditional evaluation methods, a functional battery of tests is commonly performed on the patient, allowing observation of any aberrant wrist, finger, gross grasp, or prehensile patterns. These functional tests also add objectivity to the evaluation process and assist with measurement of the patient's progress. Examples of these functional tests include VAL-PAR Work Samples or similar standardized tests and other prehensile tests such as the Purdue Pegboard or Minnesota Rate of Manipulation (Fig. 7-8).

Simulation and re-creation in the clinic of job or leisure tasks that provoke pain also assist in determining the specific tissue of dysfunction and the formulation of a course of treatment. Once the possible causative factors and the tissues involved are identified, treatment for each is instituted.

A

Figure 7-8. Functional tests. A) Minnesota Rate of Manipulation. B) VAL PAR Upper Extremity Range of Motion Work Samples.

B

A general consideration for the treatment of wrist and hand dysfunction is the maintenance of tissue homeostasis, achieved by resting from function only those tissues necessary, while allowing adjacent tissues to continue their freedom of motion. Since all biologic tissue responds to stress, the maintenance of strength through resistive therapeutic exercises, be they isometric, isotonic, or isokinetic, assists in achieving tissue homeostasis while preventing stiffness and dysfunction. Allowing functional use of the upper extremity also prevents the development of a divorcing phenomenon by repeated introduction of the extremity into the environment.

PERIPHERAL NERVE LESIONS

Certainly one of the most challenging areas of dysfunction is that of the peripheral nervous system caused by laceration or compression and/or entrapment. In the case of laceration, rehabilitation begins immediately after surgical repair with immobilization through splinting or casting of the appropriate tissues to alleviate tension across the neural repair. Depending upon preference, full motion of the immobilized tissues is often instituted anywhere from 10 days to 4 weeks after repair has been performed. In the interim, maintenance of joint and soft tissue mobility and muscle strength is achieved through active and passive exercises, massage, and joint mobilization. During the early phases of repair, the patient is educated with regard to caring for insensate areas. The patient must frequently observe the involved area for any lesions, institute proper hygiene and skin care, avoid temperature extremes, and continually visualize the affected area when near moving machinery. Although controversial, maintenance of denervated muscle is at times achieved through stimulation of denervated muscle by direct current and passive, active, and resistive exercises of the involved and surrounding tissues. This regimen can play a part in retarding atrophy and permanent damage to the muscle parenchyma.[16,41]

Since the most important function of the hand is sensibility, proper care of cutaneous nerve injuries is of primary concern. As the axons regenerate and progress distally through the neurotubule, hypersensitivity or dysesthesia may develop. This state of extreme local discomfort often responds to a desensitization program in the clinic and at home, consisting of stroking with various textures, percussion with different pressures, deep pressure or massage, and several forms of vibration.[3] Each modality or stimulus is introduced into the hypersensitive area initially at its periphery for a period of 10–20 minutes at a time. The patient must work at a low level of discomfort for the treatment to be effective in alleviating hypersensitivity. As the condition improves, more coarse textures, deeper pressure, and centralization of the vibratory stimulus all help to further decrease the level of sensitivity.

As the nerve continues to regenerate, treatment is concentrated on sensory re-education of quickly and slowly adapting fibers, as advocated by Dellon,[19] which is believed to be effective in improving the patient's rate and level of sensory recovery. During phase I, before the perception of moving or constant touch over the involved area returns, pressure with a blunt object is applied and moved across the area. Phase II commences when moving or constant touch is perceived and includes object identification and size differentiation statically and while moving across the sensory deprived area. The patient is also educated in development of protective and discriminative sensation, as described by Callahan.[11]

Loss of muscle function requires splinting for the nerve-injured patient. The prevention of deformity and pain can be accomplished by splinting at night, which maintains the hand in a safe position while avoiding ligament shortening or stretching. During the day, static or dynamic splinting assists absent or weakened motions, further improving function. Splinting should be performed only when necessary, with consideration for the following principles.

1. Splinting is performed only to improve function and should not prevent the use of the injured extremity or any existing functional tenodesis.
2. No attempt should be made to improve one function at the expense of another.
3. The splint should minimize involvement of the functional contact areas of the palm or fingertips.
4. Pressure is avoided over insensate areas and disbursed over as much surface area as possible.
5. The size of the splint is minimized as much as possible while supporting only those joints that require support.

Splinting for median nerve palsy emphasizes the preservation of the thumb web space in order to prevent contracture. By positioning the thumb in palmar abduction with an opponens hand-based splint, the patient is able to use the flexor tenodesis to perform functional prehensile and gross grasp (Fig. 7-9).

A secondary consideration for splinting with high median nerve palsy is the prevention of digital flexor tightness. Splinting in the presence of an anterior interosseous nerve palsy frequently requires a dorsally based static interphalangeal extension splint to prevent hyperextension.

In the high or low ulnar nerve palsy patient, a hand-based dorsal splint maintains the metacarpophalangeal joints of the ulnar two digits in approximately 20–30 degrees of flexion while allowing full metacarpophalangeal flexion and interphalangeal joint extension (Fig. 7-10). This design improves function by preventing the "claw" deformity and improves digital extension. The position assists the extrinsic flexors and extensors to compensate for the loss of the intrinsic muscle balance.

With a high radial nerve palsy, the patient frequently uses extensor tenodesis to create digital extension. Splinting of this patient might involve only the fabrication of a wrist cock-up splint or some form of dynamic extension for the digits, as described by Colditz.[13] Functional splinting for the posterior interosseous nerve palsy may not be necessary. A positional splint at night to hold the hand in the protective position might prevent deformity or soft tissue tightness.

Once muscle reinnervation is occurring, EMG biofeedback, in addition to electrical stimulation, is effective in retraining the reinnervated muscle. This regimen is commonly performed in conjunction with both active and resistive exercises for strengthening. During the reinnervation period, functional activities that usually incorporate the use of the involved muscles are also performed.

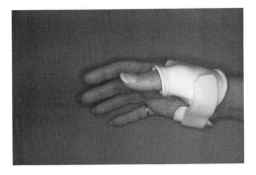

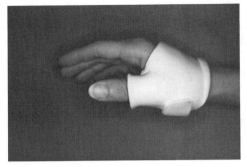

Figure 7-9. Opponens splint for low median nerve palsy.

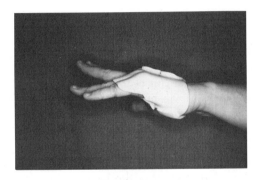

Figure 7-10. Splint for low ulnar nerve palsy maintaining metacarpophalangeal joints in flexion.

The entrapment of the median nerve beneath the flexor retinaculum and its collective signs and symptoms have been referred to as carpal tunnel syndrome. The causes of the carpal tunnel syndrome have been well documented in literature and include vascular, metabolic, and physical trauma. Entrapment of the ulnar nerve at the level of the wrist in Guyon's canal is possible. Compression or entrapment of the common or proper digital nerves in the palm or digit may also occur. Damage to both the median palmar cutaneous nerve as well as entrapment of the radial cutaneous nerve at the level of the brachioradialis has been reported.[19]

Nonsurgical treatment of carpal tunnel syndrome has been advocated by Gelberman,[25] who had a high success level, reporting a 40% recurrence in patients demonstrating mild symptoms with duration of 1 year or less, when treated with injection and splinting. Patients with severe symptoms and a duration greater than 1 year, however, have an 89% recurrence rate with this conservative treatment. The recurrence rate with injection alone was 90%. Technique utilizes injection with an anti-inflammatory agent and/or splinting for 3 weeks in approximately 20 degrees of wrist extension. Removing the causative factor is also necessary if conservative management is to be successful. Those motions that have been identified as causative include repetitive gross grasp or prehension, isometric gripping with the wrist in flexion, combined wrist flexion and ulnar deviation, and wrist flexion with combined prehension or repetitive digital motion. In addition, vibration, such as that which occurs while utilizing a hand tool, is often a cause. Treatment involves supportive therapeutic modalities, including the application of moist heat, ultrasound, paraffin, and ice to reduce active inflammation, and strengthening of those muscle groups involved while avoiding stress across the volar wrist and on inflamed tissues. Treatment for common or proper digital nerve entrapment includes static splinting, which maintains the involved joints in a position of rest to decrease tension across the site of entrapment. Additionally, therapeutic modalities and an exercise program, which includes active and resistive exercises, are instituted to preserve motion, muscle strength, and endurance. Despite aggressive care, however, surgical release may still be required in some cases.

MUSCULOTENDINOUS INJURY

Involvement of the musculotendinous unit and tendon sheaths primarily revolves around traumatic injury, such as laceration, contusion, or repetitive trauma, which results from work, activities of daily living, or recreational habits and activities. Formulating a treatment program for laceration of the flexor or extensor mechanism of the wrist requires a sound foundation in anatomy, physiology, kinesiology, and the healing process. Consideration is given to the individual tendon involved and the excursion over each of the joints it passes, its vascularity, and the presence or absence of a synovial sheath, such as the wrist flexor or extensor retinaculum or the digital flexor sheath. The primary objectives in treating the injury are to minimize the effect on other tissues, prevent adhesion formation, and maintain motion of all the musculoskeletal tissues of the hand and wrist.

Tendon Laceration

The most commonly accepted method of postsurgical treatment of digital flexor tendon laceration is the controlled early motion program as advocated by Lister, et al.,[35] Duran,[21] and Schneider.[49] The goals of this treatment are initially the avoidance of tension on the repaired tendon, promotion and restoration of flexor tendon glide, and the promotion of flexor tendon healing through intrinsic means. During the first 3–4^1/$_2$ weeks, a dorsal hood splint with rubber band traction of the involved digit(s) is required (Fig. 7-11). The wrist is placed in 25–30 degrees of flexion, the metacarpophalangeal joints are blocked in 70 degrees of flexion, and the interphalangeal joints are maintained in the neutral position. Passive flexion and active extension in this protected position are initiated, ten repetitions four times each day. The splint is removed, depending upon the course of healing and the attending surgeon's preference, between 3 and 6 weeks postoperatively, while rubber band traction in conjunction with a wristlet is continued to protect the repaired tendon. Active exercises can be begun at 3–6 weeks postoperatively, dependent upon the flexor tendon excursion and the surgeon's discretion. In general, the better the tendon glide, the greater the risk of tendon rupture, and the less resistance necessary to maximize tendon excursion. Therefore, progression through the rehabilitative program should be delayed each step of the way.

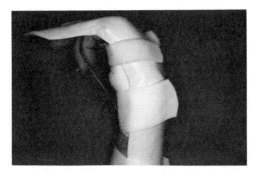

Figure 7-11. Dorsal hood splint maintaining the wrist in 20–30 degrees of flexion. Metacarpophalangeal joints in 70 degrees of flexion, interphalangeal joints in neutral, and rubber band traction on each involved digit allowing full IP extension against rubber band resistance.

Unlike the flexor tendons, traditionally, extensor tendon lacerations have been treated with immobilization, avoiding tension on the repair while preserving mobility of the noninvolved structures. Mobilization of the repaired extensor tendon is predicated on the zone in which the laceration occurs. It is beyond the scope of this chapter to discuss specific care of extensor injuries in each zone. In general, however, most digital extensor tendon lacerations are held in a static extended position for a minimum of 4 weeks (5–6 weeks for wrist extensor tendon lacerations), depending upon the zone in which the injury has occurred. This schedule is followed by an active range of motion program while gradually discontinuing the extensor splinting over a 2-week period, starting when mobilization is initiated. The patient is closely monitored for the development of any extensor lag or adhesion. Resistive range of motion specifically for extensor tendon mechanism is rarely indicated, unless there is an extensor lag or adhesion. Resistive motion is primarily aimed at increasing functional strength for gross grasp and prehension, initiating at approximately 6–8 weeks postoperatively.

Despite all efforts, some patients develop adhesions secondary to scar formation. The adhesions formed depend upon the location of the tendon injury and the extent of damage to the surrounding tissues in cases with concomitant crushing or fractures of the phalanges or metacarpals. Correction of flexion or extension contractures and aiding in the scar remodelling process can be started at 4–6 weeks postoperatively if there is sufficient tensile strength of the extensor tendon to tolerate dynamic forces. Alleviating contractures through remodelling of the scar tissue is frequently facilitated by the application of various heating modalities, such as moist heat, paraffin, and ultrasound, and by massage, while stressing the scar tissue in the direction of the desired motion via splinting and exercise.[34] The use of ultrasound prior to 4 weeks postoperatively with tendon laceration repairs remains controversial and should be reserved until after this period of time. Finally, if an extensor lag should develop, continued static extension splinting and specific strengthening should be pursued for an extended period before resorting to other methods of correction.

Tendinitis/Tenosynovitis

Tendinitis, or stenosing tenosynovitis, is probably more common than tendon laceration. The patient presents with soft tissue dysfunction involving the musculotendinous unit or its sheath, including tendinitis of the wrist, tenosynovitis of the dorsal compartment of the wrist, de Quervain's disease, or stenosing tenosynovitis of the flexor tendon sheath ("trigger finger").[59] Medical treatment of these conditions includes nonsalicylate anti-inflammatory drugs, injection with local steroids,[33] and static splinting to rest the involved tissues.[48] Rhoades, et al.[48] found conservative treatment of stenosing tenosynovitis to be most effective when a single digit is involved, after less than 4 months of symptomatology. These cases responded most favorably to the conservative regime of injection and immobilization for 3 weeks. Immobilization for de Quervain's disease can be accomplished with a custom-molded radial gutter splint. Tendinitis involving the extensors or flexors of the wrist can be immobilized with a wrist cock-up splint, while tendinitis of the extensor digitorum communis requires a functional resting splint which includes the metacarpophalangeal joints of the involved digits.

Treatment of repetitive tendon strain injuries has been separated into three stages by Browne.[9] Stage I is most reversible, and involves

1. Rest from function
2. Identification and avoidance of causative factors, such as repetitive motion or trauma-producing equipment in the work environment
3. Reduction of the workday
4. Alteration of repetitive with nonrepetitive tasks at regular intervals

During stages II and III, treatment includes rest from function and job or work task modifications, as well as splinting to support the involved structures. At this time, physical modalities are applied to reduce inflammation, including thermal agents, cryotherapy, transverse friction massage,[12] and phonophoresis. During these last two stages, corticosteroid injections as well as surgical intervention are sometimes necessary. Rehabilitation should include a balance between rest and activity. In addition to supportive physical modalities, splinting and removal of causative factors, as well as a conditioning program that includes isometrics, eccentric exercises, and flexibility exercises, should be instituted.[51]

The conservative treatment of the stenosing tenosynovitis of the flexor tendon sheath, or "trigger finger," has been proposed by Evans.[22] Treatment combines a metacarpophalangeal joint extension blocking splint, which allows full interphalangeal joint motion, flexor tendon gliding exercises, and the avoidance of causative factors. Preliminary results indicate this treatment is most effective when symptoms are of less than 4–5 months duration and without severe "triggering" and limitation of flexion. This treatment program has not proved as effective for the thumb as it has for the other four digits.

OSSEOUS LESIONS

Treatment of the skeletal system might be required as a result of trauma, such as fracture or dislocation, cumulative repetitive trauma, or arthritis (either traumatic, degenerative, or systematic). Treatment of these conditions follows the same basic principles mentioned earlier, including rest from function, maintenance of tissue homeostasis, elimination of causative factors, and treatment of symptoms with various physical modalities and techniques.

Therapeutic intervention by the rehabilitation team in fracture management most frequently occurs after the immobilization period. Therapeutic intervention should occur, however, prior to termination of immobilization for preventive purposes, introduction of the patient to the rehabilitation process, and early patient education and instruction. Early institution of an exercise program assists with preservation of noninvolved joint motion, tissue mobility, and edema control, which helps prevent further morbidity as a result of the fracture. Immobilization should restrict only those tissues and joints necessary to maintain the fracture in proper position and alignment, while permitting motion of noninvolved structures.

In treatment of restricted joint motion and reduced soft tissue mobility after the discontinuation of immobilization, careful consideration is given to the status of the healing fracture as well as to the condition of the articular cartilage. Immobilization has been shown to cause either temporary or permanent structural changes in the articular cartilage.[39] Treatment consists of the application of thermal agents, such as ultrasound, over the contracted tissue, and joint mobilization to assist in the restoration of arthrokinematics. In addition to modalities and joint mobilization, an active range of motion program is instituted as soon as possible to minimize the deleterious effects of immobilization on articular cartilage.[39] Providing the fracture is stable, as both the motion and the condition of the articular cartilage improve, resistive and passive exercises and static or dynamic splinting are instituted to assist in tissue remodelling. Static or dynamic splinting should be of low load and long duration, which has been found the most effective means by which to remodel scar tissue.[2]

Rehabilitation should not exacerbate pain or provoke inflammation, which inevitably retards overall progress. The use of other physical agents, including high volt galvanic stimulation, continuous passive motion devices, interferential current, or laser are also thought to be beneficial. Their effectiveness, for the most part, however, remains theoretical, lacking controlled investigations.

Serial casting is a particularly effective treatment for flexion contractures of the proximal interphalangeal joint.[5] Digital casting is also applied to the treatment of distal phalangeal fractures or mallet deformities. Direct application of the finger cast to the skin contours of the digit is nonremovable by the patient and is less bulky than other methods of immobilization.

Although the rehabilitation team's primary involvement is in treatment during postoperative fracture management, advances in fracture bracing places the team in a position to take a more active role in the care of hand and digital fractures. The fabrication of a metacarpal fracture brace, for example, allows for the early use of the uninjured digits and joints.[23] The brace minimizes joint stiffness and the requirement for postfracture rehabilitation.

Joint Injuries

The treatment of acute joint injuries depends upon the level of ligamentous damage, the most severe of which require surgical intervention. Those cases that do not require surgery should be placed at rest from function in a position that allows for full contact of the injured structures. Mild sprains without ligamentous disruption usually require support of the injured ligament while allowing protective active motion. In the digits, protection and support can be achieved through buddy taping or strapping. In the more severe injuries, preservation of joint stability is of paramount importance and takes precedence over mobility.[6] The more serious injuries that present with instability and ligamentous disruption commonly need 3–6 weeks of immobilization, followed by protected active motion. Dynamic or static splinting for correction of contracture and the institution of passive and resistive exercises should be delayed until 6–8 weeks after the injury.

Arthritis

Just as with joint stiffness, the treatment of arthritis requires special considerations. Medical treatment of arthritis includes intra-articular steroid injections and nonsalicylate antiinflammatory drugs.[46] Surgical intervention is often applied in the later stages of arthritic deformity. The rehabilitation specialists' primary task is patient education: instructing the patient to avoid work, recreation, and daily living activities that can place undue stress across the involved joints and surrounding soft tissue. During periods of exacerbation, the involved joints should be rested from function with static splinting. Cryotherapy is at times beneficial in alleviating pain and inflammation, provided vasculitis is not present. Conversely, patients might benefit from thermal agents such as moist heat or paraffin, which have been shown to raise joint capsule and soft tissue temperatures in the hand and foot.[4] Whirlpool can increase edema,[56] and active exercising while in the whirlpool has not been shown to significantly reduce it.[50]

During periods when arthritis is inactive, treatment is aimed at increasing strength and maintaining motion and function, accomplished through light resistive exercises, gentle passive motion, and isometric exercises. Isometric squeezing of dowels of successively decreasing diameter is an example of an exercise that aids in improving digital flexion. Various prehensile and functional activities, such as Val-Par work samples, Purdue Pegboard, and Minnesota Rate of Manipulation, for example, assist in increasing motion and hand-eye coordination, and restoring muscle balance.

REPETITIVE TRAUMA

Of increasing importance is the care of cumulative or repetitive trauma disorders of the hand and wrist. The medical community is just beginning to recognize the effect of the working environment on humans. Repetitive trauma affects neurologic as well as musculotendous systems in the hand. Occupational and environmental factors that must be considered include repetitive force, mechanical stress, posture, temperature, and vibration.[1] Accordingly, recognition of cumulative or repetitive trauma is of primary importance in preventing progression of the disease. Treatment of the initial phase includes rest from function, static splinting, thermal agents, and a flexibility and strengthening program. Later phases require permanent job alteration, continued use of external supports, and prolonged periods of rehabilitation.

SUMMARY

The treatment of hand and wrist dysfunction requires accurate diagnosis along with a thorough functional evaluation to identify specific structures involved, causative factors, aberrant motor patterns, and the establishment of a baseline by which to measure progress and response to treatment. Although specific conditions require the formulation of specialized and individualized treatment programs, a general understanding of the hand's anatomy, kinesiology, arthrokinematics, ergonomics, and their interaction with the healing tissue is essential in formulating a treatment program. The timely use of splinting, rest, thermal and electrophysiological modalities, therapeutic exercise, and joint mobilization plays a key role in the rehabilitation program, as well as the incorporation of functional activities that simulate either work, daily activities, or recreational motions.

REFERENCES

1. Armstrong TJ: Ergonomics and cumulative trauma disorders. *Hand Clin* 1986; 2:553.
2. Arom, Madden J: Effects of stress on healing wounds. *J Surg Res* 1976; 20:93.
3. Barber, LM: Desensitization of the traumatized nerve. In Hunter J (editors): *Rehabilitation of the Hand*, 2nd ed. St. Louis, C. V. Mosby Co., 1984.
4. Barrel R, et al: Comparison of in vivo temperatures produced by hydrotherapy, paraffin wax treatment, and fluidotherapy. *Phys Ther* 1980; 60:1273.
5. Bell JA: Plaster cylinder casting for contractures of the interphalangeal joints. In Hunter J, et al (editors): *Rehabilitation of the Hand*, 2nd ed. St. Louis, C.V. Mosby Co., 1984.
6. Bettinger S: Sprains and joint injuries: Therapeutic management. *Hand Clin* 1986; 2:99.
7. Bonnel F: Histologic structure of the ulnar nerve in the hand. *J Hand Surg* 1985; 10A:264.
8. Bonnel F., Vila RM: Anatomical study of the ulnar nerve in the hand. *J Hand Surg* 1985; 10B:165.
9. Browne CD, et al: Occupational repetition strain injuries: Guidelines for diagnosis and management. *Med J Aust* 1984; 140:329.
10. Brumfield RH, et al: Joint motion in wrist flexion and extension. *South Med J* 1966; 59:909.
11. Callahan A: Methods of compensation and re-education for sensory dysfunction. In Hunter J, et al (editors): *Rehabilitation of the Hand*, 2nd ed. St. Louis, C.V. Mosby Co., 1984.
12. Chamberlain GJ: Cyriax's friction massage: A review. *J O S P T* 1982–3, 4:16.
13. Colditz JC: Splinting for radial nerve palsy. *J Hand Ther* 1987; 1:18.
14. Cone RO, et al: Computed tomography of

the normal radioulnar joints. *Invest Radiol* 1983; 18:541.

15. Cooper A, Bransby B: *Treatise of Dislocations,* 5th ed. Philadelphia, Lea & Blanchard, 1844.

16. Cummings J: Conservative management of peripheral nerve injuries utilizing selective electrical stimulation of denervated muscle with exponential progressive current forms. *Am J Occup Ther* 1985; 7:1.

17. Dellon AL: Clinical use of vibratory stimuli to evaluate peripheral nerve injury and compression neuropathy. *Plas Reconstr Surg* 1980; 65:466.

18. Dellon AL: The moving two point discrimination test: clinical evaluation of the quickly adapting fiber/receptor system. *J Hand Surg* 1978; 3:474.

19. Dellon AL: Susceptability of the radial sensory nerve to entrapment and neuroma formation. Presented at *Americal Physical Therapy Association*—Combined Section Meeting Hand Rehabilitation, Atlanta, GA, Feb. 1987.

20. Dellon AL, et al: Re-education of sensations in the hand after nerve injury and repair. *Plast Reconstr Surg* 1974; 53:279.

21. Duran RJ and Jouser RG: Controlled passive motion following flexor tendon repair in zones 2 and 3. In *American Academy of Orthopedic Surgeons' Symposium on Flexor Tendon Surgery in the Hand.* St. Louis, C.V. Mosby Co., 1975.

22. Evan RB, et al: Conservative management of the trigger finger—a new approach. Read at *Tenth Annual Meeting American Society of Hand Therapists,* San Antonio, TX, Sept. 1987.

23. Ferraro MC, et al: Closed functional bracing of metacarpal fractures. *Ortho Rev* 1983; 12:49.

24. Fisk GR: The wrist. *J Bone Jt Surg* 1984; 66B:396.

25. Gelberman R, et al: Carpal tunnel syndrome, results of a prospective trial of steroid injection and splinting. *J Bone Jt Surg* 1978; 62:1181.

26. Gilula LA, et al: Roentgenographic diagnosis of the painful wrist. *Clin Ortho Rel Res* 1984; 187:52.

27. Goodman JJ, et al: Arthroplasty of the rheumatoid wrist with silicone rubber; an early evaluation. *J Hand Surg* 1980; 5:114.

28. Kaplan PE: Hemiplegia: rehabilitation of the lower extremity. In Kaplan, P and Cerullo, L (editors): *Stroke Rehabilitation.* Butterworth Publ., Boston, 1986.

29. Kaplan PE: Sensory and motor residual latency measurements in healthy patients with neuropathy, Part 1. *J Neurol Neurosurg Psych* 1976; 39:338.

30. Kaplan PE, Sahgal V: Residual latency: New applications of old technique. *Arch Phys Med Rehabil* 1978; 59:24.

31. Kraft GH, Halvorson GA: Median nerve residual latency: Normal value and use in diagnosis of carpal tunnel syndrome. *Arch Phys Med Rehabil* 1983; 64:221.

32. Lamberta FJ, et al: Volz total wrist arthroplasty in rheumatoid arthritis: A preliminary report. *J Hand Surg* 1980; 5:245.

33. Lapidus PW, Guidotti FP: Stenosing tenosynovitis of the wrist and fingers. *Clin Orthop Rel Res* 1972; 83:87.

34. Lehman J, et al: Effect of therapeutic temperature of tendon extensibility. *Arch Phys Med Rehabil* 1970; 51:481.

35. Lister GD, et al: Primary flexor tendon repair followed by immediate controlled mobilizations. *J Hand Surg* 1977; 2:441.

36. Ludwig G: Median ulnar nerve communications and carpal tunnel syndrome. *J Neurol Neurosurg Psychiat* 1977; 40:982.

37. Mackinnon SE, Dellon AL: Overlap pattern of the lateral antebrachial cutaneous and radial sensory nerve. *J Hand Surg* 1985; 10A:522.

38. Mathiowetz V, et al: Reliability and validity of grip and pinch strength evaluations. *J Hand Surg* 1984; 9:222.

39. McDonough AL: Effects of immobilization and exercise on articular cartilage—a review of literature. *Am J Occup Ther* 1981–2; 3:2.

40. Meyer RA, et al: Neural activity originating from a neuroma in the baboon. *Brain Res* 1985; 325:255.

41. Michon J, Moberg E (editors): *Traumatic Nerve Lesions of the Upper Limb*. New York, Churchill Livingstone, 1975.

42. Milch H: So-called dislocations of the lower end of the ulna. *Ann Surg* 1942; 116:282.

43. Mino DE, et al: Computed tomography of the radioulnar joint. *J Hand Surg* 1983; 8:23.

44. Morrey BF, et al: Biomechanical study of the elbow following excision of the radial head. *J Bone Surg* 1979; 61:63.

45. Palmer AK, Weiner FW: Biomechanics of the distal radioulnar joint. *Clin Orthop Rel Res* 1984; 187:26.

46. Palmieri TJ, et al: Treatment of osteoarthritis in the hand and wrist. *Hand Clin* 1987; 3:371.

47. Ray RD, et al: Rotation of the forearm—an experimental study of pronation and supination. *J Bone Joint Surg* 1951; 33A:993.

48. Rhoades CE, et al: Stenosing tenosynovitis of the fingers and thumb—results of a prospective trial of steroid injections and splinting. *Clin Orthop Rel Res* 1984; 190:236.

49 Schneider L: *Postoperative Management in Flexor Tendon Repair in Flexor Tendon Injuries*. New York, Little, Brown & Co., 1985.

50. Schultz KS: The Effect of Active Exercise During Whirlpool Treatment. Thesis, San Jose State University, Department of Occupational Therapy, San Jose, 1982.

51. Stanish WD, et al: Eccentric exercise in chronic tendonitis. *Clin Orthop Rel Res* 1986; 208:65.

52. Swanson AB: Flexible implant arthroplasty for arthritic disabilities of the radiocarpal joint. *Orthop Clin North Am* 1973; 4:383.

53. Tess EE, Moran CA: Clinical assessment recommendations. American Society of Hand Therapists, 1981.

54. Thomas JE, et al: Electrodiagnostic aspects of carpal tunnel syndrome. *Arch Neurol* 1967; 16:635.

55. Volz RG: Total wrist arthroplasty. *Clin Orthop* 1977; 128:180.

56. Walsh MT: Relationship of Hand Edema to Upper Extremity Position and Water Temperature During Whirlpool Treatments in Normals. Thesis, Temple University College of Allied Health Professions, Philadelphia, 1983.

57. Waylett J, Seibly D: A study to determine the average deviation accuracy of a commercially available volumeter. Fourth Annual Meeting American Society of Hand Therapists, Las Vegas, 1980.

58. Werner JL, Oner GE: Evaluating cutaneous pressure sensation of the hand. *A J O T* 1970; 24:347.

59. Wood MB, Dobyns JH: Sports related extra-articular wrist syndromes. *Clin Orthop Rel Res* 1986; 202:93.

8

Hip Joint Dysfunction

Paul E. Kaplan and
Ellen D. Tanner

One appendicular arthralgia that often renders a patient thoroughly and quickly disabled is hip pain. Hip joint dysfunction makes each step torture and each minute of walking seem like a whole year of exercise. Our largest and most powerful peripheral joint is extremely sensitive to the effect of inflammation, acute and chronic, and is strategically placed.

HIP STRESS

Stress forces on the hip joint are huge. During walking upon level ground, compressive forces across the hip are four times body weight.[19,23,26] During climbing stairs, these same compressive forces increase to seven times body weight.[1,23] If a 12.8 kg box is to be lifted with flexed knees to a table, the compressive forces are over three times body weight.[20] If the trunk is inclined backward during the lift, the load moment significantly increases.[19] If the knee is straight instead of bent the load moment also increases.[20] The magnitude of the weight transferred across the hip along with the relatively long lever arm built into the structure of the hip combine to guarantee a vast mechanical advantage to any muscular effort. With this mechanical advantage, any slight change in positioning would have an enormous impact in the direction of the generated torque (Fig. 8-1).

The hip's large structural mechanical advantage makes possible the controlled weight transfer necessary for walking and running. Nevertheless, small changes in position result in a vast amount of augmented stress and strain across the joint. Functional activities such as twisting, turning, stooping, or being confined in unusual body positions at small spaces place enormous loads upon the hip. If the patient is also obese, these activities will be stressful enough to cause hip dysfunction—the question is not "if" but "when" such an episode will occur. These functional activities are especially common to blue collar occupations in heavy industry, and consequently, hip dysfunction is more common in that setting (Table 8-1).

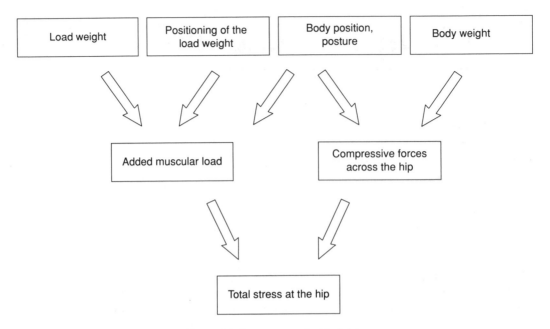

Figure 8-1. Stress across the hip joint.

HIP MOTOR CONTROL

Torques created through the built-in mechanical advantage of the hip are directed and therefore controlled through the actions of motor units. These units are themselves organized through the older, primary action of the lumbosacral plexus, through anterior and posterior divisions. For every spinal segmental anterior division peripheral nerve axis there is a posterior division correlate (Table 8-2). This anterior/posterior organization more nearly controls flexor/extensor muscular action in the hind legs of lower mammals. In primates, however, torsion of the femur on the hip and of the tibia on the femur has produced odd effects. In the human thigh, the old anterior compartment is now antero-

TABLE 8-1. FUNCTIONAL ETIOLOGY OF ELEVATED HIP STRESS

Weight or load-related stress
Obesity
Lifting, pushing, pulling loads greater than 50 lb

Position-related stress
Twisting, turning motions
Stooping or being confined in unusual positions in small spaces

TABLE 8-2. POSTERIOR ORGANIZATION OF THE LUMBOSACRAL PLEXUS

Division	Nerve	Spinal Cord Segment
Anterior	Obturator	L2–4
	Tibial	L4–S2
	Nerve to the quadratus femoris	L4–S1
	Nerve to the superior gemellus	L5–S2
Posterior	Femoral	L2–4
	Common peroneal	L4–S1
	Superior gluteal	L4–S1
	Inferior gluteal	L5–S2

medial in location while the posterior compartment is now lateral; the old anterior compartment of the leg is now posteromedial and the old posterior leg compartment is anteriorly located. These modifications insure the following.

1. The leg and thigh wind and unwind during ambulation. Consequently, no one spinal nerve segment controls any specific function—for example, as the C6 segment controls external rotation of the shoulder. The multisegmental organization maintains economic ambulation control because the loss of only one spinal segment will not affect hip function. Since at least three of the six gait disturbances are centered in the hip, gait modulation is usually intact.

2. The leg primarily moves anteroposteriorly and not medial-laterally. The knee, in lower animals, a medial-laterally oriented joint, in primates is thus rotated 90 degrees anteriorly (toward a narrow range of functional alignment). Any abnormality, therefore, that rotates, abducts, or adducts the leg and thigh past this specific window effectively obliterates knee function. Consequently, hip external rotation and abduction is always far more important than other hip activities since these motions maintain the functional window.

3. Any slight adjustment of hip function profoundly affects knee action. The knee is already on the end of a lever arm if the fulcrum of action remains at the hip. Winding and unwinding converts this motion from linear to sinusoidal orientation through space.

4. The anteroposterior organization of knee function means that pelvic tilt at the hip always depends upon at least 45 degrees of knee flexion. Thus, the knee has a feedback influence. If hip function misalignment destroys knee function, pelvic tilt is also eliminated. In this instance, one extremity is lengthened, producing a circumducted gait.

5. As the hip rotates, muscle layers controlled by anteroposterior-oriented motor control provide differential feedback to each spinal cord segment. The net effect is to give an accurate picture of exactly where the hip is in space at each moment of movement in the normal individual. This sensitive proprioceptive feedback mechanism is also disrupted with hip joint maladjustment. A good example is the

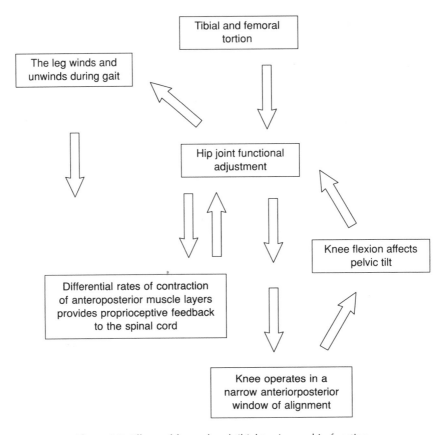

Figure 8-2. Effects of femoral and tibial torsion on hip function.

consistent hip external rotation, flexion, and abduction seen in patients with above-knee amputations. Disrupted proprioceptive feedback frequently contributes to the perceived muscular imbalance (Fig. 8-2).

DIFFERENTIAL DIAGNOSIS

The differential diagnosis of hip dysfunction used to depend largely upon physical examination.[4,14,17,18,31] "Snapping" hips could be reproduced by the patient while standing or riding a bicycle. The often accompanying trochanteric bursitis was confirmed by the patient's generating pain while trying to abduct the hip when it was already abducted and against resistance or through tenderness over the greater trochanter. Meralgia paresthetica could be diagnosed through a Tinel's-like thumping over the lateral femoral cutaneous nerve of the thigh at the level of the anterior superior iliac spine. Finally, a patient with avascular necrosis of the hip would try to limit hip abduction and internal rotation during gait. Physical examination, assisted by radiographic examination of the hip and thigh, is still important. But the whole issue of differential diagnosis has been revolutionized by noninvasive testing.

Meralgia paresthetica is an excellent example. Sensory nerve conduction techniques have been applied to the evaluation of the lateral femoral cutaneous nerve of the thigh. A reliable, reproducible nerve latency has been used to evaluate patients suspected of having meralgia paresthetica.[6,27,30] This electrodiagnostic examination is safe, sensitive, and specific. As a result, more patients can be accurately diagnosed with this condition. In the process, many patients with low back pain as well as hip dysfunction who were originally thought to have lumbar radiculitis were found to have meralgia paresthetica. Some were overweight executives with brief-type undershorts. Some, however, did not fit this pattern and still had meralgia paresthetica instead of lumbar radiculitis. Accurate diagnosis can prevent useless and unnecessary spinal surgery.

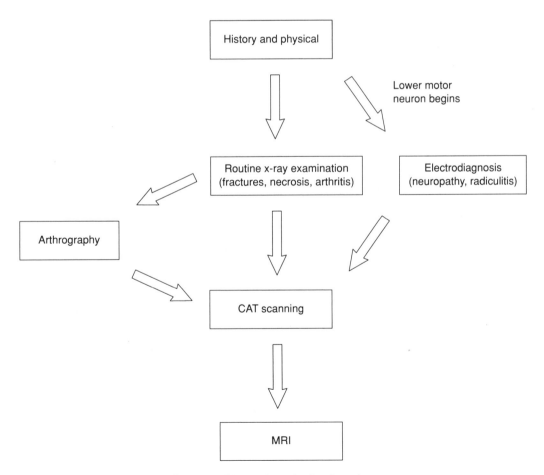

Figure 8-3. Diagnostic evaluation flow chart.

CAT scanning and MRI have been applied to the diagnosis of a wide variety of hip disorders.[3,5,25] MRI and CAT scanning are highly sensitive and relatively specific for fractures, arthritides, ischemic necrosis, tumors, and infections. Although they are expensive, they are effective after a screening physical examination and radiography. Many nonspecific and evanescent symptoms can now be correlated with specific anatomic abnormalities. Stress fractures with or without ischemic necrosis have been associated with both low back and hip failure symptoms. Moreover, MRI and CAT scanning can be combined with other evaluations. CAT scanning can be produced with a three-dimensional analysis of hip morphology.[28] CAT scanning can be applied to generate data on mineral analysis in osteoporotic patients.[9] Additionally, results of CAT scanning can be compared with those of ultrasonography to determine the presence of hip joint effusion.[8,32] No complaint of hip pain, however vague, should be without a full physical and diagnostic evaluation. Active disease processes are in this way accurately characterized, even at a relatively early stage (Fig. 8-3).

OSTEOARTHRITIS

There is some controversy about osteoarthritis as a cause of hip dysfunction. One proposal emphasizes the distinction between primary and secondary osteoarthritis.[12] Osteoarthritis, according to this hypothesis, exists as a secondary process—a reaction to deformities caused by a slipped capital femoral epiphysis, Legg-Perthes disease, dysplasia, etc. For the clinician, these fine academic distinctions are usually moot. The overweight, overworked patient with hip pain and a limp must be treated for osteoarthritis. Whether the osteoarthritis is primary or secondary, it is still the major contributor to hip symptomatology. Pain, stiffness, and physical dysfunction are so common in osteoarthritis of the hip that an effort has been made to standardize patient responses.[2] Difficulty with bending, strenuous exercise, sitting, and rising from a bed are nearly universal. Associ-

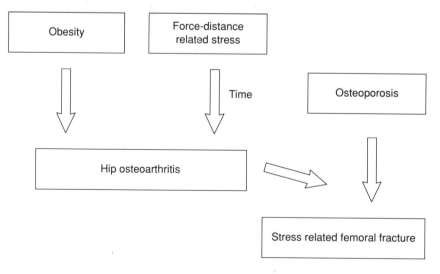

Figure 8-4. Osteoporosis and osteoarthritis of the hip.

ated psychological and social complications such as difficulty sleeping, poor relations with spouse, and boredom are also frequent problems. Wasted human potential and enormous functional morbidity consequently produce frustration and despair in both patients and clinicians.

The therapeutic dilemma is deepened by the coexistence of osteoarthritis and osteoporosis in elderly patients.[13] Osteoarthritis can contribute to the risk of femoral stress fractures in osteoporotic individuals. Reciprocally, osteoporosis does not protect an individual against osteoarthritis. The risk of iatrogenic stress fractures during therapy is high in these patients. Appropriate baseline radiographs and scanning must be obtained before therapy begins, and the patients's clinical response to the stress of therapy must be closely monitored in elderly patients (Fig. 8-4).

Medical management has been enhanced with the application of several anti-inflammatory agents to help control the osteoarthritis.[7] Surgical management is beyond the scope of this book, but certain ramifications of surgical therapy must be noted. Conservative surgical management offers a wide choice, from osteotomy to transfer of the

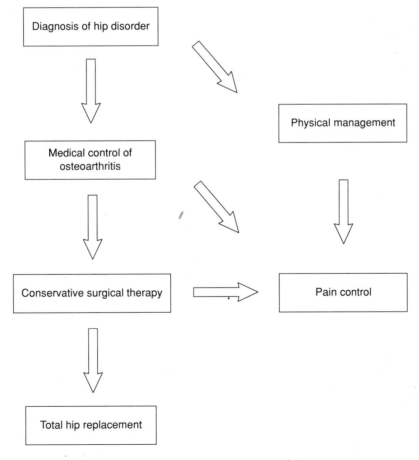

Figure 8-5. Management of hip osteoarthritis.

greater trochanter.[16,22] These procedures successfully relieve local and focal hip pain. Because CAT scan and MRI studies can provide an early and accurate diagnosis, appropriate conservative surgical management has been made much more effective. On the other hand, total hip replacement loosening can be associated with deterioration of the cement as well as formation of a synovial-like membrane at the bone-cement interface.[10] Because loosening, along with infection, is a feared complication of replacement and leads to salvage-type procedures, swifter diagnosis and the application of more conservative and medical measures should effectively manage the only indication for hip replacement in osteoarthritis, i.e., pain (Fig. 8-5).

CONSERVATIVE TREATMENT

Because pain at the hip is so disabling, appropriate treatment can mean the difference between disability and function when the hip is involved in a pathologic process. In situations in which the underlying arthritis can only be controlled and not cured, therapeutic goals are to relieve the symptoms and manage effects of progression of the pathologic process. Since putting the hip "at rest" is difficult, these goals are often thought unachievable. It is possible, however, to lower pressure across the hip and relax the hip joint with ambulation aids. A number of methods are effective in restoring strength and mobility across a moderately involved hip. When the condition is severe, surgery is often necessary to provide relief of symptomatology.

The acute soft tissue injury at the hip that is seen frequently in sports medicine is treated the same as any other injury of a similar nature—with rest, ice, and compression and elevation when possible. Rest is provided through the use of crutches, cane, or walker, when ambulation is necessary. As soon as the acute symptoms subside, heat replaces ice. Both superficial and deep heat are helpful. Gentle active-assistive exercises are begun, progressing to active and later resistive as healing progresses. In the hip, the muscles masses are large, and stresses can be enormous. For this reason it is important that strength and flexibility are conscientiously restored and that return to normal activities is carefully graded and monitored. The muscles around the hip are notorious for tightness, so thorough evaluation of the muscles' length and appropriate stretching exercises are essential. Muscle length is particularly important, for example, in trochanteric bursitis, where the shortened or tight tensor fascia lata is implicated as a causative factor. Muscular tightness around the hip is also implicated in numerous cases of low back pain.

BURSITIS

Inflammation of the bursa over the greater trochanter can be caused by a direct injury or blow to the greater trochanter,[21] by rheumatoid arthritis,[24] and by irritation by the ilictibial band.[11] The pain is often so severe that an antalgic gait is caused, and the Trendelenburg gait is seen in chronic cases. While treatment by injection might provide dramatic relief and even cure the bursitis, it is necessary to determine the cause in order to prevent recurrence. Posture, flexibility, leg length discrepancy, and gait should be evaluated, and the causative factors identified and corrected. Response to treatment by ultrasound is often dramatic in as few as three to six treatments.[15] Stretching of tight structures and strengthening of weakened muscles should accompany any treatment.

Raman and Haslock[24] found 15 out of 100 rheumatoid arthritis patients who suffered from trochanteric bursitis. Symptoms were relieved with one or two injections, even when the pain had been of long duration (1–5 years). Patients with rheumatoid arthritis who complain of hip pain could be evaluated for bursitis. Pain in the hip should not be assumed to be centered in the joint or untreatable.

In patients who do not have rheumatoid arthritis, the iliotibial band is most frequently the culprit in the absence of direct trauma. Bursitis is most common in women, because the broader pelvis, increased Q angle, imbalance between the adductors and abductors of the hip, and tight iliotibial band cause repetitive trauma to the bursa by sliding over it repeatedly[11] (Fig. 8-6). Other causative factors include leg length discrepancy, residual weakness or contracture of the gluteus medius, abnormal running mechanics, and running on banked surfaces.[21] Sometimes an audible or palpable snap can be detected when the tensor fascia lata slides over the greater trochanter (Fig. 8-7).

The ischial and the iliopectineal bursae are less frequently affected but are quite painful. Ischial bursitis can result from a direct blow, as in a fall, or from prolonged compression while sitting on a hard surface, while iliopectineal bursitis is often associated with an antalgic gait and pain in the anterior aspect of the hip.[21] A tight iliopsoas muscle or osteoarthritis can cause iliopectineal bursitis.[11] Treatment of bursitis is discussed in Chapter 1 (Fig. 8-8).

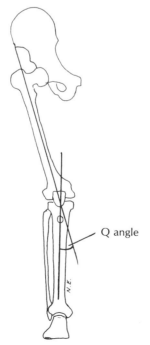

Q angle

Figure 8-6. Q Angle. The Q angle is formed by lines drawn from the anterior superior iliac crest through the patella and from the patella through the tibial tubercle.

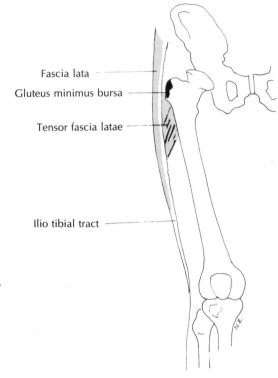

Fascia lata

Gluteus minimus bursa

Tensor fascia latae

Ilio tibial tract

Figure 8-7. Iliotibial Tract Overlying Greater Trochanter. The iliotibial tract/ fascia lata overlies the greater trochanter. During ambulation, a tight iliotibial tract moves forward and back over the trochanter, which can cause irritation and bursitis.

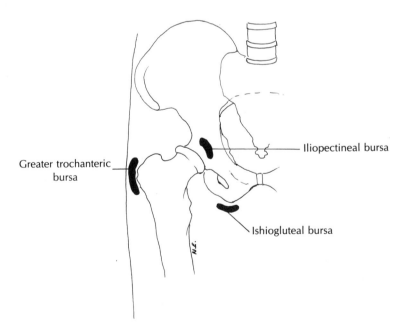

Iliopectineal bursa

Greater trochanteric bursa

Ishiogluteal bursa

Figure 8-8. Bursae of the Hip—Anterior View. Location of the three major bursae of the hip.

MANAGEMENT OF DEGENERATIVE JOINT DISEASE (OSTEOARTHRITIS)

Symptomatic degenerative joint disease is said to be the most common disease process affecting the hip.[15] The pain is associated with capsular or ligament tightness or erosion of the joint surfaces.[29] Restriction of movement is often greatest in abduction and internal rotation, causing shortening of those muscle groups.

When patients are told that their disease is not severe enough to require surgery, they are often referred for symptomatic treatment. No treatment can reverse the damage to the articular cartilage, but the nutrition of the joint can probably be improved, much of the lost range of motion can be restored, and the weakened muscles around the joint can be strengthened to help support and relieve stresses from the joint (Fig. 8-9 through 8-21). Inflammation might be reduced, pain relieved, and ambulatory function restored. Further, the joint capsule, which is implicated as the primary pain source, can be stretched, decreasing the loading on the joint and increasing the effective weight-bearing surface area of cartilage, which retards the degenerative process.[15,29] Ultrasound has been shown to be extremely effective in reduction of joint contracture in the hip and is a valuable adjunct in treatment of the joint capsule. Manual therapy techniques are useful as well in stretching the joint capsule.

In addition to direct treatment of the joint and its surrounding structures, other interventions can improve the quality of life for these patients. Correct use of the appropriate ambulation aid is taught. Numerous adaptive devices, such as raised toilet seats and other assistive devices to compensate for lost mobility or to prevent pain, are available. Instruction in energy conservation techniques is also available for these patients.

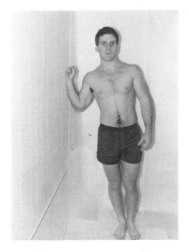 

Figure 8-9. Standing Stretch of Iliotibial Tract and Abductors of the Hip. A tight iliotibial tract can be stretched in the standing position as shown.

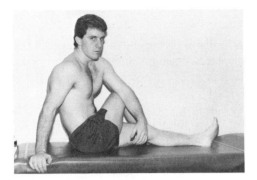

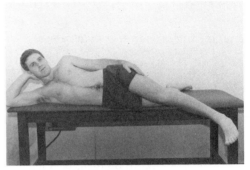

Figure 8-10. Sitting Stretch of Iliotibial Band. The iliotibial tract can be stretched in sitting, as shown.

Figure 8-11. Iliotibial Stretching Side-lying. The iliotibial tract can also be stretched in the side-lying position with the shoulders held horizontal to the mat. The most effective stretch, however, is probably in the same position as the Ober test, which determines whether the iliotibial tract is shortened. It is more difficult to stretch in this manner, however.

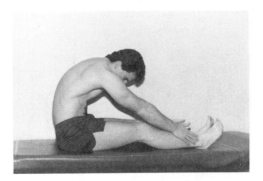

Figure 8-12. Sitting Hamstring Stretch. The hamstrings are effectively stretched in the sitting position, as shown. However, this exercise might place undue strain on the low back.

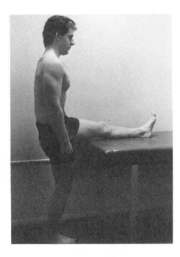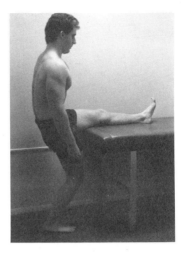

Figure 8-13. Standing Hamstring Stretch. With the leg to be stretched on a high surface, squatting with the other leg causes a strong stretch on the hamstrings. When flexibility is severely reduced, lower support surfaces are utilized initially.

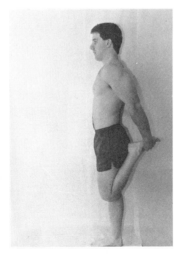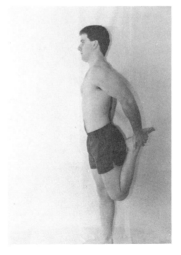

Figure 8-14. Anterior Thigh Stretch. The anterior thigh musculature is stretched in the position shown here. The same exercise can also be accomplished in the prone position.

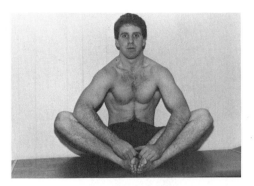

Figure 8-15. Adductor Stretch. Sitting with the soles of the feet together and the hips and knees fully flexed, pressure from the arms downward on the knees causes a strong stretch to the adductors of the hip.

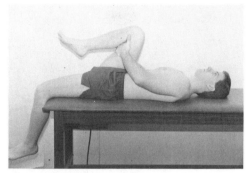

Figure 8-16. Hip Extensor Stretch. With the contralateral leg over the edge of the table, a strong stretch to the hip extensors is provided by pulling on the thigh, increasing hip flexion.

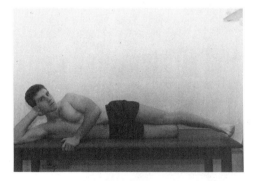

Figure 8-17. Hip Abductor Strengthening Exercise. The hip abductors are strengthened when the upper let's hip and knee are maintained in full extension, the lower leg is slightly flexed at the hip and knee for stability, and the body is in sidelying, perpendicular to the table. Lifting the foot from the table with no rotation exercises the hip abductors.

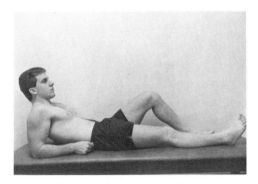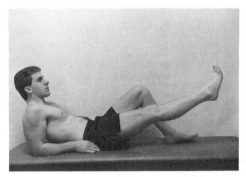

Figure 8-18. Hip Flexion Strengthening Exercise. With the contralateral leg flexed at the hip and knee, the leg is lifted from the table, with no knee flexion and no hip rotation. Weights are added at the ankle as strength improves.

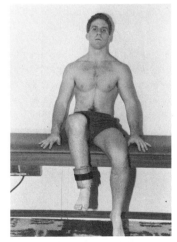

Figure 8-19. Iliopsoas Strengthening Exercise in Sitting. Strengthening of the iliopsoas is accomplished by flexion of the knee, which eliminates much of the strength of the rectus femorus. Weight is added to the ankle as the patient gains strength.

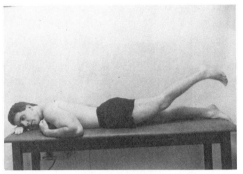

Figure 8-20. Hip Adductors Strengthening Exercise. The adductors of the hip are effectively strengthened by isometric contractions against a pillow between the legs. Little or no rotation of the hips is allowed, or the exercise strengthens the internal rotators of the hip.

Figure 8-21. Hip Extension Strengthening Exercise. In the prone position, the patient is instructed to lift the affected leg off the bed. When the knee is flexed, the glutei are working hardest. With the knee straight, the hamstrings are more active.

REFERENCES

1. Andriacchi PT, et al: A Study of lower limb mechanics during stair-climbing. *J Bone Joint Surg* 1980; 62A:749.
2. Bellamy N, Buchanan WW: A preliminary evaluation of the dimensionality and clinical importance of pain and disability in osteoarthritis of the hip and knee. *Clin Rheumatol* 1986; 5:231.
3. Berquest TH: Magnetic resonance imagery: Preliminary experience in orthopedic radiology. *Magn Reson Imaging* 1984; 2:41.
4. Blickenstaff LD, Monis JM: Fatigue fracture of the femoral neck. *J Bone Joint Surg* 1966; 48A:1031.
5. Burkus JK: CAT scan evaluation of hip dislocations with associated acetabular fractures. *Orthopedics* 1983; 6:1443.
6. Butler ET, et al: Normal nerve conduction velocity in the lateral femoral cutaneous nerve. *Arch Phys Med Rehabil* 1974; 55:31.
7. Dresner AJ: Multicenter studies with sodium meclofenatmate (Meclomen) in the United States and Canada. *Curr Thes Res* 1978; 23:590.
8. Egund N, et al: Computed tomography and ultrasonography for diagnosis of hip joint effusion in children. *Acta Orthop Scand* 1986; 57:211.
9. Gilsang V, et al: Quantitative spinal moneral analysis in children. *Ann Radiol* 1986; 29:380.
10. Goldring SR, et al: The synovial-like membrane at the bone-cement interface in loose total hip replacements and its proposed role in bone lysis. *J Bone Joint Surg* 1983; 65A:575.
11. Gould JA, Davies GJ (editors): *Orthopedic and Sports Physical Therapy.* St. Louis, C.V. Mosby Co., 1985.
12. Harris WH: Etiology of osteoarthritis of the hip. *Clin Orthop* 1986; 213:20.
13. Healey JH, et al: The coexistence and characteristics of osteoarthritis and osteoporosis. *J Bone Joint Surg* 1985; 67A:586.
14. Hungerford DS: Bone marrow pressure, venography and core decompression in ischemic necrosis of the femoral head. In *Proceedings of the Hip Society*, Vol. 7, St. Louis, C.V. Mosby Co., 1979, p. 218.
15. Kessler RM, Hertling D: *Management of Common Musculoskeletal Disorders.* Philadelphia, Harper & Row, 1983.

16. Lennox D: Osteoarthritis VI: The hip—surgical management. *Maryland Med J* 1985; 34:267.

17. Marcus ND, et al: The silent hip in avascular necrosis. *J Bone Joint Surg* 1985; 55A: 1351.

18. McBeath AA: Some common causes of hip pain. *Postgrad Med* 1985; 77:189.

19. Nemeth G, et al: Hip joint load and muscular activation during rising exercises. *Scand J Rehab Med* 1984; 16:93.

20. Nemeth G, et al: Hip load moments and muscular activity during lifting. *Scand J Rehabil Med* 1984; 16:103.

21. Nicholas JA, Hershman EB (editors): *The Lower Extremity and Spine in Sports Medicine*. St. Louis, C.V. Mosby Co., 1986.

22. Papavasiliou VA: Lateral and distal transfer of the greater trochanter. *Clin Orthop* 1986; 207:198.

23. Paul JP: Approaches to design. *Proc R Soc Lond (Biol)* 1976; 192:163.

24. Raman D, Haslock I: Trochanteric bursitis—a frequent cause of "hip" pain in rheumatoid arthritis. *Ann Rheum Dis* 1982;

25. Rick K, et al: Computerized tomography in evaluation and classification of fractures of the acetabulum. *Clin Orthop* 1984; 188: 231.

26. Rydell NW: Forces acting on the femoral head prosthesis. *Acta Orthop Scand* 1966; Suppl 88.

27. Sarcala PK, et al: Meralgia paresthetica: Electrophysiologic study. *Arch Phys Med Rehabil* 1979; 60:30.

28. Sartoris DJ, et al: A technique for multiplanar reformation and three-dimensional analysis of computed tomographic data. *J Can Assoc Radiol* 1986; 37:69.

29. Saunders HD: *Evaluation, Treatment and Prevention of Musculoskeletal Disorders*. Minneapolis, Viking Press, Inc., 1985.

30. Stevens A, Rosselle N: Sensory nerve conduction velocity of N. cutaneous femoralis lateralis. *Electromyography* 1970; 10:397.

31. Teng P: Meralgia paresthetica. *Bull Los Angeles Neurol Soc* 1972; 37:75.

32. Wilson DJ, et al: Arthrosonography of the painful hip. *Clin Radiol* 1984; 35:17.

41:602.

9

Knee Pain and Dysfunction

Paul E. Kaplan and
Ellen D. Tanner

Much information on knee lesions, their consequences, and their management, is already published. To review the clinicopathophysiology in the orthodox academic fashion would involve a great deal of repetition for only a slight chance of presenting new data. Instead, this chapter begins with a review of some of the numerous and significant advances in arthroscopy, since this procedure is frequently indicated in diagnosing and treating knee disorders, then covers management of knee disorders.

CONTEXT OF ARTHROSCOPY

Arthroscopy is used as a diagnostic tool in the clinical presentation of knee disorders.[27] Pain and tenderness are usually poorly focused but can be regular and can be characterized. Usually, the knee is stiff and the patient guards it while walking. Internal and external rotation are often limited, making any limp that much more pronounced. If stiffness becomes pronounced enough, the knee might even lock. More frequently, the knee will lock only during certain activities. These usually involve sudden and forceful internal or external rotation. Locking is not necessarily specific to meniscal disease. Swelling is the single most stressful symptom or sign to many patients. An effusion can be sampled or drained and examined for inflammation, infection, crystals, or bleeding.

Of all the items so far mentioned, the characterization of the effusion is the first item leading to specific diagnosis. Tenderness over the medial or lateral aspects of the knee or patella subluxation may be present, but might or might not be prominent on the day of evaluation. Knee instability is another valuable clue, especially with rotary or anteroposterior stress. Painful anterolateral instability is commonly associated with cruciate ligamental tears. Ligamental tears in other combinations often generate patella subluxation. Quadriceps strength will maintain a functional joint even in the presence of extensive meniscal disease (Table 9-1).

TABLE 9-1. SIGNS AND SYMPTOMS OF KNEE DISEASE

Usually General	More Specific
Pain	Effusion characteristics
Tenderness	Instability characteristics
Patella subluxation	
Quadriceps weakness	
Stiffness	
Swelling	
Intermediate	
Locking—usually, but not invariably associated with meniscal disease.	

HISTORY OF ARTHROSCOPY

To evaluate and manage tuberculosis of the knee, Professor K. Takagi adapted an endoscope as an arthroscope in the first few decades of this century. These instruments had adaptive lenses for a wide variety of photographic techniques.[39] Bircher also studied the knee with arthroscopic examinations, and he was followed by Kreuscher and then by Burman. Burman's instrument could also be used to study the elbow, ankle, and shoulder.[29] Therefore, by 1940, the basis of this diagnostic procedure had been presented.

Dr. Masaki Watanabe followed Takagi's work with an *Atlas of Arthroscopy* in 1957. He also used the arthroscope that serves as the basis of most modern instruments 21. In Toronto, in 1963, I. Macnab applied a pediatric cystoscope to arthroscopic examination of the knee.[5] R. W. Jackson went to Japan, learned arthroscopic techniques from Dr. Watanabe, and reintroduced the procedure to North America by 1965.[18,19] Arthroscopy was a skill explored and preserved in Japan but for most of its history relatively unknown to Western medicine. This lack of tradition has helped and hurt. No vested Western interest has obstructed the application of newer technology, instrumentation, or application. On the other hand, the West was insensitive and hostile to the potential of this procedure for a long time.

Meanwhile arthroscopic surgery has become a common treatment for knee disorders. It has prolonged the careers of numerous athletes and has an enormous impact upon the treatment of many joint disorders. It is particularly useful for small tears or radiolucent loose bodies.[10]

METHODOLOGY

Arthroscopy is an operative procedure performed with the patient under anesthesia. As such, the procedure is performed in accordance with the general surgical guidelines for asepsis, gentleness, and hemostasis.[20] The joint must be distended to be visualized. This is often accomplished using saline, and gas has also been used.[18,19] The scope itself fills one entry to the joint. Supplementary openings are frequently required for other instruments. In this way, tissue to be removed can be correctly positioned and cut under direct visualization. Closed circuit video systems have made it possible for the entire surgical team to view the progress of the procedure. Partial or total closed meniscectomies can be performed as well as synovectomies and chondrectomies.[16,18,19] Procedures to treat ligamental tears have also been developed.[18] National and international seminars have exponentially increased opportunities in the past decade to train new surgeons in these techniques and to trade information about new procedures.

INDICATIONS AND CONTRAINDICATIONS

We have intentionally reversed the usual order of procedure and indication to focus attention upon central issues of indications and contraindications. Generally, arthroscopy is ordered after history, physical examination, radiography, and ultrasonography have failed to clarify diagnosis of the knee's internal disorder.[20] The need for surgical intervention of some type usually has become apparent at this point. Arthroscopic surgery should be considered only for disorders known to be responsive to this intervention. Acute knee injuries, with hemarthrosis, have been an indication.[5]

Contraindications involve considerations of disease processes that affect the risk-benefit ratio. These usually include the presence in the area of active infection, inflammation, thrombophlebitis, and soft tissue or bony growth filling the knee joint. Such entities as hemarthrosis, tumor, or chronic infections can be difficult to evaluate and treat with arthroscopy and require careful evaluation by experienced personnel (Table 9-2).

TABLE 9-2. INDICATIONS AND CONTRAINDICATIONS FOR ARTHROSCOPY

Indications	Contraindications
Preliminary and less invasive evaluations have been completed	None absolutely
Surgery within the joint is probably indicated	Active arthritis, infection, hemorrhage, tumor, thrombophlebitis in the general area increases risk
Arthroscopic surgery could treat the presumed knee disorder	Wide knee capsular disruption and extravasation of joint fluid increases risk

RESULTS

The choice in treating traumatic meniscal damage is often whether to open the joint surgically or attempt closed repair through arthroscopic surgery. Closed partial meniscectomy is well tolerated.[5,6,11,12,26,32,36] Often the length of hospital stay after arthroscopy is less than 3 days. Many workers and athletes are back at work in less than 2 months. While open partial or complete meniscectomies have generated deep venous thrombosis, wound dehiscence, or hematomas, complications from arthroscopy are notable by their scarcity. Arthroscopic partial meniscectomy lends itself to the treatment of elderly patients with knee disease.[30] Nondegenerative (especially bucket-handle) meniscal tears respond better, however, than do degenerative meniscal tears. Long-term varus knee deformity might be less prevalent with the closed technique. This conclusion supports other data that symptomatic degenerative meniscal tears can be effectively managed by partial meniscectomy as long as degenerative arthritis is absent.[24]

Trends seem to indicate a slightly better statistical result after medial partial meniscectomy than after lateral partial meniscectomy.[32] Results also support the hypothesis that the generation of chondromalacia is largely age-related rather than dependent upon meniscal disease.[8,32] Thus, age controls two of the variables contributing to the results of arthroscopy. Nonetheless, elderly patients generally tolerate this procedure well and many have good or excellent results. Arthroscopy has also been used to evaluate and treat the discoid meniscus with good results (Fig. 9-1).

Arthroscopy has been performed on hemarthritic knees, and in three-fourths of the cases the hemarthrosis was associated with cruciate ligament lesions.[11,33] Early removal of blood means that the knee can be treated earlier with therapeutic exercises. This regimen limits ligament reconstruction to those knees not responding to the intensive early physical therapy.[6] With arthroscopy not only can blood be removed but the extent of cru-

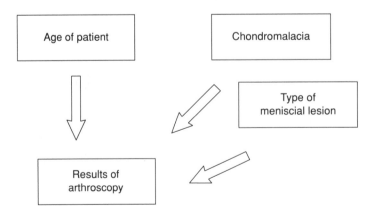

Figure 9-1. Factors contributing to results in arthroscopy of the knee.

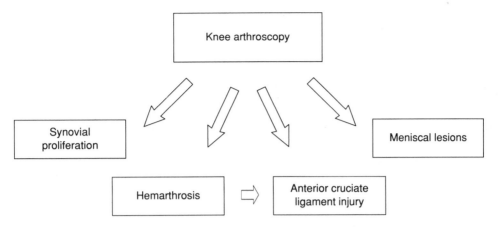

Figure 9-2. Knee arthroscopy—what can be examined and treated.

ciate ligament disruption can be estimated in the context of damage to other knee joint structures[19] (Fig. 9-2).

Once in a while, a new procedure is devised that really does change the way medical care is given. Arthroscopy is one of these procedures. Table 9-3 lists the characteristics that make it uniquely serviceable. As a result of arthroscopy, athletes' and workers' careers have been prolonged and the quality of their lives enhanced.

This procedure must be coordinated with rehabilitative services so its full power can be utilized. Such coordination is complex and detailed but not difficult and should be possible for any full-service medical center. The lack of such coordination is not due to any problems inherent in arthroscopy. The cause lies rather in poor coordination between the arthroscopist and rehabilitation medicine. Enlightened multidiciplinary conference coordination in these cases is far superior to shotgun, routine orders, since each case has its own characteristic aspects. Interdiciplinary efforts are usually the best choice; arthroscopy is not the only answer, but as part of a comprehensive effort, it is very effective.

TABLE 9-3. ADVANTAGES OF ARTHROSCOPY

Safety

Provides accurate visualization of internal joint destruction

Adds to available surgical options

Augments the chances of early vigorous physiotherapy

TREATMENT OVERVIEW

As a complex and stable weight-bearing structure, the knee poses problems not found in the elbow, shoulder, or hip. The motion that occurs at the knee is described by Cailliet[2] as a combination of rolling, rocking, and gliding. There are strong, stabilizing ligaments both inside and outside the joint capsule. Any of the stabilizing structures, as well as the usual bursae, tendons, and articular cartilage, can be injured, causing pain and dysfunction. Arthritis at the knee causes disability of a different quality and magnitude than in the upper extremity. Additionally, multiple problems are associated with the patellofemoral joint, a unique structure in the human body. The muscles around the knee are large and strong, having significant impact upon the joint itself. The retinaculum is often implicated in patellar tracking problems, as are muscular imbalances, bony abnormalities, and alignment of the extremity.[13] The menisci are frequently damaged, leading to pain and dysfunction.

The goal of treatment, as with all other injuries, is restoration of normal function. Sometimes rehabilitation improves function to a point that is better than the preinjury level, which can prevent future injuries and enhance fitness. Normal function includes multiple factors—endurance, power, strength, flexibility, and coordination—all of which are taken into account during rehabilitation.

The most important rehabilitation technique for any joint is prevention. Strength and flexibility conditioning prior to strenuous activity is well known to prevent injuries. After an injury has occurred, appropriate rehabilitation can prevent reinjury. Thorough evaluation is essential in providing effective intervention and must include determining predisposing factors to injury as well as the status of the components of normal function mentioned above.

With the increase in sports-associated injuries comes an increased demand for specific, appropriate rehabilitation so that athletes can resume their activities as quickly as possible. The arthroscope, which makes surgery possible without opening the knee joint, is the most impressive advance. The rehabilitation time has been dramatically reduced, from 3–6 months after meniscectomy in the past to a few weeks in many cases. No longer does meniscectomy consign the patient to a knee immobilizer and crutches for 6 weeks and the concomitant pain and disability.

As with the vast majority of musculoskeletal pain and dysfunction, patient education is of great importance. Not only does the patient thereby understand the problem and how to avoid further pain and injury, but compliance with the treatment program is greatly enhanced.

Treatment of the knee cannot be accomplished without taking into account the other joints of the lower extremity and their interrelations with the knee. Strength, flexibility, and alignment of the foot, ankle, and hip are considered during rehabilitation of the knee.[38] A few of the numerous causes of knee dysfunction include pronation of the foot, tibial torsion, altered Q angle, leg length discrepancy, abnormal insertion of the quadriceps on the patella, and various anomalies of the articulations of the joints of the knee.

Once knee injury has taken place, immobilization is best avoided and mobility restored as soon as possible. Muscle atrophies quickly, and the collagen fibers of ligaments show dramatic changes as a result of immobilization.[31] The joint itself might be affected by scarring, adhesions, pannus, and reduced lubrication.[31] Many patients have atrophy of the thigh muscles, which is related to ligamentous instability[40] and to patella tracking abnormalities. Therefore, intensive strengthening can make the difference between a functional knee and one that requires surgery, and strengthening should be done, even if surgery is to be performed later, to provide a better overall outcome.

PATELLOFEMORAL PAIN

Henry[15] has noted that patellofemoral complaints are one of the most, if not the most, common knee problem in athletes. Cox[3] calls patellofemoral joint pain the most common stress-related injury that afflicts young runners, seen most frequently in teenagers and young adults. Symptoms are caused by failure of any of the stabilizing structures, thus permitting inappropriate patellofemoral articulation and increasing the load on the articular surfaces.[3] Pain is felt in either the subchondral bone or the synovium, since the articular cartilage has no pain receptors. This syndrome should not be confused with chondromalacia patellae (Fig. 9-3).

The primary goals of rehabilitation are recovery of motion, power, and stability.[15] Nonoperative management, successful in 75%–85% of patients, includes alteration in activity, exercise, and anti-inflammatory medication.[9,13] Changes in activity reduce patellofemoral load up to three times, thereby decreasing the effects of the tracking problem and thus the symptoms.[13] Activities such as swimming, running in a pool, cycling, cross-country skiing, or walking should be substituted for running[25] until symptoms abate and tracking problems have been corrected. Then gradual resumption of running is al-

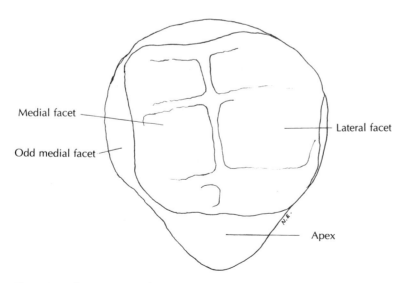

Figure 9-3. The posterior surface of the normal right patella is faceted and smooth.

lowed in a guided progression. Climbing, jumping, squatting, and kneeling should be avoided.[9]

Exercises reduce patellofemoral contact force and decrease lateral tracking via quadriceps strengthening and hamstring stretching.[13] Stretching of the iliotibial band is indicated when it is tight.[3,15,25] Straight leg raising exercises are most effective in strengthening the quadriceps without increasing the load and thereby the articular cartilage damage.[3] Isometrics are often used for patients with acute pain, changing to progressive resistive exercises with the knee in extension.[9] Henry[15] recommends hip abduction and adduction and hip flexion exercises with straight leg raising, using free weights. Electrical stimulation of the quadriceps is helpful in muscular strengthening[31] and re-education when the patient is unable to progress.[9] Exercises should be continued two or three times per week to maintain muscle tone and prevent a recurrence of misalignment and symptomatology.[3]

Orthotic devices that hold the patella medially and decrease lateral tracking decrease contact forces on the lateral patellofemoral joint are recommended.[13] These devices need not be complicated to provide stabilization. Stabilizing sleeves, pads, and braces have been found effective in reducing or relieving symptoms in 64%–93% of patients.[3]

Ice packs are used after exercise to relieve inflammation and pain.[3,25] Heat is applied prior to exercise to help increase circulation to the patellofemoral joint.[3] TENS is helpful in pain control in some cases of patellofemoral pain.[3,9]

CHONDROMALACIA PATELLAE AND OSTEOARTHRITIS OF THE PATELLA

Chondromalacia means "abnormal softening of cartilage."[7] The term is often used inappropriately as a diagnosis in cases of patellofemoral pain[4] or to describe pain behind the patella or roughening of the patellar articular cartilage seen radiographically, neither of which is correlated with patellofemoral pain.[3] There are, sometimes, clefts deep in the softened articular cartilage as a result of trauma, not obvious on the surface, that can lead to destruction of the joint in middle-aged or older people.[4] As a result of a complex biochemical "cascade," synovitis and progressive inflammation characteristic of osteoarthritis ensues.[4] Chondromalacia has been attributed to trauma and weakness of the vastus medialis.[35]

Treatment for patellofemoral osteoarthritis is similar to that for patellofemoral pain syndrome. Exercises to strengthen the quadriceps: avoidance of kneeling, jumping, prolonged standing, running, and excessive stair climbing; and a knee support are recommended.[4] Treatment with modalities as described in Chapter 2 often provides relief of pain and reduction of inflammation. Shields, et al.[35] recommends ice and electrical stimulation during the rehabilitation stage.

PLICA SYNDROME

This syndrome is caused by a synovial fold that runs along the medial and superior borders of the patella and can become thickened,[3] often because of chronic inflammation.

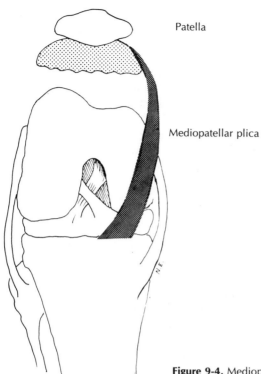

Patella

Mediopatellar plica

Figure 9-4. Mediopatellar plica with knee flexed to 90 degrees.

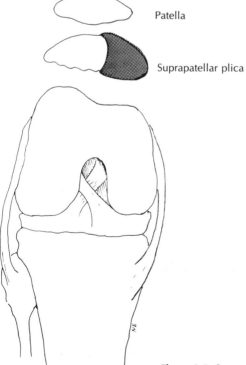

Patella

Suprapatellar plica

Figure 9-5. Suprapatellar plica overlying the patella.

This fold is present in 20%–60% of the population, and three plicae have been described.[34] This condition mimics symptoms of patellofemoral syndrome or chondromalacia patellae.[3] Pain is usually felt along the medial patella and increases with activities requiring knee flexion.[34] Arthroscopy is considered the best means of identifying abnormalities in the plicae[1] (Figs. 9-4, 9-5).

Treatment in the acute phase is directed at reducing inflammation, through medication and modalities, and strengthening the quadriceps with little or no flexion of the knee, as in short arc quadriceps exercises or isometrics.[1,34] Surgical removal of the tissue is indicated if symptoms do not remit in a reasonable period.

MENISCUS INJURIES

Atrophy of the quadriceps occurs quickly after meniscal injury, especially in the vastus medialis, where it may be noted within days of the injury.[2] Therefore, following meniscal injury strength of the quadriceps must be carefully restored or preserved. Even if the meniscus is surgically removed in part or altogether, strengthening of the quadriceps prior to surgery has a beneficial effect upon the outcome (Figs. 9-6, 9-7). With the advent of arthroscopic surgery for meniscal tears, full range of motion is restored very quickly, and long-term dysfunction avoided. In a study by Hamberg et al,[14] open meniscectomy caused 70% loss of quadriceps torque in the first week, and after 8 weeks of intensive rehabilitation, quadriceps function was not yet normal. Patients who underwent arthroscopy or arthroscopic surgery had no significant loss of torque at 1, 4 and 8 weeks postoperatively.

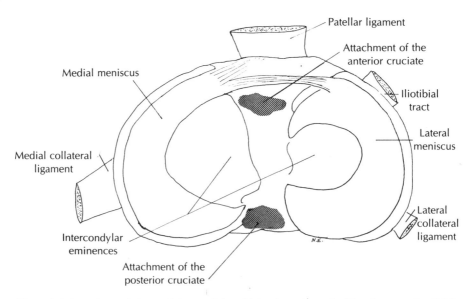

Figure 9-6. Location and shape of the medial and lateral meniscus, looking down on the tibial plateau. Also shown are the attachments of the ligaments of the knee.

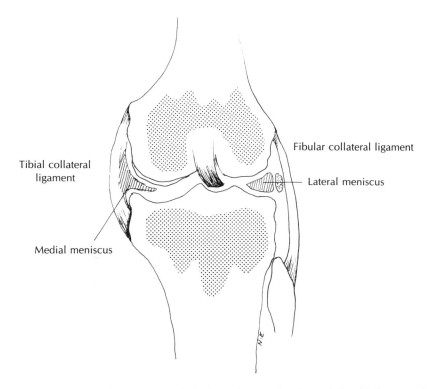

Fibular collateral ligament

Tibial collateral ligament

Lateral meniscus

Medial meniscus

Figure 9-7. Longitudinal section through the knee showing the close relationship between the medial meniscus and the tibial (medial) collateral ligament, and the lateral meniscus and the fibular (lateral) collateral ligament.

LIGAMENTOUS INJURIES

At times, ligament injuries require surgical intervention. For injuries requiring nonsurgical treatment, several options are available. Treatment depends upon the structure damaged and the severity of the damage. No hard and fast rules can be followed, but knowledge of the extent of the damage and the deleterious effects of immobilization is invaluable. Even limited motion, in a protected fashion and without pain, will help forestall or prevent the formation of restrictive fibrosis, allowing rehabilitation to proceed without complications (Fig. 9-8).

Initially, ligament injuries are treated with joint protection to allow healing of these protectors of knee stability. Acute injuries are treated for control of pain and swelling, usually with ice, compression, elevation, and rest. TENS in addition to or instead of medication is helpful in pain control. Once pain and inflammation are under control, isometric exercises are begun for hamstrings and quadriceps. Montgomery and Steadman advocate a combination of bracing and dynamic mobilization within a safe range, which can be begun immediately after injury.[31]

In anterior cruciate ligament involvement, greatest tension on the ligament from 30

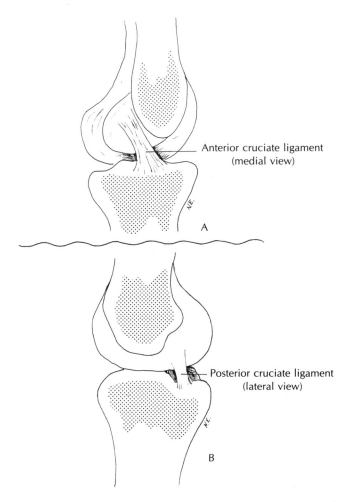

Anterior cruciate ligament
(medial view)

A

Posterior cruciate ligament
(lateral view)

B

Figure 9-8. Longitudinal section showing the medial view of the anterior cruciate ligament and the lateral view of the posterior cruciate ligament.

degrees of knee flexion to full extension, is avoided. In posterior cruciate ligament injuries, primary attention is paid to strengthening of the quadriceps to prevent posterior displacement of the tibia on the femur, and hamstring strengthening is avoided until the quadriceps strength has returned to nearly normal.[35]

The medial collateral ligament is one of the most frequently damaged ligaments.[21] There is little agreement on management of grade III injuries of the medial collateral ligament, with some practitioners advocating surgery and others nonsurgical treatment.[21] Many clinicians prefer immobilization, although there is evidence that medial collateral ligament injuries can be effectively treated without immobilization,[35] which permits the rehabilitation process to progress as fast as is tolerated by the patient. Attention is paid to restoring range of motion, which can be restricted by adhesion formation to the collateral ligaments. During early rehabilitation, extension of the knee past −30 degrees of full extension is avoided, since the collateral ligaments are stretched in extension.

When mobility is restored as rapidly as permitted by pain in a well-motivated patient, there are fewer complications, and the outcome is often good. When anxiety, low pain threshold, or prolonged immobilization are present, adhesions will form and restoration of knee mobility will be partial and prolonged. In cases in which there is already some loss of mobility, numerous physical modalities are useful in the process of remobilization. Heat, both superficial and deep (ultrasound), causes increased extensibility of soft tissue, increased blood flow, relaxation of spasm, and reduction of pain. Applied prior to other modalities, such as friction massage, manual therapy, passive stretch, electrical stimulation, or exercise in many forms, heat can be of great benefit. TENS aids in pain relief before, during, and after treatment sessions and can be applied by the patient at home as required. Iontophoresis or phonophoresis with various medications, such as hydrocortisone, xylocaine, iodine, calcium, and chloride can be used to provide further relief of symptoms and help stretch out or break up the adhesions[22] (Fig. 9-9). Electrical stimulation helps restore strength and muscle control lost as a result of knee injury. The vastus medialis loses strength first and usually gains it back last. This muscle is necessary for maintaining patellar tracking, and further injury can result if correct tracking is lost.

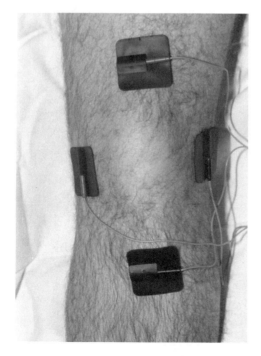

Figure 9-9. One TENS placement at the knee. Effective placement of TENS around the knee. The superior and inferior electrodes make one circuit, while the medial and lateral electrodes make the other, effectively crossing the circuits.

TENDINITIS

Martens, et al.[28] describes patellar or quadriceps tendinitis as a chronic overload lesion in the tendon near its insertion, which occurs in athletes involved in repetitive activities such as running, jumping, bicycling, or kicking. This problem is sometimes called "jumper's knee" and presents with pain and tenderness in the affected area, sometimes accompanied by local edema. Pain is usually felt at the superior aspect of the patella, at the attachment of the quadriceps mechanism, or most commonly, at the inferior pole of the patella[1] (Fig. 9-10).

Treatment is directed at reduction of inflammation and pain and at encouraging healing of the tendon, as discussed in Chapter 1. Rest is important, but immobilization is not necessary in most cases. Usually, avoiding activities that exacerbate pain is sufficient, for a period of up to 14 days, during which the stages of tendon healing described by Stanish, et al.[37] take place. Once symptoms are relieved, strengthening begins in carefully controlled increments, gradually increasing tensile forces. The tendon will then be able to withstand functional stresses placed on it by normal activities. Pain should be the limiting factor in treatment by exercise once healing is under way.

The popliteus tendon can develop tendinitis, especially during severely stressful activities such as running downhill,[1] and pain is felt along the posterolateral aspect of the knee. Treatment is the same as for other tendons (Fig. 9-11).

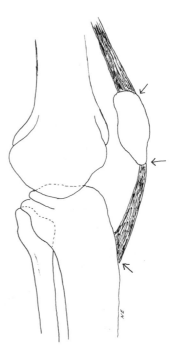

Figure 9-10. Locations of patellar tendinitis pain. The most common location of pain is at the inferior border of the patella. Other locations, marked with arrows, may also be painful.

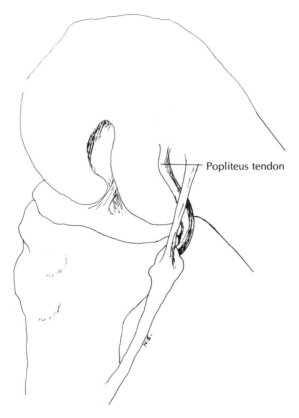

Figure 9-11. Location of popliteus tendon.

Figure 9-11. Location of popliteus tendon. Although sometimes difficult to distinguish, the popliteus tendon is a common site of tendinitis at the knee, especially suspect when pain is felt in downhill running.

BURSITIS

There are numerous bursae about the knee, which may become inflamed as a result of trauma or repetitive activities. Figure 9-12 demonstrates their locations. Bursitis may be acute, or, as in patients who suffer recurrent trauma to the bursa, chronic.

Prepatellar bursitis is one of the most common types and may be associated with other injuries of the knee. If swelling is not extreme, ice, compression, rest, and elevation are indicated. In cases of extreme swelling, aspiration is helpful in relieving pressure and reducing recovery time.[1]

Pes anserinus bursitis is commonly associated with repetitive activities, such as running long distances.[1] Training errors, tight hamstrings, and anatomic deviations, such as genu valgum and tibial torsion, can increase the pressure on the bursa.[1] Correcting the

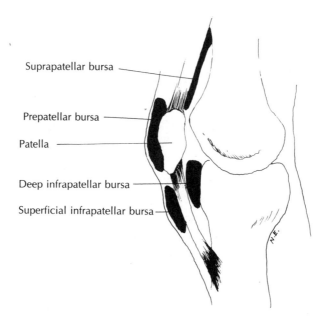

Suprapatellar bursa

Prepatellar bursa

Patella

Deep infrapatellar bursa

Superficial infrapatellar bursa

Figure 9-12. Locations of bursae. The most commonly affected bursae around the knee are the suprapatellar, prepatellar, deep infrapatellar, and superficial infrapatellar bursae.

precipitating factors is an important part of the treatment, which also includes the program described in Chapter 1.

ARTHRITIS

Because it is an important weight-bearing joint, the knee presents problems not encountered in the joints of the upper extremities. As discussed in Chapter 2, one of the most important aspects of treatment of arthritic flare-ups is rest. Resting the knee normally means that the patient does no walking. While it is useful to reduce the weight-bearing stress on the knee, preventing the patient from walking is not reasonable or desirable under most circumstances. Weight-bearing stresses can be significantly reduced, or even eliminated, with judicious use of ambulation aids, such as a walker, crutches, or cane. The entire load can be removed from a single knee by a non–weight-bearing or toe-touch weight-bearing gait with either a walker or crutches. The applicability of these devices depends, though, upon the noninvolvement of the joints that will "take up the slack." If the other knee is similarly affected, or the upper extremities are unable to bear the body's weight, these devices cannot be effectively employed. Patients suffering arthritic flare-ups should be encouraged to reduce their activity levels, however, until the acute phase has passed.

Treatment of the rheumatoid knee is aimed at prevention of deformity and maintenance of strength. Unfortunately, the position of greatest comfort is usually in flexion, which increases the likelihood that flexion contractures, quadriceps atrophy, and fibrous adhesions will result.[2] The patient should not place pillows under the knees and can benefit from resting splints. Isometric quadriceps exercises are done frequently, as are gentle active-assistive range of motion exercises within the limits of pain. Heat or ice is helpful in reducing the discomfort and thereby facilitating range of motion exercise. As soon as the acute inflammation subsides, active exercises are begun and gradually progressed to resistive exercises to restore strength. Patients are instructed carefully in a home exercise program and cautioned of the possible effects of noncompliance.

Osteoarthritic knees are treated with heat or ice for pain and spasm relief, by weight loss programs, and by strengthening muscular imbalances. One of the most common changes in the osteoarthritic knee is knee flexion contracture, so patients should be taught early how to avoid allowing contracture to occur. Strengthening of the vastus medialis is important, as is stretching of tight hamstrings. EMG biofeedback can be of great value in specifically strengthening the vastus medialis.[23] Tight gastrocnemius muscles can also be detrimental, especially in women who wear only shoes with heels higher than 1 in. Although it can be difficult to accomplish, stretching of the gastrocnemius is often helpful in preventing ankle plantarflexion or knee flexion contractures (Figs. 9-13 and 9-14).

Figure 9-13. Quadriceps Setting. Usually the first exercise because it is the least traumatic, is an isometric exercise for strengthening the quadriceps. The knee is in the extended position and the quadriceps tensed and held. Please see Chapter 8 for more exercises of muscles around the knee.

Figure 9-14. Short Arc Quadriceps Exercise. With a towel roll or other support under the knee, the foot is lifted from the table until the knee is fully extended, held, and lowered slowly. This exercise minimizes the stresses on the patella while providing strengthening for the quadriceps in the most important range. Weight is added to the ankle as indicated.

REFERENCES

1. Boland AL: Soft tissue injuries of the knee. In Nicholas JA, Hershman EB (editors): *The Lower Extremity and Spine in Sports Medicine*. St. Louis, C.V. Mosby Co., 1986, p 983.
2. Cailliet R: *Knee Pain and Disability*. Philadelphia, F.A. Davis Co., 1973.
3. Chrisman OD: The role of articular cartilage in patellofemoral pain. *Orthop Clin North Am* 1986; 17:231.
4. Cox JS: Patellofemoral problems in runners. *Clin Sports Med* 1985; 4:699.
5. Dandy DJ: *Arthroscopic Surgery of the Knee*. New York, Churchill Livingstone, 1981.
6. Dandy DJ, et al: Arthroscopy and the management of the ruptured anterior cruciate ligament. *Clin Orthop Rel Res* 1982; 167:43.
7. *Dorland's Pocket Medical Dictionary*, 21st ed. Philadelphia, W.B. Saunders Co., 1968.
8. Fahmy NRM, et al: Relationship between meniscal tears and osteoarthrosis of the knee. Presented to the British Orthopaedic Association, April 8, 1981.
9. Fisher RL: Conservative treatment of patellofemoral pain. *Orthop Clin North Am* 1986; 17:269.
10. Freeman MAR: *Arthritis of the Knee*. New York, Springer-Verlag, 1980.
11. Gelquist J, et al: Arthroscopy in acute injuries of the knee joint. *Acta Orthop Scand* 1977, 48:190.
12. Goodfellow JW: Closed meniscectomy. *J Bone Joint Surg* 1983; 65B:373.
13. Grana WA, Kriegshauser LA: Scientific basis of extensor mechanism disorders. *Clin Sports Med* 1985; 4:247.
14. Hamberg P, et al: The effect of diagnostic and operative arthroscopy and open meniscectomy on muscle strength in the thigh. *Am J Sports Med* 1983; 11:289.
15. Henry JH: The patellofemoral joint. In Nicholas JA, Hershman EB (editors): *The Lower Extremity and Spine in Sports Medicine*. St. Louis, C.V. Mosby Co., 1986, p 1013.
16. Highgenboten VL: Arthroscopic synovectomy. *Orthop Clin North Am* 1982; 13:399.
17. Ikeuchi H: Arthroscopic treatment of the discoid lateral meniscus. *Clin Orthop Rel Res* 1982; 167:19.
18. Jackson RW: Anterior cruciate ligament injuries. In Casscells SW (editor): *Arthroscopy: Diagnostic and Surgical Practice*. Philadelphia, Lea & Febiger, 1984, p 52.
19. Jackson RW: The scope of arthroscopy. *Clin Orthop Rel Res* 1986; 208:69.
20. Johnson LL: *Arthroscopic Surgery*. St. Louis, C.V. Mosby Co., 1986.
21. Jones RE, et al: Nonoperative management of isolated Grade III collateral ligament injury in high school football players. *Clin Orthop* 1986; 213:137.
22. Kahn J: *Low Volt Technique (Clinical Electrotherapy)*. Syosset, NY, Self, 1985.
23. King AC, et al: EMG biofeedback-controlled exercise in chronic arthritic knee pain. *Arch Phys Med Rehabil* 1984; 65:341.
24. Lotke PA, et al: Late results following medial meniscectomy in an older population. *J Bone Joint Surg* 1981; 63A:115.
25. Lutter LD: The knee and running. *Clin Sports Med* 1985; 4:685.
26. Lysholm J, Gillquist J: Endoscopic meniscectomy. *Int J Orthop* 1981; 5:265.
27. Macnicol MF: *The Problem Knee*. Rockville, MD, Aspen Publishers, Inc., 1986.
28. Martens M, et al: Patellar tendinitis: Pathology and results of treatment. *Acta Orthop Scand* 1982; 53:445.
29. Mayer L, Burman MS: Arthroscopy in the diagnosis of meniscal lesions of the knee joint. *Am J Surg* 1939; 43:501.
30. McBride GG, et al: Arthroscopic partial medial meniscectomy. *J Bone Joint Surg* 1984; 66A:547.
31. Montgomery JB, Steadman JR: Rehabilitation of the injured knee. *Clin Sports Med* 1985; 4:333.
32. Northmore-Ball MD, Dandy DJ: Long

term results of arthroscopic partial meniscectomy. *Clin Orthop Rel Res* 1982; 167:34.

33. Noyes ER, et al: Arthroscopy in acute traumatic hemarthrosis of the knee. *J Bone
Joint Surg* 1980; 62A:687.

34. Saunders HD: *Evaluation, Treatment and
Prevention of Musculoskeletal Disorders*.
Minneapolis, Viking Press, 1985.

35. Shields CL, et al: Rehabilitation of the knee
in athletes. In Nicholas JA, Hershman EB
(editors): *The Lower Extremity and Spine in
Sports Medicine*. St. Louis, C.V. Mosby
Co., 1986, p 1055.

36. Simpson DA: Open and closed meniscectomy. *J Bond Joint Surg* 1986; 68B:301.

37. Stanish WD, et al: Tendinitis: The analysis
and treatment for running. *Clin Sports Med*
1985; 4:593.

38. Stanitski CL: Rehabilitation following knee
injury. *Clin Sports Med* 1985; 4:495.

39. Takagi K: The arthroscope. *J Jap Orthop
Assn* 1939; 14:359.

40. Tegner Y, et al: Two-year follow-up of conservative treatment of knee ligament injuries. *Acta Orthop Scand* 1984; 55:176.

41. Watanabe M, Takeda S: The Number 21
arthroscope. *J Jap Orthop Assn* 1960;
34:1041.

10

Foot and Ankle Dysfunction

Gerard J. Jablonowski

Following an injury or any painful condition which has compromised normal foot function, proper clinical management is essential for the return of efficient painless locomotion. Providing the stable base for proximal support as well as the flexibility for adaptation to and shock absorption from the ground surface during gait, the human foot must perform distinct synchronous movements, much like a finely tuned, highly efficient engine. If a single component of the structure is altered by restriction, weakness, overstress or malalignment, the entire mechanism will be affected. Simple walking may become a painful, labored, fatiguing event and proximal joint function in the lower extremity will be compromised by abnormal stresses and eventual pathologies.

There are many different problems that can affect the foot and ankle complex. Some traumatic and others non-traumatic, resulting from structural abnormality or faulty biomechanics. A need to accurately identify the structure at fault and assess the etiological factors contributing to the symptoms is the key element in establishing the proper course of management. As noted by Root,[51] the symptoms of many foot and ankle pathologies have been accepted as an entity in themselves and treatment has been based upon an attempt to erradicate the symptoms without consideration of their cause. Such a shortsighted approach will not lead to the ultimate resolution of the patient's problem. An understanding of common lesions, implications of faulty structure and normal and abnormal biomechanics of the foot, ankle and lower leg during gait will provide the clinician with the insight into the etiological factors causing or contributing to the local lesions identified. Such insight enhances the practitioners ability to achieve satisfactory outcomes in the management of patients presenting with painful foot and ankle pathologies. (See Table 10-1.)

TREATMENT MODEL

A suggested treatment model for foot and ankle dysfunction begins with the identification of the lesion through history taking, proper diagnostic studies and clinical examination. A treatment model can be followed to maximize the effectiveness of clinical intervention depending on the origin of the dysfunction, traumatic or non-traumatic. The major differ-

**TABLE 10-1. PATHOLOGIES COMMONLY
ASSOCIATED WITH ABNORMAL SUBTALAR
PRONATION**

1. Plantarfasciitis
2. Painful bunion
3. Metatarsalgia
4. Painful corns and calluses
5. Painful neuromas
6. "Shin splints"
7. Stress fractures
8. Medial knee pain
9. Patello-femoral syndromes
10. Ilio-tibial band syndrome
11. Diffuse hip pain
12. Sacro-iliac dysfunction
13. Low back pain

ence is the early incorporation of a biomechanical evaluation and structural assessment of
the foot and lower leg during acute treatment of nontraumatic disorders (See Figures
10-1, 10-2).

Traumatic disorders of the ankle require a thorough evaluation to determine the
structure(s) involved. A history of the injury, selective tissue tension, palpation, neurological exam, stability tests, and x-rays typically comprise the musculoskeletal examination. Once the primary tissue defect has been determined and isolated, a period of immobilization (depending on degree of instability) and rest from weight bearing is
indicated to allow the early phase of tissue healing to progress. Modalities given for pain
relief can also aid in the acceleration of healing by increasing blood flow, and ridding the
area inflammatory debris. In this acute phase of healing, the patient's proximal limb
strength and mobility of non-involved tissues are maintained through an active exercise
program.

As tissue healing progresses into the sub-acute and chronic stages, a program of
manual therapy to remobilize the affected joints is implemented. The goal is to restore
normal arthrokinematics and full osteokinematic range of motion. An exercise program
progresses to restore premorbid functional strength and flexibility, according to the patient's tolerance. In addition, coordination and proprioception must be continually evaluated and addressed if deficiencies are noted.

As gait becomes normalized and the affected joints remobilize, a thorough biomechanical evaluation of the structure of the foot, ankle, and lower leg should be conducted
so as to rule out the possibility of faulty biomechanics as a contributor to the traumatic
event. (This is of particular importance in the treatment of repeated similar injuries to
the ankle.)

During the functional activity testing of running, jumping, cutting, squatting, etc. a
correction of any faulty biomechanics, structural (through orthotics or accommodators) or

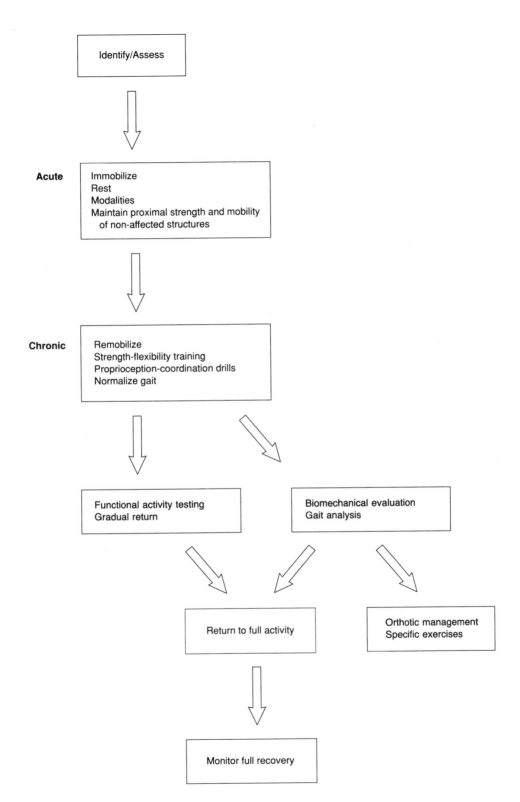

Figure 10-1. Treatment model for traumatic origin.

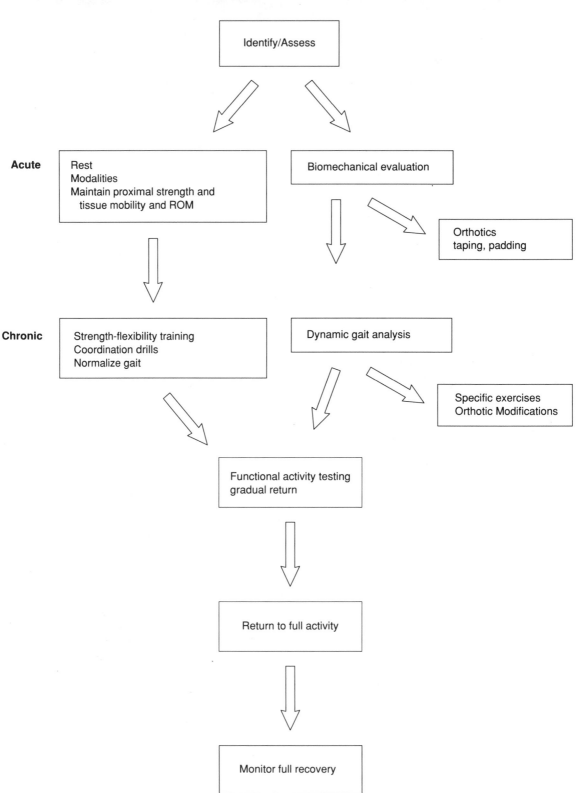

Acute

Identify/Assess

Rest
Modalities
Maintain proximal strength and
    tissue mobility and ROM

Biomechanical evaluation

Orthotics
taping, padding

Chronic

Strength-flexibility training
Coordination drills
Normalize gait

Dynamic gait analysis

Specific exercises
Orthotic Modifications

Functional activity testing
gradual return

Return to full activity

Monitor full recovery

206

Figure 10-2. Treatment model for non-traumatic origin.

functional (flexibility and strengthening exercises), is made before the patient can return to full activity.

During the return to full activity, periodic monitoring is carried out to be sure of normalization of gait without proximal musculoskeletal compromise. Those disorders of the foot and ankle not of a traumatic origin, but rather of an insidious onset, require a slightly different clinical approach. As the primary lesion is identified through thorough clinical examination, a primary etiology is discovered. Just as in traumatic origin lesion, a period of rest of the irritated tissue is provided through taping, strapping or limitation of weight bearing activity. Modalities for pain relief and acceleration of healing can be offered, as well as a program of maintaining mobility of non-affected tissue, joint range of motion, and proximal limb strength.

Simultaneous with this acute care treatment, a thorough structural biomechanical evaluation is completed to determine if primary etiology rests in faulty biomechanics. If so, the structural defect should be addressed in the acute phase. Tissue stress reduction could be aided by the addition of orthotics or alterations to the patient's footwear.

As tissue healing progresses to the sub-acute and chronic stages, treatment goals are restoration of premorbid functional strength, flexibility, and restoration of normal arthrokinematic and full osteokinematic range of motion. Also, a dynamic gait analysis is performed to evaluate for functional deformity which can be addressed through advanced limb, foot, and ankle flexibility and strengthening exercises, as well as orthotic modification.

With continued progression of tissue healing and symptom diminuation, functional activity testing can be performed with an eye on normalization. As full activity levels are attained, periodic monitoring is indicated to ensure the recovery process and prevent re-occurrence.

BASIC EXAMINATION SCHEME

Clinical examination of foot and ankle pathology includes observation in weight bearing and non-weight bearing, as well as with the patient walking (if tolerated or indicated). A Cyriax evaluation scheme of selective tissue tension is best accomplished in non-weight bearing, thus allowing the examiner to distinguish between a contractile tissue lesion (muscle, tendon) versus an inert structure lesion (ligament, capsule). Active and passive motions of the talocural, subtalar, midtarsal, metatarsophalangeal and interphalangeal joints along with resisted isometrics of the anterior and posterior tibialis, gastroc-soleus, peroneals, extensor hallicus longus, flexor hallicus longus, extensor digitorum brevis, and flexor digitorum longus constitute the selective tissue tension exam. Gross lesions of the static stabilizers and dynamic mobilizers are thus isolated. More subtle lesions may require the greater forces of weight bearing to sufficiently stress the tissue and produce clinical symptoms.[11] Joint mobility (restriction versus hypermobility) is noted through assessing the "joint play" available at key joints. Palpation of suspected areas can reveal local tenderness. Visual inspection of the feet might note structural abnormality, edema, corns and/or calluses. An inspection of the patient's well worn shoes will often help in determining their weight bearing pattern, and provide several clues as to the cause of the primary dysfunction.

Further examination should focus on the neurological status of the foot and lower extremity. Areas of decreased sensitivity and/or hypersensitivity should be specifically noted. If Tarsal Tunnel Syndrome is suspected, a Tinel sign can be sought by tapping the posterior tibial nerve behind the medial malleolus. (An electromyogram [EMG] and nerve conduction velocity [NCV] test confirms this entrapment syndrome.)

The circulatory status of the foot can be assessed by palpating the dorsalis pedis and posterior tibial pulses, noting their presence and intensity, and by evaluating the efficiency of the venous return from the foot. Pooling, discoloration or visible varicosities are examined.

Having evaluated the primary lesion by identifying the tissue at fault, and establishing a baseline of the patient's foot status, thoroughness dictates the performance of a biomechanical evaluation of the foot and lower extremity. Such an evaluation will augment the assessment and enhance the treatment program, if biomechanical abnormalities are discovered and determined to be the cause of, or a complicating factor to, the primary lesion. The biomechanical evaluation of the lower limb will be discussed later in this chapter.

COMMON FOOT AND ANKLE DISORDERS

Traumatic injuries are defined as those sustained in a single event which damaged a specific structure(s). Non-traumatic disorders are those which develop insidiously, usually as a result of overuse and eventual tissue breakdown or inflammation.

In discussing these pathologies, focus will be on noting key evaluation techniques and findings, clinical signs and symptoms, and conservative and surgical treatment options. Not to be construed as a "cook book" approach to ankle and foot pathology, these are accepted treatment methods, determined effective by experience and documented in the scientific literature. There remains allowance for the practitioner to develop his/her own conclusions and insight regarding treatment planning on a individual basis. Treatment regimes should always be modified and tailored to fit the needs of the individual patient.

Traumatic Disorders

Ankle Sprain. Any of the ligaments which support the ankle may be partially or completely torn by an extraordinary force which exerts excessive tension on its fibers. The inversion ankle sprain, with tearing of the anterior talofibular ligament, is the most commonly sustained. Forced ankle inversion combined with plantarflexion, which places the anterior talofibular ligament in its most vulnerable position, is described as the typical mechanism of injury. The next most frequently sprained ligaments at the ankle are the calcaneocuboid (combined supination and adduction of the forefoot) and calcaneofibular ligaments (inversion of the hindfoot in neutral plantarflexed-dorsiflexed position). Portions of the deltoid ligament might also be sprained (through combined plantarflexion, eversion, and abduction of the hindfoot), but more often a forced eversion stress will result in an avulsion of the medial malleolus rather than tearing of the ligament itself.[36]

Ankle sprains are classified according to severity and subsequent joint instability. A classification system developed by O'Donoghue[48] notes specific clinical features for each degree. Treatment of ankle sprains is dependent upon the degree. A mild or first degree sprain is one in which there is a partial tear of fibers of one of the ligaments without any functional weakening of the ligament as a whole. Examination reveals mild to moderate pain upon application of the injury mechanism stress and no joint instability.

Treatment of first degree sprains is relatively uncomplicated and primarily symptomatic. Often the patient will not even seek professional help. Any mild swelling is best managed with ice, compression and elevation, and the patient may be instructed to rest (partial weight-bearing) until symptoms subside. Since there is no disruption in the integrity of the joint, external support is usually not indicated. In the subacute stage, joint movements should be reassessed for stability, since the clinician can more easily determine the integrity of the ligaments in this stage, while swelling is reduced and the protective muscle spasm relieved. Strengthening and mobility exercises should address any deficiencies found in the chronic phase.

Moderate or second degree sprains demonstrate an actual tear of a portion of the ligament, yet the structural integrity of the ligament as a whole remains. No gross abnormal motion is noted, but some ligament laxity may be apparent due to overstretch and thinning of intact fibers. There is characteristic severe pain, moderate diffuse swelling, ecchymosis, and tenderness over the lateral aspect of the foot and ankle, particularly at the site of the damaged ligament.

In addition to the management of the pain and edema through rest, ice, compression and elevation, goals of treatment should be to protect the ankle against further injury while healing progresses and to maintain an ideal environment in which tissue repair mechanisms can effectively engineer reconstruction of the damaged portion of the ligament. Stirrup splints or strapping is utilized to limit subtalar motion and subsequently limit stress on the lateral ligaments. A functional degree of plantarflexion and dorsiflexion is maintained, which deters atrophy of the extrinsic controls of foot motion while promoting adequate circulatory mechanisms.

As the acute phase resolves, weight bearing can progress as tolerated provided the ankle has protective support. A Gibney basket-weave adhesive-tape strapping gives very effective support and when properly applied, effectively limits inversion and eversion of the ankle. The tape is applied beginning from the medial side of the leg, (for inversion sprain) passed down and under the heel with the foot held in eversion, and brought up the lateral side of the leg. A cut-out foam or felt horseshoe pad can be applied around the lateral malleolus to help control swelling. Alternate strips of stirrups and anchors are applied to complete the weave. The taping procedure is usually ended with figure eights and heel locks to assure the restriction of subtalar motion. (The interested reader would be advised to consult leading sports medicine texts for complete instructions and variations in taping methods for ankle sprains.) Early joint motion within a painfree range should be performed and encouraged as the symptoms subside. Gradual return to activity must be closely monitored and a graded, progressive level of intensity should be established which tests the strength, flexibility, and integrity of the healed ankle structures.

A severe or third degree sprain demonstrates a total loss of the functional stability of the ankle due to a complete rupture of one or more of the ligaments. Swelling is extreme and pain is severe. In some cases, however, abnormal motion can be clinically reproduced without undue pain due to total disruption of the ligaments along with their innervation of pain fibers. Rupture of the anterior talofibular ligament and the calcaneofibular ligament will result in gross lateral ankle instability. Inversion stress x-rays of the ankle reveal an abnormal degree of talar tilting within the mortise joint as compared to the uninvolved side. A positive anterior drawer sign (forward slide of the talus on the tibia) indicates a total tear of the anterior talofibular ligament, since this is the only structure preventing forward subluxation of the talus.[23]

The immediate treatment aim for the third degree sprain is to restore the integrity of the ligament and permit it to heal.[48] Much debate exists as to whether or not surgical repair for ruptures of the anterior talofibular ligament is the treatment of choice. Successful treatment of such tears has been achieved both surgically as well as by immobilization without surgical repair. When at least two of the lateral ligaments, the anterior talofibular and the calcaneofibular, are torn, there is support for surgical repair.[61] If surgical repair is chosen, the best results have been reported if the procedure is done promptly after the injury and when the ligament ends can still be defined and sutured. In delayed surgical procedures, two to three weeks post injury, it is virtually impossible to define ligament ends and approximate them for adequate suturing. Late procedures are usually more successful when a reconstruction, rather than a repair, is performed.[48]

Following prolonged immobilization either post-op or nonoperative, rehabilitation of the third degree ankle sprain progresses in a similar fashion as the second degree, but with a slower progression of treatment phases, and protective measures continued longer. More efforts at muscle strengthening and joint mobilization might be needed to address the soft tissue changes which may result from the prolonged period of immobilization. Return to normal vigorous activity should be monitored closely and be more gradual. Even though new collagen is laid down within the first few weeks following injury, it takes several months for this new tissue to regain normal tensile strength, and it remains highly susceptible to re-injury during this period.[36]

In both second and third degree sprain rehabilitation programs, joint proprioception and lower limb coordination should be addressed through balance board ankle exercises. These balance drills enhance the ankle's ability to quickly react to changing position by contraction of support musculature and reestablishment of the neurological mechanism innervation needed for the reflexive detection of joint position to preserve stability and maximize function.

For the athlete or active individual, return to full activity must include gradual training in straight line running, figure eights, and progressive cutting. As the running distance approaches one mile and the cutting can be performed at 90 degree turns with no pain and without limping, the patient is ready for full competitive play or heavy work activity.

Fractures. Ankle and foot fractures are common traumatic injuries often sustained in motor vehicle accidents, athletic competition, and falls. The mechanisms of fracture are usually external rotation, abduction, adduction, vertical compression, or various combinations of these forces.[18] Fractures about the foot and ankle are diagnosed via radiographic studies and are best treated with precise repositioning of the bone fragments either through closed manipulative reduction or open reduction with internal fixation. For most fractures, the open reduction method provides better alignment and normalization of joint surfaces and is preferred; as even slight malpositioning of the healed bones can lead to chronic foot dysfunction and rapid onset of degenerative joint disease.[58]

There are many types of ankle and foot fractures documented in the scientific literature. Ankle mortise fractures are classified along anatomic lines as monomalleolar bimalleolar, or trimalleolar.[58] The trimalleolar, or cotton fracture, includes a fracture of the posterior lip of the articular surface of the tibia in addition to fractures of the medial and lateral malleoli. Other common fractures of the foot include calcaneal fractures, transchondral fractures of the talus, metatarsal fractures and fractures of the phalanges. It is not within the scope of this text to describe the operative reduction procedures for these fractures, but rather to inform the practitioner as to their occurrence and underscoring the importance of radiographic studies for each traumatic ankle and foot injury encountered.

The immediate treatment of the fractured or suspected fractured ankle or foot involves careful handling of the part to attempt gentle correction of any gross deformity. This will help prevent compounding the fracture and causing impairment of the circulation. Splinting the ankle, applying ice packs and firm compression dressing, and elevation will minimize the subsequent swelling accompanying the fracture and reduce the chances for impairment of circulation.[49]

After operative or closed reduction procedures have taken place, the patient will be immobilized for a period of time depending on the procedure and the original type of fracture. Some procedures call for early bivalving of the cast to allow for active range of motion exercises to be performed daily during the period of bone healing. Non-weight bearing and/or protective weight bearing with crutches or a walking cast is advocated until satisfactory union has occured.

Specific therapy management protocols are often developed by practitioners for each procedure. Close communication between therapist and surgeon assures proper, safe, and effective rehabilitation. General management of the post fracture ankle involves:

1. Graded mobilization of stiff joints (initiation of procedures dependent upon amount of healing and the presence of any intra-articular fixation devices).
2. Stretching of soft tissue to ensure full range of motion.
3. Strengthening of extrinsic and intrinsic musculature which has atrophied or weakened as a result of the original injury and subsequent period of immobilization. Treatment will usually progress as with second and third degree sprains, as balance and coordination drills test the proprioceptive ability of the ankle, and functional activities are gradually incorporated into the rehabilitation program. Fabrication of an orthotic or accommodative device may be necessary, which is worn if the fracture reduction has caused a change in the normal biomechanics of the foot, and these changes are expressed in the late chronic phase.

Peroneal Tendon Dislocations. The peroneal tendons pass behind and beneath the lateral malleolus. If the ankle encounters a force which suddenly places it in a dorsiflexed and everted position, and there is a simultaneous forceful contraction of the peroneals, the peroneal retinaculum can be torn, allowing the tendons to dislocate anteriorly over the lateral malleolus.[73] Recurrent dislocations of this nature causes repeated ankle sprains. They tend to destabilize the lateral aspect of the ankle, minimizing counterbalance forces against ankle inversion. Also, chronic peroneal tendonitis may develop with repeated episodes.

Examination usually reveals a stable ankle and subtalar joint with no signs of abnormal movement. The patient might be able to demonstrate subluxation of the tendons with active dorsiflexion and eversion, and reduction with plantarflexion. There may be swelling evident posterior to the lateral malleolus, in the acute phase. Radiographic studies are usually normal unless there is an associated fracture of the lateral malleolus.[33]

Conservative treatment involves managing the acute symptoms of swelling, pain and inflammation with rest, cold modalities, compression, and elevation. A small felt compression pad may be taped over the tendons in the region of the posterior lateral malleolus to help keep them in place. This, however, should be considered only as a temporary measure, with the understanding that surgical intervention will be needed for long term resolution of the problem if it interferes with normal activity. The peroneal retinaculum must be repaired along with a deepening of the peroneal groove behind the lateral malleolus.[73] Peroneal strengthening exercises are cautiously progressed after satisfactory healing of the retinaculum.

Achilles Rupture. Rupture of the achilles tendon is a low incidence injury, but extremely disabling, occurring most often in men between the ages of 30 and 50.[7] Most often, this sudden injury occurs to men with normally sedentary lifestyles who engage in "weekend athletics," mainly basketball and racquet sports. There is uncertainty as to the exact causes for the rupture. One theory suggests chronic degeneration of the tendon due to compromised blood supply with increasing age.[15] Another suggests the failure of the normal inhibitory mechanism of the musculotendinous unit.[26] Achilles tendon rupture following repeated steroid injections, most often in treating a chronic Achilles tendonitis, has also been described.[37] Common mechanisms for Achilles tendon rupture include a sudden uncontrolled overstretch on a planted foot, or a forceful contraction of a relaxed or deconditioned gastrocnemius.[9]

Examination of the calf reveals a palpable gap in the Achilles tendon and the belly of the gastrocnemius and is retracted into the upper portion of the calf. Ecchymosis is noted around the heel and there is inability to perform toe raises. A positive Thompson test is demonstrated. With the patient prone to having both feet over the edge of the examining table, the calf muscles on the injured side are squeezed by the examiner. If the tendon is intact, the foot will plantarflex. If the tendon is ruptured, the foot will not plantarflex.[30]

Currently, both operative and non-operative treatment is employed in the management of Achilles rupture. Both techniques involve casting for six to nine weeks with the foot in a plantarflexed posture to promote healing of the approximated tendon fragment ends.[9] Upon removal of the cast, an ankle-foot orthosis (AFO) with a dorsiflexion lock and/or a one inch heel lift is worn by the patient to prevent undue stress on the tendon. Gradually, the thickness of the lift is reduced and the splint wearing discontinued when the patient demonstrates a normal painfree gait with a minimum of 10 degrees of ankle dorsiflexion. Early therapeutic intervention includes passive range of motion in the direction of plantarflexion and active exercises into dorsiflexion within a protected range. Progressive strengthening of the gastrocs is safely undertaken through isokinetic exercises

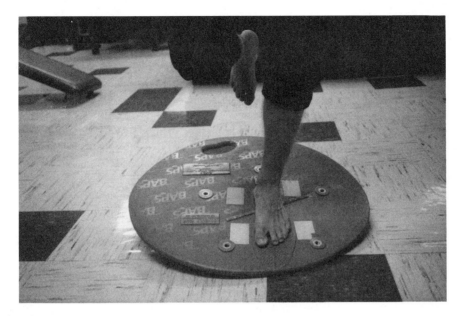

Figure 10-3. Exercising on BAPS board for return to full normal biomechanical motion of the ankle.

and toe raises. Other ankle support musculature weakened by the period of immobilization must also be strengthened. Selected joint mobilization techniques are often employed to loosen stiff joint capsules. Ankle range of motion exercises can be performed on the multiplaner BAPS board (Biomechanical Ankle Platform System Camp /Medipedic Jackson, Michigan) for the return of normal biomechanical ankle function (See Figure 10-3). Preceding these activities with a warm whirlpool frequently relaxes any spasm which could inhibit mobility. Finishing the exercise session with a 10 minute ice application prevents any post-exercise induced swelling.

Stone bruises. An often encountered injury sustained by athletes and joggers, the "stone bruise" is a painful contusion to the plantar surface of the heel. The pathology is very common in active elderly individuals, where the fibro-fatty pad of the heel breaks down and thins with age. A single event of the heel striking a hard object (a stone, for example) or repeated events of heel strike on hard surfaces, can produce the painful heel condition.

Examination usually reveals a swollen and tender area of the calcaneal fat pad. Therapeutic intervention is to recommend rest and local heat or ice applications to control pain and swelling. Heel pads or "donut" pads usually provide no significant relief. A molded heel cup which compresses available fibro-fatty tissue from the sides of the heel to the plantar surface and prevents its displacement, often provides satisfactory relief of symptoms and will allow renewed activity without discomfort.[10] Patients with chronic plantar heel pain must wear the molded heel cup during all weight bearing activity, as it provides a mechanical solution to their problem.

Common Overuse Pathologies

Certain pathologies develop as a result of chronic repeated microstresses to otherwise normal, healthy tissue. Most often symptoms occur either in the extreme abuse associated with strenuous athletic competition, overuse associated with training and recreational exercise, or with cumulative effects of many years' microdamage to tissue from normal occupational and daily activities as middle age is reached. Often, these extrinsic factors compound intrinsic factors such as insufficient tissue elasticity or poor biomechanical function. Pain and inflammation eventually result in the foot, ankle and/or lower leg. Ultimately, limited mobility ensues and degenerative changes may occur.

Chronic cases of tendinitis, bursitis and periostitis are not uncommon in the foot, due to long term repeated microtraumas. These pathologies are generally classified as "overuse syndromes." In addition to identification of the other associated symptoms, it is paramount that the clinician recognize overuse microstress as an etiological factor and discover the ultimate reasons for the abnormal tissue stresses, both intrinsic and extrinsic. Measures must be taken which address the patient's activity patterns and habits as well as any biomechanical disturbance in the foot and lower limb. Then the clinician will have provided the complete measures to not only control present sypmtoms, but effectively prevent their reoccurrence.

Plantar Calcaneal Spurs and Fasciitis. Plantar calcaneal spurs (heel spurs) and plantar fasciitis are common pathologies often considered together. They both involve basically the same symptoms, develop via similar mechanisms, and are treated in the same manner.

Plantar fasciitis is an inflammation of the plantar fascia of the foot at or near its insertion at the plantar surface of the calcaneus. Heel spur is a bony prominence which develops on the anterior plantar aspect of the calcaneus, just anterior to the tuberosity where the plantar fascia inserts.

Frequently, heel spurs are not symptomatic. This has been shown to be the case with the discovery of calcaneal spurs on x-rays of non-symptomatic heels.[62] Leading theories suggest they develop as a response to normal stress traction of the plantar fascia on the periosteum, chronic inflammation of the plantar fascia at its origin, or secondary to traumatic periostitis.[67]

Plantar fascia helps stabilize the medial longitudinal arch and stabilizes the foot at the terminal phases of propulsion. It is stretched excessively when the medial longitudinal arch is depressed and the hallux and other toes passively dorsiflex for the push-off phase of gait.[10]

Plantar calcaneal spurs and fasciitis may present as pain on the medial undersurface of the heel occurring on standing, walking, and running. Some local swelling and tenderness are often evident upon palpation. Passive dorsiflexion of the toes may also increase the local discomfort. Radiographic studies may or may not reveal a spur. Typically, the patient reports the pain being at its' worst when first getting up in the morning during weight bearing. The pain then subsides to a more tolerable level after several minutes of walking.

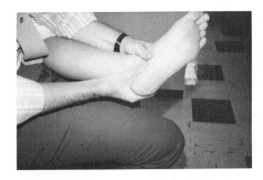

Figure 10-4. Optimal therapist hand positioning for effective mobilization of the subtalar joint.

Plantar fasciitis could occur with certain foot deformities such as restricted subtalar joint motion (rigid foot) and with excessive pronation related to pes planus (flat foot).

The treatment goals of plantar calcaneal spur and fasciitis are to control the inflammation at the insertion of the fascia into the calcaneus, and to relieve undue stress in the planar fascia itself. A period of non-weight bearing is recommended until symptoms subside. If a primary foot deformity exists which leads to excessive subtalar pronation, a biomechanical orthotic device with an added arch support should be fabricated. A heel pad with a relief area for the tender heel region or a heel cup can be incorporated into the device. Also, strapping the arch for added support has shown to be helpful for symptom control in the acute phase, pending completion of the orthotic.

With a rigid foot-type, exhibiting limited subtalar range of motion, mobilization of the subtalar joint along with controlled stretching of the plantar fascia will help decrease symptoms by lessening tissue tension[10] (See Figure 10-4). Ultrasound and phonophoresis with 10% hydrocortisone has been used with limited success to control acute pain and inflammation. Medical management usually involves anti-inflammatory drugs and/or corticosteroid injections at two to three week intervals until symptoms subside. Rarely is surgery necessary.

Metatarsalgia. Metatarsalgia is a syndrome describing pain over the metatarsal heads and/or in the metatarsophalangeal joints. Scranton[53] classifies metatarsalgia according to whether a weight bearing imbalance exists between metatarsals (Primary) or whether the forefoot pain is due to other factors such as rheumatoid arthritis or stress fractures (Secondary). Both primary and secondary metatarsalgia have characteristic associated reactive keratosis. A third classification of forefoot pain is that without reactive keratosis, such as neuromas, gout, and plantar fasciitis.

Many theories have been developed to attempt to explain the etiology of primary metatarsalgia. One theory suggests that a short first metatarsal will cause the second to have to bear a disproportionate amount of weight as the body propels forward over the foot, and thereby result in ultimate joint dysfunction and the development of a painful callus. Another theory suggests failure of the transverse metatarsal arch to be maintained by the transverse metatarsal ligaments and the transverse head of the adductor hallucis muscle. This can be compounded by weakness of the intrinsic toe flexors during propulsion, causing only minimal elevation of the central metatarsal heads, and thereby increasing weight bearing forces on them.[67]

Extrinsic factors such as excessive weight gain and wearing high heeled shoes have been suggested as causes for, or aggrevating circumstances of the condition. Pes planus foot type, with assicated plantar subluxation of the metatarsal heads and pes cavus, with associated prolonged forefoot weight bearing periods, have also been implicated. Any tendency toward excessive pronation will lead to hypermobile functioning of the first and fifth ray and thereby cause a relative depression of the transverse metatarsal arch.

Metatarsalgia is common in middle-aged persons with a pronation tendency.[9] Upon examination, plantar calluses are usually evident over the second and third metatarsal heads. Direct pressure over the involved metatarsal heads will elicit pain. Patients normally only report pain upon weight bearing. Night pain is not common. Typical findings on gait analysis include diminished push-off with a decreased propulsion phase, prolonged double limb support, prolonged midstance, and decreased stride.[14] Some patients severely alter their gait to weight bear excessively on the lateral border to prevent any weight shift across the metatarsal heads.

Conservative treatment includes removal of calluses, and shoe modifications. A metatarsal pad just proximal to the metatarsal heads two, three, and four or an external metatarsal bar, shock alleviating material under the metatarsal heads, and decreasing heel height are helpful. Warm foot soaks or local cold applications, depending on the acuity of the symptoms, frequently help control the symptoms. Chronic management must include exercises to strengthen intrinsic toe flexors and stretch the heel cord. Weight reduction is a key factor also. Orthotic fabrication is essential to correct excessive pronation tendency, if such a biomechanical imbalance exists.

In the event of failure of conservative approach with severe symptoms persisting and limiting work or recreational activity, surgical intervention may be necessary. An operative procedure to shorten the involved metatarsal will relieve pain from the plantar callosity. Contraindications to any surgical procedures include pes cavus foot type, neurological problems, impaired circulation, and diabetes.[62]

Stress Fractures. Stress fractures occur when bone fails to adapt to new or unusual loading patterns. Sensitive, painful periosteum is the first warning of a possible impending failure of bone to accept stresses placed upon it. Under normal conditions, microdamage stimulates new bone where needed. However, when the microdamage outpaces the development of new bone, a stress fracture results.[17] Avid exercise, running, and intense training are common provocative activities.

Two factors are often implicated in the development of stress fractures—muscle fatigue and biomechanical imbalance. The function of well rested, prepared muscle is to absorb shock from the bones. As muscles fatigue, more shock waves are transmitted into the bones, causing repeated microdamage during weight bearing. Biomechanical imbalances which either cause a rigid, poorly adapting foot or promote hypermobility, which causes muscle overuse and eventual fatigue, will also lead to stress fractures.[42] Hard running surfaces and improper shoes can also be factors contributing to the cause of stress fractures.

Stress fractures are most commonly seen in the metatarsals (March fracture) but are also evidenced in other tarsal bones, as well as the tibia and fibula. On rare occasions they are detected in the femur. A typical finding is localized sharp pain reported by the patient without any history of acute injury. Usually, however, a highly repetitive or stressful activity will be reported in the not-too-distant past or continues to the present, during which the pain is felt.

Examination findings may include palpable tenderness at the involved site. X-rays are commonly negative initially, but at a two or three week follow-up, radiographic studies show an increased area of density where callus formation is occuring at the site of the stress fracture.[48]

Initial therapeutic management is to reduce the symptom-causing activity to comfortable limits, or stop it altogether. With metatarsal involvement, O'Donoghue[48] recommends a walking cast to protect the lesion while healing progresses. Return to full activity should only be attempted when the callus mass is completely consolidated.

Chronic management of a stress fracture must include a functional orthotic device for control the biomechanical imbalances. Added shock absorption material to the shoe of the rigid foot type is necessary. Intrinsic and extrinsic muscle strength, endurance and flexibility re-training is carried out during rehabilitation for safe and effective return to original activities. If not properly treated, stress fracture symptoms could persist 6–12 months and return to desired recreational or work activities might be impossible.

Shin Splints. It is commonly agreed that "shin splints" should not be a diagnostic entity, but rather a catch-all term to describe any inflammatory condition of the lower leg often associated with running. This overuse syndrome typically presents as a dull ache which increases gradually after running, and progresses to pain during and after the run. It is very common in poorly conditioned athletes as they begin a training program.

There are three distinct pathologies which can constitute "shin splints." The first involves irritation of the posterior tibialis muscle along its origin at the posterior aspect of the tibia. Overuse irritation of the posterior tibialis occurs during running when the subtalar joint hyperpronates either due to a structural deformity or functional compensation for a rearfoot or forefoot malalignment. The posterior tibialis functions to control pronation. This added eccentric loading of the muscle, compounded by an added functional demand to assist in resupination of the foot from a disadvantageous position eventually causes microtearing at the tendo-periosteal attachment and resultant periostitis along the posterior aspect of the tibia.

A second distinct pathology may be irritation of the interosseous membrane between the tibia and fibula. That portion of the membrane serving as an attachment site for the posterior tibialis tendon becomes inflamed and irritated due to abnormal traction stresses from the tendon as it responds to hyperpronation of the foot.

A third distinct pathological entity is periostitis, which results from overuse stresses of the anterior tibialis at its origin attachment along the anterior lateral aspect of the tibia. Acting as a decelerator for plantarflexion at heel strike, the anterior tibialis is eccentricly overworked when the heel cord is tight or shortened, producing more rapid forefoot loading and heel rise. This situation can be further complicated by the foot with a pronation tendency. The anterior tibialis is involved in pronation control and accelerates supination from a disadvantagous position.

Other pathologies are evidenced in the lower leg and are often classified and diagnosed specifically, but are sometimes included in the catch-all of shin splints. They are stress fractures and anterior compartment syndrome. Stress fractures, as mentioned earlier, can be a progression of the periostitis resulting from poor shock absorption by these readily fatigued muscles.

Inflammation from anterior tibialis microdamage might, in rare instances, go on to an anterior compartment syndrome.[48] This is a potentially serious condition involving compression of the neurovascular structures of the anterior portion of the lower leg. This syndrome is identified by measuring the fluid pressure in the anterior compartment while the patient is symptomatic, usually after a long run on a treadmill. Pressures can reach near or above diastolic blood pressure.[42] Swelling is evident along the anterior shin, and the patient typically reports diffuse, non-specific tenderness and some numbness of the foot. To prevent catastrophic eventual neurological deficit and/or tissue necrosis, a surgical procedure of fasciotomy and compartmental release should be done after diagnosis is ascertained.

Treatment for those pathologies classified as shin splints is diverse. For the acute pain symptoms, studies have shown that ultrasound, phonophoresis, iontophoresis, and ice massage are all equally effective.[59] Contrast baths and compression taping of the lower leg during activity are also helpful in decreasing symptoms in some cases. A decrease in activity level is recommended while tissue heals.

Ultimately, resolution will only come about if the reason for the tissue overstress is corrected. Controlling compensatory hyperpronation through biomechanical orthotic devices along with exercises to stretch and strengthen the anterior and posterior tibialis as well as stretch the heel cords, are essential elements of the treatment program. Gradual return to full activity should be closely monitored for pain.

Tarsal Tunnel Syndrome. Tarsal tunnel syndrome is an entrapment neuropathy of the posterior tibial nerve as it passes beneath the flexor retinaculum under the medial malleolus, a region anatomically referred to as the "tarsal tunnel." Much like its counterpart the carpal tunnel, the tarsal tunnel is a space limited area with several structures passing through it. The roof of the tunnel is the laciniate ligament. The floor is the bony depression on the posterior aspect of the medial malleolus.[2] In addition to the posterior tibial nerve, the occupants of the tarsal tunnel are the tendons of the tibialis posterior, flexor digitorum longus and flexor hallucis longus muscles, and the posterior tibial artery. Any irritation of these structures passing through the tunnel, or any abnormality of the structures comprising the tunnel, can lead to an inflammatory response or narrowing of the tunnel. As the neurovascular bundle is compressed, the rich blood supply which normally feeds the nerve is compromised and the nerve begins to loose its normal elasticity which is needed during ankle motion.[9] The pathology of the entrapment neuropathy ensues.

Causes of tarsal tunnel syndrome include structural and functional abnormalities of the foot leading to excessive pronation, pes planus deformity, tenosynovitis of the posterior tibialis tendon, work or activity posture stressing the medial compartment, direct trauma, or medial ligament sprain.

Symptoms of the neuropathy include local edema, tenderness, and paresthesias either into the plantar surface of the foot (medial/lateral plantar branches) or into the heel region (calcaneal branch). Pain will be intensified with full passive eversion and dorsiflexion. The patient could report the symptoms either with activity or at rest, with a common complaint of night "burning" or "tingling" of the toes which interferes with sleep.

A positive Tinel sign can be elicited by tapping over the posterior inferior region of the medial malleolus. Sensory exam may reveal deficits in the specific dermatomal distribution, and intrinsic muscle weakness is often evident in the toe flexors and abductor hallucis. Confirmation of the syndrome is obtained through EMG/NCV studies, showing a decreased nerve conduction velocity across the tunnel and ultimately signs of muscle denervation.

Treatment is typically aimed at controlling the symptoms and correcting any primary structural abnormality. TENS for pain control and ankle immobilization in a neutral position appear to be helpful. A functional orthotic might be indicated to balance the excessively pronating foot type.

Medical management to supplement physical measures often includes corticosteroid injection and oral anti-inflammatory medication. In the event of the failure of conservative treatment, operative section of the flexor retinaculum and release of the compressed nerve is necessary for symptom relief.[27]

Long term management includes restoring the normal flexibility and range of motion of the ankle as well as restrengthening any intrinsic foot muscles which have become weakened due to neural interruption. A similar entrapment neuropathy of the medial plantar branch of the posterior tibialis nerve exists as the nerve passes through an opening in the abductor hallucis muscle. This neuropathy may have similar etiologies and symptoms as tarsal tunnel syndrome. Treatment approaches to the problem are basically identical.[10]

Morton's Neuroma. The most lateral branch of the medial plantar nerve may become impinged and irritated between the third and fourth or second and third metatarsal heads, forming a painful "neuroma" or thickening of part of the nerve.[62] This condition is most prevalent in middle-aged females, and the wearing of tight fitting and high heeled shoes has been implicated as a possible cause for the development of the neuroma. Excessive pressures on hypermobile metatarsals as well as activities which promote repeated or prolonoged extension of the toes have been identified as possible etiologies.[10] Symptoms include local pain and paresthesias, which the patient reports is temporarily relieved by removing the shoes and massaging the ball of the foot and toes.[9] Pain may be present upon walking and persist into the night. Upon examination there is pain with compressing the metatarsal heads together and with direct pressure exerted between the metatarsal heads (See Figure 10-5). Some sensory changes may be noted in the toes in extreme chronic cases.

Conservative therapeutic management includes resting from any provacative activity and being sure that the footwear is wide enough at the toe box to lessen any intermetatarsal pressure. A metatarsal relief pad can be incorporated into the shoe to lessen weight bearing over the tender neuroma site. Ultrasound and phonophoresis with anti-inflammatory or analgesic suspension may be helpful in providing symptom relief. The entire foot should be examined so that any structural or dynamic deformity contributing to excessive forefoot weight bearing or abnormal pronation leading to metatarsal hyper-

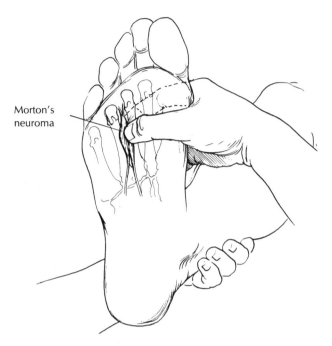

Morton's
neuroma

Figure 10-5. Palpating for Morton's Neuroma (Reproduced with permission from Hoppenfeld, *Physical Examination of the Spine and Extremities,* Appleton-Century-Crofts, Pub.).

mobility late in stance can be corrected with a functional orthotic. A cavus foot type can be accommodated with an arch support.

Medical management will usually include the direct injection of corticosteroids. Very often, conservative management will not yield satisfactory long term results. A surgical procedure to excise the swollen nerve segment is a highly effective treatment.[62] After the surgery, attention must be paid to the biomechanics of the foot and the footware habits of the patient in order to prevent a reoccurrence of the neuroma.

Achilles Tendinitis. Achilles tendinitis is an inflammation occurring in the loose connective tissue about the tendon near its insertion into the calcaneus.[9] It occurs most often as a result of repeated overstress on the tendon, but can also develop secondary to direct trauma to the area. Tightness in the heel cords is often implicated as the reason for the excessive stress during repetitive activities. Acute symptoms include local tenderness, heat, redness, and swelling over the Achilles tendon near its calcaneal insertion. Pain can be clinically elicited with forced passive ankle dorsiflexion and active resisted plantarflexion. Palpable crepitus is detected over the tendon during motion.[11]

Acute therapeutic intervention includes avoiding the provocative activities, resting the tendon from motion, elevating the foot, and the local application of ice. Phonophoresis with hydrocortisone is also an effective modality for symptom control in this stage.

As the acute symptoms begin to subside, a period of protective immobilization, either by use of a walking cast boot or other splint which prevents ankle joint motion, is essential to provide adequate tissue healing.[48] Upon removal of the cast or splint in the chronic stage, strapping the ankle to prevent dorsiflexion beyond neutral, and/or having the patient wear a one-half inch heel lift in the shoe will take pressure off the tendon. This will assist in the resumption of normal activity.

The patient's return to work or recreational activity should be closely monitored. "Bursts" of activity which put excessive demands on the gastrocs should be avoided. Long term exercises of heel cord stretching and gastroc strengthening are carried out on a regular basis. (Heel cord stretching should be performed with the foot in a supinated position, so that the dorsiflxion achieved is all at the ankle and not at the midtarsal, thereby maximizing the effort to effectively lengthen the heel cord.) Toe raising calf strengthening exercises are carefully monitored for any return of symptoms in the posterior heel region.

In addition, medical management includes the direct injection of long acting local anesthetics into the tissue surrounding the Achilles tendon. As symptoms subside, supportive measures such as a heel lift and activity restriction should still be implemented. The patient should not be allowed to develop a false sense of security. Sufficient time for tissue healing must be allowed and the rehabilitation of strengthening and stretching carried out. Also, any foot deformity which can contribute to the overstress of the Achilles tendon should be addressed. An equinus foot should be supported with an accommodative device.

Posterior Calcaneal Bursitis.

Posterior Calcaneal Bursitis. Two small bursae are located in the posterior region of the heel. The retrocalcaneal bursa lies between the anterior surface of the Achilles tendon and the posterior superior angle of the calcaneus. The calcaneal bursa lies between the insertion of the Achilles tendon and the overlying skin.[23] Either or both of these bursae can become inflamed and painful secondary to Achilles tendinitis, or primarily as a result of excessive pressure upon the area caused by friction from tight, ill-fitting shoes. Acute symptoms include tenderness to palpation at the site of the inflammation. Swelling is often just beneath the skin with calcaneal bursitis, and behind the Achilles tendon with retrocalcaneal bursitis. The overlying skin and bursa walls may feel thickened from the irritation of chronic pressure. Pain will also be intensified with passive dorsiflexion of the ankle, as the bursa is compressed by the taut Achilles tendon.

Therapy in the acute phase may include rest and elevation of the foot, and local ice applications. For calcaneal bursitis, phonophoresis with hydrocortisone preparation is beneficial. Removal of any external pressure source which could have precipitated the bursitis is essential. Often the shoe heel counter is cut away or trimmed down, or an open back shoe is worn. If primary Achilles tendinitis was the precipitating factor in the development of the bursitis, management of the tendon lesion can be carried out as mentioned earlier. As full painfree range of motion is restored in the ankle, the patient can resume normal activities.

In addition to the above physical measures to control symptoms, medical management frequently includes needle aspiration of the swollen bursa, followed by a corticosteriod injection. Surgical excision of the bursa is not useful. Nonetheless, excision of any exostosis could be effective, especially for men.[67]

COMMON FOOT DEFORMITIES

Hallux Valgus

Hallux valgus, or bunion, is a deformity in which the great toe is deflected laterally and a bony prominence develops secondarily over the medial aspect of the first metatarsal head and neck (See Figure 10-6). There exist many theoretical explanations for the etiology behind the development of hallux valgus, including heredity, mechanical overpressure of narrow shoes, arthritic changes of the first metatarsophalangeal joint, flatfeet and excessive pronation.[28] Regardless of the etiology, the deformity is maintained and heightened by the bowstring effect of the tendon of the extensor hallucis longus and the inability of the abductor hallucis muscle, due to the plantar displacement of its tendon, to resist the progressive valgus deformity aggravated by the adductor hallucis.[62] Pain from the deformity often arises from continued pressure over the medial bony prominence resulting in a bursa which frequently becomes inflamed and thickened. Also, degenerative arthritic changes develop in the metatarsophalanageal joint.

Conservative treatment is often sufficient for mild cases. Relief of pressure over the medial bony prominence by a felt ring and the wearing of a shoe with an enlarged toe box will relieve the bursitic pain. Arthritic pain usually requires intra-articular injection of corticosteroids, modalities of heat or cold as indicated, rest and salicylates.[67] In selected feet where it can be determined that excessive pronation has been a prime etiological factor in the development of the deformity, treatment must also include the use of a biomechanical orthotic device to rectify the pronation. Only through an attempt to correct the primary etiology will the chances of successful long term relief be favorable.[28]

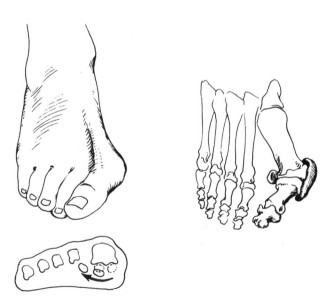

Figure 10-6. Hallux valgus deformity showing bunion development along with lateral displacement of sesamoids. (Reproduced with permission from Hoppenfeld, *Physical Examination of the Spine and Extremities,* Appleton-Century-Crofts, Pub.)

Surgical correction may be indicated for the relief of symptoms in the failure of conservative methods. Many surgical techniques have been developed to correct the deformity and relieve the patient's symptoms. Although it is not within the scope of this text to fully describe these techniques, common goals for most procedures have been stated by Stewart:[62]

1. Correction of the valgus deformity of the proximal phalanx.
2. Removal of any exostotic bone from the medial and dorsal aspects of the first metatarsal head and, when necessary, removal of the bursa that protects it.
3. Correction of the varus deformity of the first metatarsal.
4. Correction of any excessive tightness of the extensors of the great toe.
5. Correction of any other associated deformities of the forefoot, such as corns, hammertoes, and subluxations of joints.

Ultimately, to prevent reoccurrence, it is incumbent upon the surgeon to examine the entire foot and consider the fabrication of a biomechanical orthotic device to control excessive pronation, it can be considered an etiology in the development of the hallux valgus.[28] Continued propulsion in subtalar pronation will eventually cause renewed pathological hypermobility of the first metatarsal and perhaps lead to the eventual return of the deformity along with rapid progression of degenerative changes in the newly aligned joint surfaces.

Pes Planus

A Pes Planus deformity, or flatfoot, is characterized by the depression of the medial longitudinal arch due to medial and plantar displacement of the head of the talus. If the medial longitudinal arch is absent in both weight bearing and non-weight bearing, the patient is said to have a "rigid" flatfoot. If the medial longitudinal arch is present in non-weight bearing, but depresses during weight bearing (foot flat portion of stance), it is classified as a "supple" or "hypermobile" flatfoot.

Flatfeet in and of themselves do not always cause symptoms. However, the normal biomechanical process of gait is altered and may cause several functional changes in the foot and the lower extremity. Such changes include excessive pronation of the subtalar joint and the structural consequences associated with it—shortened peroneals, tightened, tender plantar fascia, deltoid ligament laxity and overstretch weakening of the tibialis posterior muscle over time.[16] Eventually, the tarsal bones assume a configuration adapted to this altered position, producing altered biomechanics of tarsal motion which leads to degenerative arthritis and secondary deformities such as hallux valgus.[67] Symptom development over time can also invade the knees, hips and low back, which suffer the proximal effects of the closed kinetic chain response to this abnormal pronation of the foot.

A plan of conservative management of hypermobile pes planus depends on the severity of the deformity and symptoms. A basic plan of treatment should include:

1. Relieving medial foot ligament and joint capsule stress by the use of a biomechanical orthotic device.
2. Strengthening of the dynamic supporters of the medial longitudinal arch, namely the foot inverters (posterior tibialis, anterior tibialis) and the toe flexors.
3. Stretching of the heel cords.

Arch supports alone will not offer the needed balance and correction of faulty biomechanics of the foot. An orthotic device, custom molded to the patient and posted to the specific degree of the deformity, with an added medial arch support will better accomplish the goal of rectifying the problems associated with the excessive pronation of the subtalar joint. (Later in this chapter a more complete discussion of the role of orthotics in the total management of foot problems will be presented.)

Plantar toe flexor strengthening exercises are best accomplished with towel gripping. A bath towel is placed on the floor and the patient "grips" the towel repeatedly with his toes, holding each grip with a maximum force for at least 7 to 10 seconds. Foot inverter strengthening can be accomplished with the use of the weighted BAPS board (Biomechanical Ankle Platform System Camp/Medipedic Jackson, Michigan) stressing resistence to the action of subtalar inversion with the foot in both plantarflexion (posterior tibialis) and dorsiflexion (anterior tibialis). Also, repeated lateral border weight bearing, raising the medial border of the foot, will assist in these efforts. Walking barefoot in sand should also be encouraged as part of the exercise program. Rigid flatfeet often require surgical intervention when conservative treatment of symptoms fails to produce lasting results. The procedure often employed is the triple arthrodesis, where the valgus angulation of the rearfoot is corrected so that the anterior portion of the calcaneus is repositioned to align below and provide support to the head of the talus.[67]

Pes Cavus

Pes cavus deformity is characterized by an extremely high medial longitudinal arch due to the equinus position of the forefoot in relation to the hindfoot. If the high arch depresses upon weight bearing, the condition is known as "flexible" pes cavus. If the medial longitudinal arch remains high with full weight bearing, the condition is "rigid" pes cavus.[65] Accompanying characteristics of the pes cavus often include a contracture of the plantar fascia which worsens with time. A fixed varus deformity of the heel develops with shortening of the Achilles tendon and clawing of the toes.[67] Weight bearing is distributed unevenly across the metatarsal heads and along the lateral border of the foot. The cavus foot is a poor shock absorber and therefore susceptible to many shock related pathologies, including stress fractures, heel pain and metatarsalgia.

Pes cavus can be idopathic or secondary to various neurological diseases such as poliomyelitis, Friedreich's ataxia, cauda equina tumor, spina bifida, and Charcot-Marie-Tooth disease. It can also develop secondary to direct congenital talipes equinovarus (or "clubfoot").[67]

Flexible pes cavus responds well to conservative management with biomechanical orthotic devices and exercise. The orthotic device should:

1. Provide support to the plantar fascia and medial longitudinal arch.
2. Balance any forefoot to rearfoot malalignment, or calcaneus deformity.
3. Provide adequate shock absorption, particularly at the heel and along the metatarsal heads.

Any muscle weakness and/or tightness should be addressed through an aggressive exercise program as tolerated. Tight ankle inverters and plantarflexors, and weak everters often accompany pes cavus.[16] Resistent symptoms subsequent to a rigid pes cavus may not respond as well to conservative orthotic and exercise management. Some require surgical correction. The specific operative technique performed depends on the age of the patient (whether bony growth is complete), the type of pes cavus, the nature of any calcaneal and toe deformities and the amount of muscle strength.[67]

Hammer Toes

Hammer toes are fixed deformities manifested as flexion contractures at the proximal interphalangeal joints with associated dorsiflexion at the metatarsophalangeal joints. The distal phalanx may form a callus on its tip due to its downward position and a painful corn commonly develops on the dorsum of the proximal interphalangeal joint from shoe pressure. The second toe is affected most often.[9] The cause of hammer toe seems to be either congenital or acquired and usually occurs in an excessively pronated foot. A tightly fitting shoe or hallux valgus deformity can frequently initiate and enhance the deformity.

Conservative treatment of the deformity centers around protecting the protruding joint and its callus with pads and custom-made "space shoes" which relieve any crowding of the toes. Persistent symptoms secondary to the deformity require surgical intervention. One such procedure involves straightening the toe by excising the head of the proximal phalanx and the base of the distal phalanx and inserting an intramedullary pin to provide fixation while the bones fuse.[67]

Hallux Rigidus

With hallux rigidus, motion of the first metatarsophalangeal joint is painfully restricted, but no deformity is evident. Restriction is greatest in dorsiflexion. Hallux regidus is frequently bilateral and is usually due to repeated minor trauma which ultimately results in degenerative changes of the first metatarsophalangeal joint. Also, anatomic abnormalities of the foot, where there is an abnormally long first metatarsal bone and the biomechanical problem of excessive forefoot pronation, have been implicated as causes.[40]

The symptoms of hallux rigidus are progressive increasing stiffness with associated pain at the base of the first toe upon walking. The patient will demonstrate a protective gait, shortening the propulsive phase, shifting terminal weight bearing laterally, and stepping with an oblique bend to the foot. Such maneuvers avoid motion or pressure over the first metatarsophalangeal joint. Also, the patient's well worn shoes will show lateral oblique rather than the normal transverse creases over the toes[23] (See Figures 10-7, 10-8).

Therapeutic management of the acutely inflamed hallux rigidus includes rest and ice to control swelling and pain. As swelling reduces, local heat to the flexor hallucis brevis will relieve any protective muscle spasm. As acute symptoms further subside, repeated passive dorsiflexion exercises are attempted at the first metatarsophalangeal joint to achieve full dorsiflexion. The patient should be encouraged to normalize his gait and allow weight transfer over the first metatarsophalangeal joint as tolerated. In advanced cases, when osteoarthritic changes are radiographically evident, injections of corticosteroids frequently controls symptoms, but ultimately surgical correction might be necessary. A Keller arthroplasty, with resection of the proximal half of the promixal phalanx, provides good mobility and pain relief, but compromises the propulsion power in gait.[62]

   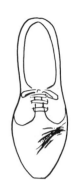

Figure 10-7. Normal toe creases on well worn shoes. (Reproduced with permission from Hoppenfeld, *Physical Examination of the Spine and Extremities,* Appleton-Century-Crofts, Pub.)

Figure 10-8. Lateral oblique toe creases of chronic Hallux Rigidus. (Reproduced with permission from Hoppenfeld, *Physical Examination of the Spine and Extremities,* Appleton-Century-Crofts, Pub.)

COMMON SKIN LESIONS

Calluses

A callus is an area of skin tissue hypertrophy occurring secondary to extreme vertical pressures of weight bearing and/or the horizontal shearing forces of function. Usually they are evident under the bony prominences of the metatarsal heads and medial or lateral ridges of the calcaneus. Calluses could also be evident over the skin at the posterior calcaneal insertion of the Achilles tendon.

This reactive hyperkeratosis is a defense mechanism of the skin to reinforce an area that might otherwise breakdown and ulcerate due to the unusual amounts of pressure exerted over it. Calluses are obvious tell-tale signs of the weight distribution pattern over the heel and forefoot. Plantar callus lesion patterns are pieces of the biomechanical puzzle leading to an assessment of functional abnormalities of the foot which could be leading to symptoms elsewhere in the foot, ankle and/or lower limb. For example, with excessive subtalar pronation secondary to compensatory rearfoot (calcaneal) varus, calluses will often be noted under the second and third metatarsal heads. This results because the first metatarsal becomes unstable and dorsiflexes from ground reactive forces. A disproportionate amount of weight bearing is then shifted more laterally at the end phase of midstance and into early propulsion. (An indepth analysis of compensatory subtalar pronation will follow later in this chapter.)

Painful calluses are best managed with methods to eliminate abnormal weight bearing patterns temporarily through the use of metatarsal pads (placed proximal to the lesion site) and ultimately by a functional orthotic. Resistive painful lesions secondary to rigid structural deformity might need an accommodative device to protect them in normal activity. The lesion itself is best treated with gradual trimming of the hypertrophic area after warm soaks. Removal of the callus can be accomplished with a scapel, pumice stone or callus file. The area is then protected with a surrounding built-up donut pad to prevent further stress on the skin. As long as abnormal forces are prohibited or redistributed, normal skin will eventually return.

Corns

There are two types of corns commonly referred to as hard corns and soft corns. Hard corns are usually found on the dorsum of deformed toes (hammertoes, clawtoes) and result from excessive pressure and friction between the phalanges and the shoe. Soft corns are found between the toes and result from excessive pressure between the toes and are kept soft by the moisture between the toes. Corns, like calluses, ultimately respond to removal of the pressure source. Shoes with adequate toe boxes are essential and correction of the toe deformity must be addressed for permanent resolution.

Corns are treated in the same manner as calluses: gradual trimming and protecting the lesion from the abnormal pressure by either donut pads or "corn pads" for hard corns, and by the use of toe spacers (often lambswool) for soft corns.

Plantar Warts

Plantar warts are vascular papillomatous growths, often with marked hyperkeratosis, found over the weight bearing areas of the foot. Their etiology remains questionable, with viral, traumatic or even psychogenic origin theorized. They occur as single lesions, in groupings, or in a clumped mosaic pattern.

Numerous treatments for plantar warts have been reported in the literature, including acid and liquid nitrogen applications, surgical removal, and ultrasound. Chemical removal commonly includes the use of 40% salicylic acid plaster pads placed over the wart and removed at weekly intervals.[6] The method of the therapeutic effect of ultrasound on plantar warts is not known. It appears, however, to have therapeutic value when applied with direct technique at dosages between 0.6 and 1.5 watts per centimeter squared for 15 minutes, once a week.[68] Ultrasound treatment can be favored because there are no other adverse tissue reactions, such as scaring and its potentially painful disability, which is often associated with acid or liquid nitrogen removal techniques.[72] The problem of reoccurrence must be considered in any treatment method chosen.

BIOMECHANICAL CONSIDERATIONS

To fully investigate the nature of foot, ankle and lower leg pathologies, the clinician must have a working knowledge of the biomechanical function of the foot and lower extremity. Being the body's only contact element with the ground, the foot responds with a series of complex interactions to the reactive forces of the supporting surface. These forces are accepted by the foot and translated to the rest of the lower limb and proximal musculoskeletal system. The nature of this foot-ground interaction must be appreciated as a possible etiological factor in foot and ankle dysfunction as well as development of abnormalities.

Recognizing normal relationships and interactions that various parts and systems have with each other to provide smooth, efficient locomotion, abnormal relationships, and applying these factors in the assessment, treatment, and prediction of pathology, are the goals of a clinical biomechanical approach to pain and dysfunction of the foot and lower leg. Consideration of this approach, along with a thorough understanding of the specific pathology, adds an extra dimension to therapeutic intervention. This dimension often provides the ultimate problem resolution.

Lower Extremity Biomechanical Evaluation

Having identified the symptom source through a complete history, selective tissue tension (Cyriax model), neurological scan and other diagnostic studies, a lower extremity biomechanical evaluation is essential to determine the primary etiology of non-traumatic disorders and certain deformities. The biomechanical examination should be a systematic approach, noting bone length and torsion, joint position, alignment, stability and range of motion, and soft tissue flexibility. Neutral positions are established and deviations from the norm should be recognized both in the static as well as dynamic state.

Proximal joint deformity and torsion of long bones and/or restriction of joint motion, leads to the development of adaptive changes in the foot, and subsequent pathology as the foot and leg complex attempt to maintain balanced locomotion. The reverse can also apply. Faulty alignment, restricted mobility, or hypermobility of foot structures, leading to abnormal mechanics of foot function during gait, can lead to joint and soft tissue changes in the lower leg, knee, hip and low back. Any of these changes may become pathological when resulting from, or exposed to, the demanding stress of athletics, the repetitive stress of work activity, or from the cumulative effects on aging joints.

The elements of a thorough lower extremity biomechanical evaluation to consider are:

a. Inspection of well worn footgear
b. Overall posture
c. Leg lengths
d. Pelvic position–crest levels, rotational assymetry
e. Hip range of motion (all planes)
f. Femoral alignment–coxa vara/valga, femoral anteversion/retroversion, femoral torsion (greater trochanter-femoral condyles)
g. Proximal soft tissue flexibility–hip flexors (Thomas Test), tensor fascia (Ober Test), hamstrings (SLR)
h. Knee range of motion
i. Tibia alignment–genu varum/valgum/recurvatum, tibial torsion (femoral condyles-malleolei), Q-angle
j. Distal soft tissue flexibility–gastroc-soleus, anterior tibialis, peroneals, posterior tibialis
k. Ankle dorsiflexion/plantarflexion
l. Subtalar neutral position (rearfoot varus/valgus)
m. Calcaneal (subtalar joint) inversion/eversion
n. Rearfoot-forefoot alignment (in subtalar neutral)
o. First ray position and mobility
p. Fifth ray mobility
q. Hallux dorsiflexion
r. Medial longitudinal arch height (weight bearing vs non-weight bearing)

These findings are all considered in the static non-weight bearing state. The response of structures to the dynamics of gait and weight bearing must also be considered. The dynamic components of the evaluation assess joint mobility and stability, muscle function, as well as proprioception. Gait analysis through force plates or the Electrodynogram (EDG) System 1184® (Langer Biomechanics Group, Inc. Deer Park, NY) examines the segmental foot responses to the floor and gives a clear picture as to the weight bearing patterns and segmental functions (See Figures 10-9—10-11). Electromyography (EMG) evaluates discrete phasic activity of muscle function. Patient visualization through video tape allows the examiner to view proximal structure motion and position, such as lower leg rotation, knee position, motion, and hip and pelvic responses to gait. Also, the use of the Biomechanical Ankle Platform System (BAPS) allows the examiner to look for correlation between the dynamics of subtalar function and the statics of subtalar measurement. Various degrees of weight bearing and range of motion can be assessed via this analysis, looking for painful arcs, and coordination of motion as well as proximal limb responses to loaded subtalar motions.

Normal Alignment

No one foot can be said to be absolutely "normal." Human variances exist that exhibit many "makes and models" of feet. Most authors, however, define a normal foot as one which is free of any visible pathology or structural abnormality. Certain relationships in the alignment of foot bones and segments are optimal to make the foot an efficient tool

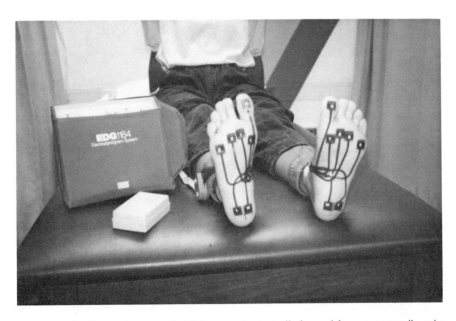

Figure 10-9. The Electrodynogram (EDG) System. Strategically located force sensors adhered to the plantar surface of the foot measure vertical forces over key weight bearing segments. Also note waist pack (force data collector) and remote activator.

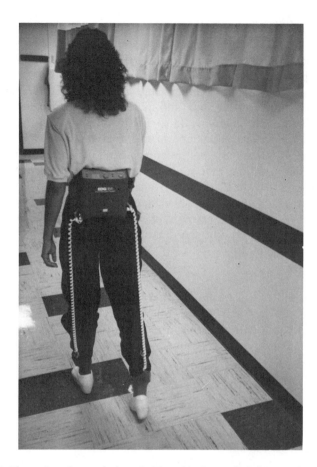

Figure 10-10. The patient then ambulates in his or her "normal" fashion as the remotely activated force data collector stores the information from the foot sensors.

Figure 10-11. Information from the force data collector is then fed into the computer for a printout of the gait analysis showing peak forces, duration of forces, as well as gait phases to determine symmetry of function and segmental stability during function.

for adapting to ground surfaces and propelling the body during locomotion. Recognizing these normal conditions will allow the clinician to note any abnormality and determine the degree of dysfunction that may be attributed to such structural deficits.

Normal static non-weight bearing alignment is based on the anatomical positioning of key foot structures—calcaneus, forefoot, and first ray segment—with the subtalar joint in its neutral position. Mertin L. Root[52] determined and defined the neutral position of the subtalar joint as that point where the joint is neither pronated nor supinated. The subtalar neutral position is found through palpation of the head of the talus medially and laterally in its articulation with the navicular. When the subtalar joint is pronated, the talus adducts. When the subtalar joint is supinated, the talus abducts. Upon pronation and supination of the subtalar joint, the examiner can palpate the medial and lateral head of the talus, respectively. The talar head bulges beyond the tuberosity of the navicular medially upon pronation and becomes prominent on the dorsolateral aspect of the foot with supination. The subtalar neutral position is recorded when the medial and lateral sides of the head of the talus do not protrude and provide a symmetrical feel in reference to the navicular. This is often referred to as the congruency grip of the talonavicular joint. The talocrural joint should be maintained neither plantarflexed nor dorsiflexed and the forefoot is locked on the rearfoot with a dorsiflexion/abduction force offered through the fourth and fifth metatarsal heads, bringing the forefoot into its maximum position of pronation. With the subtalar joint in neutral, a bisection of the posterior calcaneus should form a vertical relationship with the distal one-third of the lower leg in the frontal plane (See Figure 10-12). An inverted calcaneal relationship to the lower leg is referred to as rearfoot varus, while any everted calcaneal relationship is known as rearfoot valgus. Most feet will show a mild 2 to 3 degrees calcaneal varus angulation. This finding is therefore normal and presents little to no functional consequences in gait.[52,66]

With the forefoot locked on the rearfoot, the plane of the forefoot can be evaluated as it relates to the plane of the plantar surface of the calcaneus. Ideally, they should be in

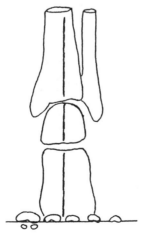

Figure 10-12. Normal foot alignment in subtalar neutral, showing vertical relationship between lower leg and calcaneus.

the same plane. If the plane of the forefoot, as noted by the plane of the metatarsal heads, is in an inverted position relative to the plantar plane of the rearfoot, a forefoot varus deformity exists. If the forefoot plane is everted relative to the rearfoot, a forefoot valgus abnormality is said to exist. By maintaining this neutral posture of the foot, the position of the first metatarsal head can be evaluated. In the normal foot, it should rest in the same plane as the metatarsal heads two through five. Positioning below that plane is referred to as a plantarflexed first ray.

Range of motion of the subtalar joint varies greatly among individuals. Inman[29] has estimated a range of between 20 to 62 degrees. The minimum range of motion required for normal ambulation is generally considered to be approximately 18 degrees with 6 degrees of pronation and 12 degrees of supination.[64] What is essential to note is that the amount of supination is twice that of pronation. Motion about the midtarsal joint is best evaluated to determine the effectiveness of the locking mechanism created by subtalar supination. The motion of the subtalar joint affects the stability of the foot through its control of the midtarsal joint. When the subtalar joint is pronated, the midtarsal joint is unlocked and the forefoot is flexible. Conversely, when the subtalar joint is supinated, the midtarsal joint is locked and the forefoot is more rigid[41] (See Figure 10-13).

The first metatarsal segment should demonstrate adequate mobility in both dorsiflexion and plantarflexion. Approximately one finger breadth of motion in each direction from the plane of the other metatarsal is commonly considered normal. With elimination of first ray dorsiflexion by the patient standing, the first metatarsophalangeal joint should exhibit a minimum of 45 degrees extension for normal hallux function in gait.[24] For biomechanically normal foot function, the ankle joint requires a minimum of

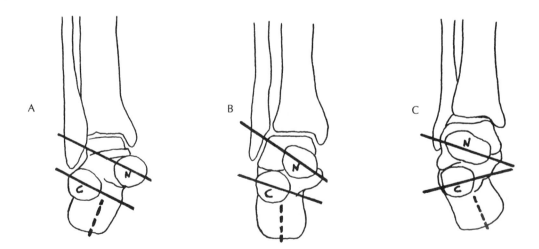

Figure 10-13. Subtalar joint position influencing axes of motion of midtarsal joint and subsequently mobility of the forefoot. A) Pronation-axes of Talonavicular and Calcaneocuboid are virtually parallel, allowing high degree of mobility. B) Neutral-axes of midtarsal begin converging to limit mobility. C) Supination-maximum convergence of midtarsal joint axes greatly limiting forefoot mobility.

20 degrees of plantarflexion and 10 degrees of dorsiflexion with the knee extended and foot in its neutral position.[52]

Dynamic weight bearing normalcy is determined by the osseus segments involved in the complex events of gait, demonstrating an ideal alignment just prior to heel-off. The subtalar joint should be in its neutral position, with the plantar surface of the calcaneus in the transverse plane. The forefoot should be locked in maximal pronation (its neutral position) with the metatarsal heads all in the transverse plane. This positional relationship provides for maximum osseus stability preparing the foot for propulsion. The knee and ankle usually lie in the transverse plane, and there is no significant rotational or torsional influences from proximal structures. The distal one-third of the lower leg should be perpendicular to the supporting surface.[52]

Open Versus Closed Kinetic Chain

Understanding the normal function of the foot requires an appreciation for the conditions under which the foot performs the complex events of gait. The segments of the foot are operating under closed kinetic chain conditions during the weight bearing phase of locomotion. The closed kinetic chain refers to that kinesiological state where several joints are linked successively and the terminal segment is not free, as when the foot is in contact with the ground during the stance phase of gait. The open kinetic chain refers to that in which the terminal segment is free, or not in contact with the ground.

Motion terms should be clearly defined when referring to either open or closed chain situations. Closed kinetic chain pronation and supination of the subtalar joint produces talus and calcaeus movement as well as motion of the knee and tibia. One of the primary functions of the subtalar joint is to serve as a torque converter allowing internal and external leg rotation to occur during the stance phase, while the foot is planted firmly on the ground.[13] Inman's[29] description of the subtalar joint as a "mitered hinge" joining the tibia and foot clearly demonstrates a rotational relationship: pronation with internal rotation of the tibia and supination with external rotation. During closed chain pronation, the calcaneus everts and the talus adducts and plantarflexes. The leg internally rotates and the knee flexes. With open chain pronation (as the leg and talus are held stable) the calcaneus and foot evert, abduct, and dorsiflex. During closed chain supination, the leg externally rotates, the knee extends, and the talus abducts, and dorsiflexes while the calcaneus inverts. Open chain supination results in calcaneus and foot inversion, adduction, and plantarflexion.

Closed chain muscle activity is expressed as its function occurring during the stance phase of gait. Examples include the function of the posterior tibialis and peroneus longus muscles. In open chain, non-weight bearing and the action of the posterior tibialis is inversion of the subtalar joint and assisting in plantarflexion of the ankle,[35] strictly an acceleration function. However, Root describes the closed chain function of the posterior tibialis as a decelerator of subtalar pronation and leg internal rotation during the contact phase, and as an accelerator of subtalar supination and leg external rotation during the midstance phase of gait.[52] In open chain, the peroneus longus is an everter of the foot and assists in plantarflexion of the ankle.[35] In closed chain, its action stabilizes the first ray and transfers body weight from the lateral to the medial side of the forefoot during propulsion.[52]

Gait Analysis

Synchronous motions of the knee, ankle and subtalar joints are essential for normal, efficient, symptom-free ambulation. As a review of the dynamics of normal gait, components of the gait cycle are considered, noting the joint motion which should be occurring.

Contact Phase. The contact phase of stance is defined as the heel strike to the foot-flat interval. At heel strike the subtalar joint is in a supinated position and immediately crosses into pronation with contact on the supporting surface. This pronation continues through the contact phase. The ankle joint is plantarflexed and the knee is flexing throughout contact.

Mid-Stance. The mid-stance phase is the interval from foot-flat to heel-rise. At the point of foot flat, the subtalar joint begins to resupinate, so that at a point just prior to heel rise it has returned to its neutral position and then continues into supination. The ankle begins dorsiflexing as the tibia moves over the talus. The knee extends during this interval to be at or near full extension at the point of heel rise.

Propulsion. Propulsion is the final interval of the stance phase, from heel-rise to toe-off. During the propulsion phase, the subtalar joint continues to supinate, locking the mid tarsal for the forefoot's role as a rigid lever, ready for toe-off. Beyond toe-off, the subtalar joint begins to move back toward pronation, preparing the heel for its next contact. The ankle plantarflexes as the heel rises to produce the power for propulsion and immediately dorsiflexes as swing begins in order to clear the toes from the ground. The knee begins flexing at heel-rise as the femur moves over the fixed tibia and continues to flex beyond toe off into the swing phase. The foot clears the ground and then extends to prepare the foot for the next heel strike in the linear progression of locomotion.

By applying these expected normal phasic sequences of gait to the pathologies discussed earlier and the biomechanical abnormalities to follow, the clinician will be able to make a clear assessment as to the nature of the biomechanical disruption and its subsequent impact on the proximal joints and their associated soft tissue. Conditions such as muscle strain, tendinitis, capsular stress, joint laxity, joint restriction, and muscle imbalance can become apparent when putting together the pieces of the biomechanically induced foot and lower leg dysfunction puzzle.

COMMON BIOMECHANICAL ABNORMALITIES

Rearfoot Varus

A subtalar neutral calcaneal inversion generally greater than 2 to 3 degrees is commonly classified as a rearfoot varus deformity. (See Figure 10-14.) This particular osseus abnormality may lead to increased lateral heel contact. In the absence of any other forefoot or mid tarsal abnormality, retrocalcaneal irritation and/or exostosis may develop.[64] In extreme cases and if sufficient subtalar pronation range of motion is available, the rearfoot undergoes excessive pronation as a compensatory motion to bring the medial condyle of the heel in contact with the supporting surface. This compensatory pronation occurs only

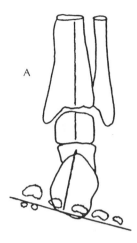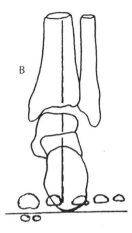

Figure 10-14. A) Rearfoot varus deformity in subtalar neutral. B) Compensated rearfoot varus leading to subtalar pronation in midstance.

when both the heel and foot are in ground contact, so at heel-rise, when the deformity leaves the ground, the subtalar joint may resupinate to make partial or full recovery for the forefoot to lock and be an efficient rigid lever for propulsion.[44] In the event of partial recovery, mildy abnormal transverse leg rotation will occur in response to this pronation and be out of phase with the normal tibia and femoral external rotations which occur with knee and hip extension. The forefoot is slightly hypermobile, leading to symptoms of metatarsalgia, knee pain, calf fatigue, and cramping.

Forefoot Varus

As mentioned previously, the forefoot inverted relative to the rearfoot in subtalar neutral is referred to as forefoot varus. This deformity is most likely resultant from incomplete derotation of the head and neck of the talus from their infantile alignment.[21] Forefoot varus leads to a great degree of compensatory subtalar pronation as the medial border of the forefoot must reach the floor during the mid-stance phase of gait. In order for the midtarsal joint to provide this extra motion, the subtalar joint must remain pronated be yond the normal 25 percent of stance and well into the later stages of mid-stance and propulsion (See Figure 10-15). This excessive propation leads to hypermobility of the forefoot segments during propulsion, rendering them very inefficient and vulnerable to breakdown. Symptoms of plantar fasciitis are often present due to the depression of the medial longitudinal arch during propulsion. Also, metatarsalgia, interdigital neuroma, and stress fractures may develop. A hallux valgus deformity is not uncommon with this foot type. This compensatory pronation could also lead to "shin splints" due to the excessive overuses and stretch of the posterior tibialis in its efforts to eccentrically control hyperpronation beyond contact into midstance.[64] Associated abnormal leg rotations might cause stress on proximal structures of the knee and hip. With increased medial leg rota-

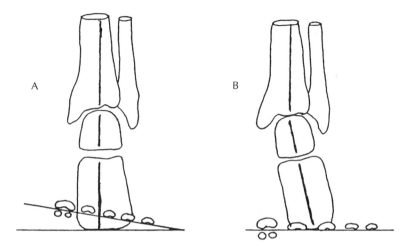

Figure 10-15. A) Forefoot varus deformity in subtalar nuetral. B) Fully compensated forefoot varus in direction of subtalar pronation in midstance.

tion, patients will tend to demonstrate tightness of the hip adductors and internal rotators, with noticeable loss of hip external rotation.[45] Medial knee pain will ensue due to either the valgus position the knee assumes with pronation and/or due to phasic disruption of normal tibia external rotation which must accompany knee extension. (Knee extension is accomplished with relative tibia external rotation achieved by abnormally increased internal rotation of the femur. This also explains the previously mentioned symptoms of tight hip internal rotators secondary to abnormal pronation.)

The abnormal leg rotations and valgus position assumed by the knee will eventually lead to tracking dysfunction of the patella due to a changing Q-angle.[16] If the forefoot varus compensation is unilateral, a functionally short limb is created on that side due to the adduction and plantarflexion of the talus, which can lead to pelvic imbalances causing sacro-iliac pathology and/or scoliosis.

If sufficient subtalar and midtarsal pronation range of motion is not available to provide compensation, a condition known as non-compensated forefoot varus exists. Characteristic of this deformity is increased lateral foot weight bearing, with excessive forces being taken by the fourth and fifth metatarsal heads. Calluses are noted and matatarsal pain ensues symptomatic of stress fractures. This uncompensated foot type is a poor shock absorber and an inefficient propulsion tool.

Forefoot Valgus

When the forefoot is in an everted position relative to the rearfoot, forefoot valgus deformity exists. Compensation for this deformity occurs in the direction of supination during the early phase of mid-stance[13] (See Figure 10-16). At heel-off, the subtalar joint pronates in response to extensive lateral ground reaction forces.[44] This is the reversed phasic activity of pronation-supination normally seen in gait. In addition to highly destructive forces

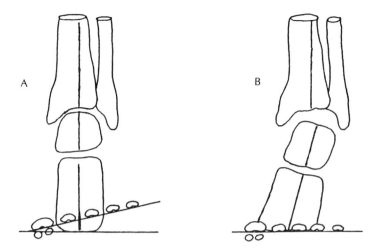

Figure 10-16. A) Forefoot valgus deformity in subtalar neutral. B) Fully compensated forefoot valgus in direction of subtalar supination in midstance.

being absorbed by the first and fifth metatarsal heads, those previously mentioned consequences of a hypermobile forefoot in propulsion can occur. Also, repeated inversion ankle sprains have been reported with forefoot valgus deformity.[66]

Similar symptoms are associated with the deformity of a rigid plantarflexed first ray. Noticeable by the marked plantarflexed position of the first metatarsal head relative to the other metatarsal heads, a rigid plantarflexed first ray causes reverse phase supination and pronation as a compensation, and extremely destructive forces are met by the hypomobile first ray segment.

Tibia Vara

Tibia vara is a frontal plane angulation deformity of the lower leg which is manifested by the distal portion of the tibia turned toward the midline of the body. This deformity, extrinsic to the foot, causes a compensatory pronation of the subtalar joint in order to bring the calcaneus into a vertical position during the contact phase of gait. Without the compensatory pronation, only lateral heel weight bearing occurs.

Tibia vara is measured as the angle formed between the ground surface and the bisection of the distal one-third of the lower leg during weight bearing.[24] The degree of this deformity must be taken into account when determining the amount of compensatory pronation the subtalar joint must undergo to bring the medial calcaneal condyle in contact with the supporting surface.

The biomechanical consequences from any significant tibia vara are very similar to compensated rearfoot varus as discussed previously. Management of associated symptoms of this deformity is by use of a functional orthotic with a medial rearfoot post which limits hyperpronation by effectively raising the supporting surface to the medial heel.

Functional Equinus

Functional equinus is a less apparent type of abnormality leading to excessive compensatory subtalar pronation. Structural equinus is often obvious in the pes cavus foot type where the plane of the forefoot is below that of the rearfoot. A functional equinus does not always demonstrate the osseous deformity and results primarily from excessive tightness of the gastroc-soleus.

To prevent a rapid heel rise during gait due to this limited dorsiflexion of the ankle, the midtarsal joint must allow for additional dorsiflexion. In order to "unlock" the midtarsal to provide this extra mobility, the subtalar joint must remain pronated well into midstance. As the talus plantarflexes with this closed chain pronation, the forefoot becomes dorsiflexed relative to it. A minimum of ten degrees of ankle dorsiflexion is required to compensate for this abnormal motion.[52]

The pathological consequences of the compensated functional equinus include those conditions secondary to excessive subtalar pronation as seen in compensated forefoot varus. Treatment is best focused on aggressively stretching the tight gastroc-soleus to allow for an increase in ankle dorsiflexion.

ROLE OF FUNCTIONAL FOOT ORTHOTICS

Foot and lower leg pathologies, secondary to structural deformity and imbalance, are effectively managed by the incorporation of proper functional orthotics into the treatment regime. The goal of orthotic management is to provide control of abnormal foot motion and to provide functional shock absorption by improving the subtalar joint position. The orthotic permits the foot to function at or near its neutral position at the end of midstance (or just prior to heel-off). Abnormal compensatory motion of the subtalar and/or midtarsal joints, due to deformity or improper alignment of the lower leg, calcaneus or forefoot, is prevented by the orthotic. The orthotic then acts to decrease or eliminate those destructive closed kinetic chain proximal limb and intrinsic foot stresses which occur in response to the poorly positioned subtalar joint during gait.

Subtalar and midtarsal compensation occurs in order to bring the normal weight bearing segments of the heel and forefoot in contact with the supporting surface to more evenly balance the weight bearing load. Non-compensating deformities, where insufficient subtalar or midtarsal range of motion is available, often produce destructive forces over small areas of the weight bearing plantar surface of the foot.

All closed kinetic chain motions, as discussed previously, should accompany each position change of the subtalar joint due to its triplaner motion. Compensatory pronation can be perceived as an asset to better balance plantar surface weight bearing, and, at the same time, be considered a liability to other intrinsic foot and proximal limb structures, which respond in an abnormal fashion, out of their normal phasic sequences of gait. Hypermobility of the forefoot during propulsion, dysphasic function and overuse of the posterior tibialis muscle, and internal rotation of the tibia late in midstance as the knee extends, can all lead to common biomechanical pathologies such as plantarfasciitis, "shin splints," knee pain, and patella tracking dysfunction.

The biomechanical orthotic device is custom fabricated from a neutral position cast slipper impression of the patient's feet (See Figures 10-17, 10-18, 10-19). This "negative"

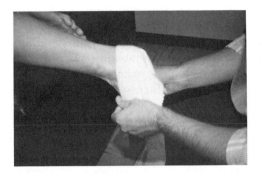

Figure 10-17. Plaster strips are molded over the patient's foot to form "slipper" cast.

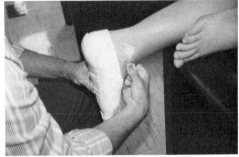

Figure 10-18. The neutral subtalar position is then palpated and maintained while the cast hardens. The forefoot is maintained pronated with a dorsiflexion force over the 4th and 5th metatarsal heads.

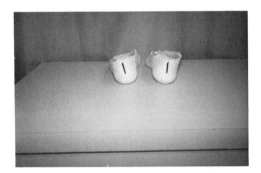

Figure 10-19. Finished neutral position slipper casts capturing the calcaneal angle and rearfoot to forefoot relationship. Casts are ready to be shipped to orthotic lab for fabrication of the biomechanical orthotic.

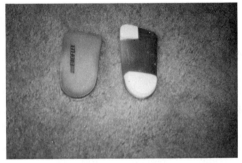

Figure 10-20. The finished biomechancical orthotic. Note rearfoot posting of the medial forefoot.

cast is then filled with plaster to form a positive mold over which the basic orthotic shell can be fabricated, capturing the contours of the foot, the calcaneal position, as well as the rearfoot to forefoot plantar relationship. To this thermoplastic shell, posting material is added at the heel and/or forefoot, depending on the specific biomechanical abnormality. Posts are wedges which fill in the angulation deformity of the forefoot and/or calcaneus as measured in the biomechanical exam and captured in the neutral position cast (See Figure 10-20).

The post functions to bring the supporting surface up to the weight bearing plantar surface of the foot, providing a ground force against which to weight bear, without the foot having to undergo excessive compensatory motions. The result is a balanced foot functioning at or near its neutral position. Muscle actions are then better sequenced, weight bearing force transmission is improved, and abnormal motions are reduced. All of this occurs without restriction of the normal degree of motion needed for shock absorption and adaptation to the supporting surface.

Orthotics are typically classified as rigid, semi-rigid, and flexible (soft), depending on the type of material used in their fabrication and the amount of mobility inherent in the device. Severe control problems, particularly in the very young where there is excessive range of motion, often require the rigid type of orthotic. Problems with more restriction of mobility and "fixed" deformities respond best to the more flexible accommodative devices which provide support and relief to painful areas of the foot. Most other problems, where sufficient joint motion is present and control of abnormal compensatory motion is required in an active patient, semi-rigid devices are indicated due to their inherent ability to control and yet better yield to normal motions required in locomotion and running. When orthotic therapy is indicated, it must be combined with activity modification and exercise to stretch and strengthen those muscles which have been compromised due to the abnormal biomechanical stresses influencing them over time. Key groups needing improved flexibility usually include the gastrocs, hamstrings, hip adductors, iliotibial band, peroneals, and hip internal and external rotators. Strengthening of the posterior tibialis, anterior tibialis, and toe flexors is usually indicated.

Functional orthotics are not a panacea for all foot problems. Careful patient evaluation and selection are the keys to successful management if orthotics are to be utilized in the overall therapeutic regime. Only the practitioner with a keen awareness of foot and lower leg biomechanics should prescribe the design of the custom molded functional orthotic.

REFERENCES

1. Alexander IJ, et al: Morton's neuroma: A review of recent concepts. *Orthopedics* 1987; 10:103.
2. Anson BJ, McVay CB: *Surgical Anatomy*, ed 6. Philadelphia, W.B. Saunders Co, 1984, vol 2, p 1283.
3. Arnoldi CC: Effecten of elastisk kompression pa stromnighastigheden i laeggens dye vener. *Ugeskr Lageger* 1976; 138:275.
4. Basmajian JV, Deluca, CJ: *Muscles Alive*, ed 5. Baltimore, Williams and Wilkins Co, 1985.
5. Benick RJ: The constrains mechanism of the human tarsus. *Acta Orth Scand* 1985; 56 (suppl):215.
6. Berkow R (ed): *The Merck Manual of Diagnoses and Therapy*, ed 13. West Point, Pa., Merck, Sharp and Dohme Research Laboratories, 1977.
7. Beskin JL, Sanders RA, Hunter SC, et al: Surgical repair of Achilles tendon ruptures. *Am J Sports Med* 1987; 15:1-8.
8. Cailliet R: *Foot and Ankle Pain*, Philadelphia, F.A. Davis Co, 1968.
9. Cailliet R: *Foot and Ankle Pain*, ed 2. Philadelphia, F.A. Davis Co, 1983.
10. Corrigan B, Maitland GD: *Practical Orthopaedic Medicine*. London, Butterworth and Co, 1983.
11. Cyriax J: *Textbook of Orthopaedic Medicine: Diagnosis of Soft Tissue Lesions*, ed 7. London, Bailliere Tindall, 1978, vol 1.
12. Debrumer HU: *Orthopadisches Diagnostikum*. G. Thieme Publ, Stuttgart, 1973, p 149.
13. DiGiovanni JE, Smith SD: Normal biomechanics of the adult rearfoot: A radiographic analysis. *J Am Podiatry Assoc* 1976; 66:812.
14. Dimonte P, Light H: Pathomechanics, gait deviations, and treatment of the rheumatoid foot: A clinical report. *J Am Phys Ther Assoc* 1982; 62:1148.
15. Fox JM, Blazina ME, Jobe FW, et al: De-

generation and rupture of the Achilles tendon. *Clin Orthop* 1975; 107:221.

16. Franco AH: Pes cavus and pes planus: Analyses and treatment. *J Am Phys Ther Assoc* 1978; 67:688.

17. Frankel VH: Fatigue fractures, biomechanical consideration. *J Bone Joint Surg Am* 1972; 54:1345.

18. Gartland JJ: *Fundamentals of Orthopaedics*, ed 3. Philadelphia, W.B. Saunders Co, 1979.

19. Glew G: Elastic band injuries. *Br Med J* 1967; 4:488.

20. Grant JCB, Basmajian JV: Grant's Method of Anatomy: By Regions Descriptive and Deductive, ed 7. Baltimore, Williams and Wilkins, 1965.

21. Hlavac HF: Compensatory forefoot varus. *J Am Podiatry Assoc* 1970; 60:229.

22. Holzegel K: Beitrag zum Kompressdionsdruck elastischer verbande am unterschenkel. Dtsch Gesundheitsu 1970; 25:887.

23. Hoppenfeld S: *Physical Examination of the Spine and Extremities*, New York, Appleton-Century-Crofts, 1976.

24. Hunt GC: Examination of the lower extremity dysfunction, in Gould JA III, Davies GJ (eds): *Orthopaedic and Sports Physical Therapy*, St. Louis, C.V. Mosby Co, 1985, vol 2.

25. Huson A: *Ein ontleedkundig-functioneel onderyoek van de voetwortel*, thesis. Univ of Leiden, Luctor et Emergo, 1961.

26. Inglis AE, Sculco TP: Surgical repair of ruptures of the tendo Achillis. *Clin Orthop* 1981; 156:160.

27. Ingram AJ: Miscellaneous affections of the nervous system, in Edmonson AS, Crenshaw AH (eds): *Campbell's Operative Orthopaedics*, ed 6. St. Louis, C.V. Mosby Co, 1980, vol 2.

28. Inman VT: Hallux valgus: A review of etiological factors. *Orthop Clin North Am* 1974; 5:59.

29. Inman VT: *The Joints of the Ankle*. Baltimore, Williams and Wilkins, 1976.

30. Justis EJ: Affections of muscles, tendons, and associated structures, in Edmonson AS, Crenshaw AH (eds): *Campbell's Operative*

Orthopaedics, ed 6. St. Louis, C.V. Mosby Co, 1980, vol 2.

31. Kaplan PE: Posterior interosseous neuropathies: Natural history. *Arch Phys Med Rehabil* 1984; 65:399.

32. Kaplan PE: Hemiplegia: Rehabilitation of the lower extremity. In Kaplan, P., Cerullo, L. *Stroke Rehabilitation*, Boston, Butterworth Publ., 1986; p 119.

33. Keene JS, Lange RH: Diagnostic dilemmas in foot and ankle injuries. *JAMA* 1986; 256:247.

34. Kelikian H, Kelikian AS: *Disorders of the Ankle*. Philadelphia, W.B. Saunders, 1985.

35. Kendall FP, McCreary EK: *Muscle Testing and Function*, ed 3, Baltimore, Williams and Wilkins, 1983.

36. Kessler RM, Hertling D: *Management of Common Musculoskeletal Disorders: Physical Therapy Principles and Methods*, Philadelphia, Harper and Row Publishers Inc, 1983.

37. Kleinnmann M, Gross AE: Achilles tendon rupture following steroid injection. *J Bone Joint Surg Am* 1983; 65:1345.

38. LeLievre J: Pathologie du pied. Masson et Cie, Editeurs, Paris, 1961.

39. Lindsjo T, et al: Measurement of the motion range in the loaded ankle. *Clin Orthop* 1985; 199:68.

40. Magee DJ: *Orthopedic Physical Assessment*. Philadelphia, W.B. Saunders Co, 1987.

41. Mann RA: Surgical implications of biomechanics of the foot and ankle. *Clin Orthop* 1980; 146:111.

42. Mann RA, Baxter DE, Lutter LD: Running Symposium. *Foot Ankle* 1981; 1:190.

43. Martens MA, et al: Recurrent dislocation of the peroneal tendons. *Amer J Sports Med* 1986; 14:148.

44. McPoil TG, Brocato RS: The foot and ankle: biomechanical evaluation and treatment, in Gould JA III, Davies GJ (eds): *Orthopaedic and Sports Physical Therapy*, St. Louis, C.V. Mosby Co, 1985, vol 2.

45. Merrifield HH: Influence of gait patterns on hip rotation and foot deviation. *J Am Podiatr Med Assoc* 1970; 60:345.

46. Nitz AJ, et al: Nerve injury and grade II

and III ankle sprains. *Am J Sports Med* 1985; 13:177.

47. Nobel W: Peroneal palsy due to hematoma in the common peroneal nerve sheath after distal torsional fractures and ankle sprains. *J Bone Joint Surg* 1966; 48A:1484.

48. O'Donoghue DH: *Treatment of Injuries To Athletes*, ed 4. Philadelphia, W.B. Saunders Co, 1984.

49. Parlasca R, et al: Effects of ligamentous injury on ankle and subtalar joints: A schematic study. *Clin Orthop Rel Res* 1979; 14.

50. Rasmussen O: Stability of the ankle joint. *Acta Ortho Scand* 1985; 56 (suppl):221.

51. Root ML: An Approach to Foot Orthopedics. *J Am Podiatr Med Assoc* 1964; 54:115.

52. Root ML Orien WP, Weed JH, et al: *Biomechanical Examination of the Foot*, Los Angeles, Clinical Biomechanic Corp, 1971.

53. Scranton PE: Metatarsalgia diagnosis and treatment. *J Bone Joint Surg Am* 1980; 62:723.

54. Segal D: Displaced ankle fractures treated surgically and postoperative management. In Segal, D., *Instructional Course Lectures*. St. Louis, C.V. Mosby Co, 1979.

55. Segal D et al: Functional bracing and rehabilitation of ankle fractures. *Clin Orthop* 1985; 199:39.

56. Segal D, et al: Clinical application of computerized axial tomography (CAT) scanning of calcaneus fractures. *Clin Orthop* 1985; 199:114.

57. Sidey JD: Weak ankles. *Br J Med* 1969; 3:623.

58. Sisk TD: Fractures, in Edmonson As, Crenshaw AH (eds): *Campbell's Operative Orthopaedics*, ed 6. St. Louis, C.V. Mosby Co, 1980, vol 1.

59. Smith W, Winn F, Parette R: Comparative study using four modalities in shin splint treatments. *J Orthop Sports Phys Ther* 1986; 8:77.

60. Solomon MA, et al: CT scanning of the foot and ankle. *AJR* 1986; 146:1204.

61. Staples OS: Ruptures of the fibular collateral ligaments of the ankle: Result study. Immediate surgical treatment. *J Bone Joint Surg Am* 1975; 47:101.

62. Stewart M: Miscellaneous affections of the foot, in Edmonson AS, Crenshaw AH (eds): *Campbell's Operative Orthopaedics*, ed 6. St. Louis, C.V. Mosby Co, 1980, vol 2.

63. Stormont DM, et al: Stability of the loaded ankle. *Amer J Sports Med* 1985; 13:295.

64. Subotnick SI: Biomechanics of the subtalar and midtarsal joints. *J Am Podiatr Med Assoc* 1975; 65:756.

65. Subotnick SI: The cavus foot. *Phys Sports Med* 1980; 8(7):53.

66. Subotnick SI: *Podiatrics Sports Medicine*, Mount Kinsco, NY, Future Publishing Co, 1975, vol 4.

67. Turek SL: *Orthopaedics: Principles and Their Application*, ed 3. Philadelphia, J.B. Lippincott Co, 1977.

68. Vaughn DT: Direct methods versus underwater method in the treatment of plantar warts with ultrasound. *Phys Ther* 1973; 53:396.

69. Viljakka T: Mechanics of knee and ankle bandages. *Acta Orthop Scand* 1986; 57:54.

70. Warwick R, Williams P: *Grays Anatomy*, ed 35 British. W.B. Saunders Co, Philadelphia, 1973.

71. Weseley N, et al: Roentgen measurements of ankle flexion-extension motion. *Clin Orthop Rel Res* 1969; 65:167.

72. Ziskin MC, Michlovitz SL: Therapeutic ultrasound, in Michlovitz SL (ed): *Thermal Agents in Rehabilitation*. Philadelphia, F.A. Davis Co, 1986.

73. Zoellner G, Clancy WG Jr: Recurrent dislocation of the peroneal tendon. *J Bone Joint Surg Am* 1979; 61:292.

11

Psychological Factors and Treatment of Pain

Susan P. Buckelew and
Robert G. Frank

This chapter first reviews mechanisms of pain perception and differences between acute and chronic pain. It then focuses on preventing chronic pain and the dysfunction associated with chronic pain, which is of major concern, given the modest expectations for outcome of treatment in chronic pain patients.

OVERVIEW

Pain is the number one reason people seek physician care. Thirty-three percent of the people in industrialized countries suffer from some form of pain. Of these, half to two thirds are partially or totally disabled for periods of days, weeks, months, or permanently. Sixty billion dollars were lost annually to health care cost, compensation, and litigation in 1980.[11] Re-estimating these figures based on 1983 costs, as much as 80 to 90 billion dollars are spent annually.[16] There is no way to estimate the degree of human suffering involved.[11]

Despite the frequency of pain and its significant impact on society, controversy remains about its assessment and treatment. A subjective experience, pain eludes precise measurement and successful treatment.

According to the International Association for the Study of Pain, pain is an unpleasant and emotional experience associated with actual or potential tissue damage or described in terms of such damage.[42,31] By definition, the relationship between pain and tissue damage is variable. "Real" pain, such as phantom limb pain, can occur in the absence of ongoing tissue damage. Pain is clearly a subjective and personal experience that includes a sensory component as well as an emotional or suffering component.

MECHANISMS OF PAIN

Three major theories have been proposed to account for the perception of pain. According to specificity theory, stimulation of nocioceptors (injury-sensitive free nerve endings) results in transmission of impulses from the peripheral to the central nervous system via the spinothalamic tracts. Stimulation of the sensory motor strip results in the perception of pain and the awareness of localization.[41] Traditionally, specificity theories imply a direct relationship between tissue damage and pain perception. Specificity models are compatible with Descartian views; specific nerve endings are stimulated, and transmission of nerve impulses follows specific pathways, resulting in pain. Specificity implies a direct relationship between fiber stimulation, tissue damage, and pain.[20] Although intuitively tempting, this theory does not account for many of the complexities of pain seen in the clinical setting. For example, the theory fails to explain differences in pain reported by people with similar injuries. One young man might report extensive pain from a broken arm yet another young man can play football for several hours with a similar injury. The specificity theory continues to be taught and provides the basis for some surgical ablation interventions. This theory assumes specific receptors, pain fibers, and a pain center.

Patterning theory, the second major pain theory, holds that the pattern of stimulation of non–pain-specific nerve endings is responsible for pain perception. Patterning theorists hypothesize that the perception of pain represents a temporal or spatial summation of the skin sensory input at the dorsal horn cells. While pressure alone is not perceived as painful, continual pressure may be quite painful. Pain also frequently occurs when the pattern of nerve impulses is disrupted by disease or injury.[41] Livingston,[36] a pattern theorist, conceptualized pain as having a cybernetic effect. Initial peripheral stimulation transmitted centrally via the dorsal column of the spinal cord becomes self-sustaining. Neuronal spinal cord firing is sustained, forming a loop or a reverberating circuit (engram) that no longer requires peripheral stimulation for pain perception. Phantom limb pain, according to patterning theory, occurs when the damage from injury or amputation initially stimulates the sequence of abnormal firing, which becomes self-sustaining. This theory, then, can account in some instances for the variability of tissue damage and pain perception.[41] Nevertheless, a cordotomy and other spinal cord ablation procedures do not relieve pain,[57] suggesting that the mechanism actually involves the brain rather than the spinal cord. Although an advancement in pain theory, patterning theories are inadequate in describing and treating the complexities of pain seen in clinical practice.

Melzack and Wall[40] proposed the gate control theory, which they subsequently modified in 1982.[41] This theory suggests that a neural mechanism in the dorsal horns of the spinal cord acts like a gate to increase and decrease the flow of nerve impulses from peripheral fibers to the central nervous system. Somatic input is subjected to modulating influence of the gate both before pain is perceived and before a response evoked. The degree to which the gate opens or closes is dependent on the relative number of large (A beta) diameter fibers and small (A delta and C) fibers and by the descending influence from the brain. If the amount of information passing through the gates exceeds the critical level, neural areas responsible for pain perceptions and response are activated. Despite some controversy regarding specific predictions, the gate control theory provides a conceptual model that accounts for the complexity of pain and the relative role of psychological factors in pain perceptions and response.

ACUTE VERSUS CHRONIC PAIN

Melzack and Wall[39,41] describe three different types of pain based on a time dimension—transient, acute, and chronic pain. Transient pain is experienced with a stubbed toe. This pain lasts briefly and is typically associated with minimal tissue damage. There are two components of transient pain: the sensory and localizing perception (stabbing pain in toe) followed by the dull, suffering component (aching, agonizing).

Acute pain, or pain that persists beyond a few minutes, is the type of pain experienced with a broken arm. In addition to the sensory and affective component, this type of pain is usually characterized by anxiety. Autonomic changes associated with acute pain include (1) tachycardia, (2) increased systolic and diastolic blood pressure, (3) pupil dilation, (4) increased striated muscle tension, (5) decrease in gut motility and salivatary flow, and (6) the release of catecholamines.[28,51] These autonomic changes are also consistent with the stress response as described by Cannon.[13] Indeed, acute pain is a significant stressor that is accompanied by significant levels of anxiety. Somatic symptoms of anxiety include the symptoms of autonomic arousal. These are often accompanied by concerns about employment and length of time for rehabilitation. The hopelessness that accompanies chronic pain is typically not seen. With acute pain, the pain is typically seen as secondary to specific tissue damage. Expectations involving the pain typically are positive, since most people realistically expect the pain to diminish over time.

Chronic benign pain on the other hand is pain that persists beyond the recovery time expected. Most people define chronic pain as pain that persists beyond 6 months; some authors contend that chronic pain begins as early as 3 months. According to Melzack and Wall,[39,41] chronic pain is frequently characterized by feelings of depression. Pain and depression have long been hypothesized as related, although the exact relationship is not understood.

Acute pain is typically considered secondary to another medical condition that requires treatment (e.g., a broken arm). Among persons suffering with chronic pain (e.g., headaches or back pain), however, the pain itself is considered the primary problem. The relationship between organic disease and pain complaint becomes less direct.

From an ecologic viewpoint, pain stimulates a need to act. Pain signals real or potential tissue damage resulting in a "fight-or-flee" response.[41] In an acute pain situation, pain requires either avoidance of the situation or some actual attempt to protect oneself through fight. In an acute pain situation, a fight or flight response may serve survival needs. Chronic pain, on the other hand, is less likely to accurately signal ongoing or new tissue damage. In this situation, attempts to stabilize through rest or fight are frequently misdirected.

For example, S. K., a 26-year-old woman with myofascial pain status post discitis, anterior and posterior fusion, first developed pain problems after a work-related injury 3 years prior to referral to a comprehensive pain management program. Severe pain left her bedridden for most of the 3 years. Her course of recovery was complicated by neurogenic symptoms, surgery, and discitis. At the time of her initial evaluation, she was extremely distressed and able to tolerate only 3 hours daily of sitting activity. Her circle of friends had narrowed. She felt as though her life was "on hold" and, in fact, had deferred decisions about school, career, and marriage. The diminishing of her social contacts and non–pain-related activities further increased her depression and deconditioning. In this example, personality factors in combination with very real physical pain resulted in an apparently irreversible behavioral cycle.

TREATMENT OUTCOME

Acute back pain affects approximately 80% of Americans at some time.[37] As many as 15% of all employees are injured while working.[3] In the United States, truck drivers, handlers, and nursing aid occupations have the highest incidence of injury. Approximately 90% of back pain naturally resolves by itself within 3 months of injury regardless of the treatment.[23,37] Although a number of treatment interventions have been recommended, long-term success of different interventions compared to no treatment is equivocal.[21]

Despite the excellent recovery experienced by most injured persons, some 10% continue to experience chronic, intractable back pain. In the last 15 years, more than 1000 pain clinics have been developed to treat these persons utilizing a variety of interventions including behavioral interventions, physical manipulations, TENS application, flexion-extension exercise programs, biofeedback, and psychotherapy. The extensive efforts of such multidisciplinary treatment approaches appear promising.[37]

Summarizing the immediate and long-term outcome of interdisciplinary pain management programs is difficult for several reasons. Improvement statistics vary depending on how each author defined successful treatment. Psychological test scores, subjective ratings, medication use, activity levels, and return to work have been used to define outcome.[4] Admission criteria are important in assessing outcome. Programs with more stringent admission criteria obtain better outcomes. Programs that treat more difficult patients have poor outcomes, although individual patients can make significant progress. Unavailability of appropriate control groups also make an adequate assessment of outcome difficult.

PREVENTION OF CHRONIC PAIN

Because effective interventions for chronic pain are limited, and given the high incidence of acute pain, prevention programs are clearly warranted.[33] Within the fields of public health and mental health, recognition of the importance of preventive models is growing.

Caplan introduced into psychiatric literature the concept of a threefold model of prevention with the community designated as the target population.[14] Applying Caplan's work and the prevention work in public health to the arena of medical care and chronic pain, prevention includes, in reverse order: tertiary prevention, secondary prevention, and primary prevention.

Tertiary Prevention

Tertiary prevention includes efforts to reduce the impact of chronic pain on the community, or reduce disability. Persons with a history of chronic pain or who are currently suffering from some form of chronic pain are seen as the target population. At the individual level, tertiary prevention is equivalent to rehabilitation efforts. The goals of rehabilitation efforts may directly impact on the pain or reduce the identified person's dysfunction. At the community level, tertiary prevention suggests the need to assess societal and cultural factors that hamper rehabilitation efforts. Prevention efforts to decrease disability would involve social policy change. In a thought-provoking article, Carron and DeGood compared disability and progress in an individual's rehabilitation efforts in New Zealand and the United States.[15] Despite equivalent pain levels between the groups of patients, 49% of the U.S. sample and only 17% of the New Zealand sample were unable to return to gainful employment and continued to receive pain-related financial compensation.

New Zealand's treatment and public response differed from that in the U. S. in many ways, including (1) the availability of worker's compensation for non–work-related injuries, (2) the absence of adversarial relationships among employer, insurer, and claimant through no-fault compensation, (3) a rapid rehabilitation intervention, and 4) substantial penalties for refusal of alternative employment. Indeed, in the United States no vocational alternatives could be readily available for poorly educated laborers who are unable to return to their jobs. In fact, for these persons, who composes the majority afflicted with chronic pain, the pursuit of social security disability benefits might appear the only possible alternative. Unfortunately, only 2% of open-ended Social Security Disability recipients were rehabilitated between 1967 and 1976. Given the substantial cost of medical care, litigation, and disability reimbursement, social policy changes to increase lump sum settlements, programs to financially encourage potential employers to hire persons with chronic pain, and a no-fault compensation system that reinforces return to work practices and decreases the need for litigation could likely serve as significant tertiary prevention methods by reducing the dysfunction associated with chronic pain.[15]

Secondary Prevention

Secondary prevention programs are designed to reduce the rate of a disorder by decreasing its prevalence. This requires reducing the number of actual cases in a given "high risk" population. With back pain, there are several possible ways of identifying high risk groups. For example, persons with acute pain are high risk for developing recurring acute pain "flare-ups" as well as for developing chronic pain. Other potential risk groups include firemen, truck drivers, nurses, and manual laborers whose jobs make them vulnerable. Use of specific psychological or physical functioning tests are often useful in identifying high risk for development of chronic pain and then implementing secondary prevention programs.

Primary Prevention

Primary prevention programs focus on lowering the incidence of new chronic pain cases in the general population. These programs reduce the incidence of new injuries by both increasing the sturdiness of the general American public and by reducing occupational hazards or other environmental risks. Introducing body mechanics courses, stress management programs, and conditioning and flexibility exercise programs into our elementary and secondary educational system to promote the hardiness of our youth are potentially effective primary prevention efforts.

Instead, psychosocial evaluations, physical examinations, and functional assessment techniques are designed to work toward comprehensive programs to increase functional activity and reduce the impact of an injury on lifestyle. They, therefore, are part of a multidisciplinary preventative rehabilitation program to reduce the incidence of chronic pain. Regardless of the intervention, approximately 90% of persons with acute back injury fully recover and return to work.[23] Nonetheless, the cost of health care and rehabilitation for the remaining 10% is substantial. In addition to appropriate medication management, programs such as a back school and stress management training, often withheld until 2 or 3 years after the injury, should be administered earlier. In this way, educational programs developed for the treatment of chronic pain could be administered early to prevent the development of a chronic pain syndrome.

Both persons employed in high risk for injury jobs as well as persons who have sustained an injury are targeted for secondary efforts to prevent the onset of chronic pain. Although all patients who sustain an injury are high risk for both a recurrence of the injury and development of chronic pain, a closer look at variables related to prognosis and recovery following the acute injury is warranted. In addition, recommendations for aggressive medical management and stress management are proposed.

VARIABLES LIKELY TO AFFECT PROGNOSIS

Factors likely affect the ability to return to work after an acute injury include physical conditioning, work environment, and psychosocial variables.

Physical Conditioning

Biering-Sorenson reported on a 1-year prospective study of 928 subjects who participated in a general health survey that included physical examination regarding low back pain.[8] One year after participating in the physical examination, 99% of the subjects completed a questionnaire reporting subsequent low back problems. Isometric endurance of back muscles was a successful predictor of prevention of a first-time injury. Those persons with a hypermobile back were most likely to develop back pain. Not surprisingly, the best predictors of recurrent or persistent low back pain included residual weakness in trunk muscles and reduced back and hamstring flexibility. Retrospectively, unequal leg length, height, and weight were positively associated with recurrent pain. The residual weakness as a predictor of recurrence of back injury is confounded by the severity of original injury and time elapsed since recovery from previous injury. These results do not necessarily indicate a premorbid predisposing factor for a first-time injury. Other studies suggest that the more physically fit firefighters had significantly fewer back injuries than the less fit individuals. These measures included flexibility, isometric lifting strength, and cardiovascular measures.[12]

Job Characteristics

In addition to individual physical fitness characteristics, one must consider the environmental factors most associated with back injury. As previously mentioned, truck drivers, handlers of heavy loads, and nurses or nursing aides constitute high risk occupations. Additionally, a number of job characteristics are associated with back injury, including heavy lifting, awkward or sustained postures, and exposure to whole body vibration. Body mechanics are often taught to reduce the incidence of injury. Unfortunately, back schools have not been particularly successful in reducing the incidence of injury. Many job and environmental demands reduce the likelihood of compliance with improved body mechanics. For example, a nursing unit that is "short staffed" commonly results in reluctance of a nurse or nursing aide to ask for assistance in a particular task. Nursing aides in such high stress jobs as a state psychiatric facility may be unable to obtain assistance with a highly volatile psychiatric patient. Jobs should be designed with better regard to the body mechanics and positions required to complete work tasks. Although definitive evidence for the effectiveness of back schools over and above a placebo effect has yet to be demonstrated,[24] to date back schools appear more effective if used immediately after an injury and less effective for chronic pain management.[34]

Psychosocial Factors

In addition to the physical functioning and the environmental and job factors, psychosocial variables, including family history, psychological distress, health attitudes, and beliefs, are important factors in predicting response to pain.

Pain is typically associated with a number of psychological and social changes. Many people with chronic pain have stopped working or lost their jobs. Loss of work generally begins a reverberating cycle of psychological and familial problems. Without work, self-esteem diminishes, leading to depression and, frequently, marital dysfunction. In turn, these changes can further depress the patient and exacerbate the chain of problems. As the problems deepen, the patient generally has difficulty in parenting responsibilities as well as marriage. Not the least of the problems is financial difficulty created or increased by the loss of work, and financial problems often lead to refusal or reduction of treatment options.

Pain, although a personal and subjective experience, affects the family, and in turn the family's response to the pain can affect the pain experience and pain behavior.[27,53] Pain is communicated to the family verbally and behaviorally (grimacing, bracing, etc.). The role of pain in the family is varied. Pain is incorporated in a dysfunctional family by providing a socially acceptable method to avoid communication and avoid attempts to resolve conflict. Pain behaviors are often inadvertently reinforced by family members who encourage the patients to rest and go to the doctor, and use the opportunity to care for the "sick" family member. Pain may serve as a family control mechanism. Decisions in these cases are made because of "pain." For example, imagine the patient who has never particularly gotten along with his or her in-laws. Although the family has planned for some time to attend a family get-together, J. has been active all day but has noticed that the back pain has been increasing. As J. thinks about a long evening in a stressful situation and thinks about the pain, J. decides that the pain is too severe and says to spouse, "I cannot go to your mother's for dinner. My back hurts too much. I suppose it will be okay if you want to go without me." At the same time the nonverbal behaviors suggest to the spouse that the pain is severe, putting the spouse in the difficult position of choosing between the parent and staying home with the "ill" spouse. Although the spouse is likely angry with this change of plans, it is difficult to vent the anger directly; after all, J. is "sick" and it is not his or her fault.

A family history of depression or chronic pain appears associated with the development of chronic pain. To the psychologist, the occurrence of chronic pain is less important than the response of the family member or family to the illness or depression.[17] Within this family context, the individual develops attitudes, beliefs, and somatic style.[5] For example, J. B., a 25-year-old married woman with severe headaches, despite reassurance following "normal CT scan findings," recalled her aunt's death following cerebral metastasis and continued to be apprehensive and fearful of the "meaning" of her own headaches. In our clinic, men with low back pain often had fathers who developed back pain secondary to work-related injuries. Despite this consistent finding, there appears a significant difference in attitude among those patients whose fathers were able to resume work responsibilities and those whose fathers were "disabled" because of the pain.

In our assessment at the University of Missouri pain management center program, we use the Symptom Checklist-90-Revised (SCL-90-R),[19] a symptom distress index to assess psychosocial distress. Although this instrument is a valid and reliable one for understanding current distress, elevated scores reflect psychopathology but do not provide assessment of the chronicity of psychological distress. During the interview, we were able to better determine whether the elevated SCL-90-R scores reflect a chronic state of poor coping or simply a response to the current stressors.[49]

Perhaps one of the most important psychological variables associated with chronic pain is the feeling of loss of control. Many people feel a loss of control after developing a painful condition. Often the total focus is on pain, which consumes all aspects of the patient's life. Traditional medical management often reinforces passivity and reliance on health care professionals. Early efforts to involve the patient in dealing with the pain by making active decisions about returning to work, choosing alternative employment, or vocational rehabilitation are critical. Increasing perceived control may be crucial in offsetting the development of depression.[2]

PSYCHOLOGICAL CONSIDERATIONS IN EARLY MANAGEMENT OF ACUTE PAIN CONDITIONS

Early medical management of acute pain conditions is the single best way to prevent the development of chronic pain problems. The prevention of chronic pain requires that the physician quickly intervene utilizing "aggressive, conservative medical management." That phrase may seem like an oxymoron, but this is precisely the type of care needed to avoid the development of chronic pain conditions. What is "aggressive, conservative medical management"?

Low back pain is most effectively managed through conservative treatment, typically consisting of rest, stretching, modalities, and anti-inflammatory drugs. These specific interventions, which have a proven history of efficacy, rightly form the first line of intervention for a person who has new, acute back injury. In addition, it is well recognized that physicians provide nonspecific interventions. Included in this category are the attention and caring the physician provides and directives to reduce stress. We typically think of a placebo as a nonspecific treatment effect, but the original Latin term means "pleasing healer." It is the role of a pleasing healer that the treating physician must undertake to provide conservative, aggressive medical management. In this role, acute pain is usually associated with anxiety. The anxiety is manifested or filtered through well-developed coping traits, but is almost always the predominating mood in the newly injured person. The physician must assess the injured person's perceptions of the injury and beliefs regarding the pain. The patient's anxiety engenders dependence upon the physician, which can be used to facilitate effective conservative management.

During the first visit, the physician must, with the same care given to the physical examination, assess the patient's expectations and beliefs regarding the pain. Once these are understood, the physician can generate a medical regimen that treats both the physical and psychological aspects of the acute pain problem. For example, S. J., a highly anxious patient, believed he would be permanently disabled by a severe muscle strain sustained 2 weeks prior to evaluation. Physical examination revealed an absence of neurologic signs and severe paraspinal tenderness. This patient will benefit from discussing the time course for his recovery and the need for extensive rest and restriction of his activities. He also is aided by a prescription for a narcotic analgesic that provides extensive pain relief and diminution of anxiety. Although narcotic analgesics are generally avoided, if given with appropriate time constraints (e.g., this medicine will provide the relief you need for the next 14 days, and after that we will use other medicines), these medications can be extremely effective in acute pain conditions.

Early management of acute pain requires an understanding of the conditioning effects associated with pain relief. If a person complains of pain and is then prescribed medications that provide relief, pain complaints become associated with relief. Because physicians tend to respond to pain complaints after the patient complains several times, the patient learns relief will come only after many complaints. Consequently, pain complaints increase. This pattern, which is called a "variable ratio learning schedule," is particularly problematic in the treatment of pain conditions. Once established, a variable ratio schedule is highly resistant to change. This type of behavior chain can become the underpinning of chronic pain syndrome.

Early medical and psychological management of acute pain conditions is the single best way to prevent the development of chronic pain problems, using education and narcotic analgesics on a time-limited, around the clock basis for appropriate patients. This medication regimen prevents the development of the conditioned complaint response and when accompanied by a thorough assessment of attitudes and beliefs and reassurance, builds an effective rehabilitation team alliance with the patient.

PSYCHOLOGICAL INTERVENTION

In addition to aggressive medical management of acute pain, psychological interventions can prevent generation and exacerbation of chronic pain.

Self-regulation strategies are based on a biopsychosocial view of pain. In this model, pain is viewed as a multifaceted experience involving a physical predisposition (weak back musculature), psychophysiologic responses involving the neck-back musculature, recent stressors, and ineffective coping with these stressors. Perhaps the diathesis-stress model best describes this approach to pain. Following an injury or as a result of poor conditioning, specific musculoskeletal problems develop. As one attempts to cope with perceived life stressors, an individual response stereotypy develops involving hyperactive back musculature. Repeated or sustained muscular hypertension of the involved area leads to ischemia, oxygen depletion at affected sites, and the release of pain-eliciting substances (such as bradykinin). Stress-related sympathetic arousal also often leads to ischemia, inducing muscle spasms and pain. Pain itself becomes a stressor leading to immobility, greater tension, and pain. A pain-tension-pain cycle develops in the person who is coping ineffectively with perceived stressors.[25,52] As predicted by the diathesis-stress model, chronic back pain patients respond with elevated paraspinal EMGs and longer recovery to personally relevant stressors as compared with a general pain sample and normal controls.[26]

There is a wide range of self-regulation techniques designed to alter this cycle by improving coping with stressors, reducing perceived stressors, and reducing hyperactivity of back musculature. Most of these techniques include some form of relaxation training.

Relaxation techniques are exercises designed to reduce muscle tension, decrease autonomic arousal, and reduce cognitive concerns, fears, or worries. Specific techniques include progressive muscle relaxation, autogenics, and imagery. Although relaxation training has been described as the aspirin of behavioral medicine, the analogy is less than perfect.[47] Like aspirin, relaxation techniques may provide some relief for a number of problems including headaches, back pain, and hypertension. Like aspirin, relaxation techniques are not a cure and, in fact, could be only one part of a much larger intervention package. Unlike an aspirin, however, relaxation training requires active participation for effective management of symptoms.

Progressive muscle relaxation (PMR) refers to a relaxation technique that focuses on training low muscle tension with the assumption that low muscle tension and anxiety are incompatible responses. Jacobson, who founded the technique in the early 1900s and published the first comprehensive book on deep muscle relaxation in 1938, outlined 15 major muscle groups involved in muscle tension and relaxation.[32] A person individually must tense one specific muscle group and then relax that same group, with the goal of reducing the tension beyond the initial tension level. Jacobson's daily program of 1–9 hours with 56 sessions of systematic training was revised, modified, and condensed by Wolpe to six lessons of 20 minutes each with twice a day 15-minute practice sessions between lessons, making training more practical.

PMR involves five basic steps. First, the subject passively focuses on the involved muscle group. At a signal from the trainer or tape, this muscle is tensed and the tension is maintained for 5–7 seconds. At the next signal from the trainer, the tension is quickly released. The subject is then instructed to continue passively observing the sensations associated with relaxation so that an awareness of the difference between a tensed muscle and a relaxed muscle can be identified. PMR is most useful for persons who can identify high levels of muscle tension before training.[32]

Autogenic training, developed by Schultz and Luthe,[48] consists of six standard exercises that combine relaxation instruction and suggestion. Autogenic training relies on repetitive phrases to reduce muscle tension and autonomic arousal. For example, to assist with muscle relaxation, one repeats the phrase "My arm is heavy" over and over again. To produce peripheral dilation this phrase is expanded to include "My arm is heavy and warm." Systematic autogenic training includes six exercises: (1) focus on feelings of heaviness, (2) focus on sensations of warmth in the limbs, (3) notice reduction in heart rate and blood pressure, (4) passive concentration via phrases suggesting that relaxation be allowed rather than forced, and (5) focus on feelings of coolness in the forehead.

Benson,[7] after a review of meditation practices and relaxation training techniques including autogenics and PMR, identified four basic principles that lead to a generalized relaxation response. These four include a quiet environment, an object to dwell upon, a passive attitude, and a comfortable position. Based on this, he recommends the following procedures to practice the relaxation response.

1. Sit quietly in a comfortable position.
2. Close your eyes.
3. Deeply relax all your muscles, beginning at your feet and progressing up to your face. Keep them relaxed.
4. Breathe through your nose. Become aware of your breathing. As you breathe out, say the word "**one**" silently to yourself. For example, breathe **in . . . out, "one"**; **in . . . out, "one"**; etc. Breathe easily and naturally.
5. Continue for 10 to 20 minutes. You may open your eyes to check the time, but do not use an alarm. When you finish, sit quietly for several minutes, at first with your eyes closed and later with your eyes opened. Do not stand up for a few minutes.
6. Do not worry about whether you are successful in achieving a deep level of relaxation. Maintain a passive attitude and permit relaxation to occur at its own pace. When distracting thoughts occur, try to ignore them by not dwelling upon them and return to repeating "one." With practice, the response should come with little effort. Practice the technique once or twice daily, but not within 2 hours after any meal, since the digestive processes seem to interfere with the elicitation of the relaxation response.[7 (pp. 162–163)]

Biofeedback refers to the use of electronic equipment to provide feedback of psychophysiologic functioning including muscle tension, heart rate, peripheral blood flow, and electrodermal responsivity for the purpose of learning to modify physiologic functioning of which we have less voluntary control. Developed in the late 1960s and early 1970s, biofeedback did not live up to the initial claims and continues to be surrounded by controversy.[29,38,43,45,46,50,58] A complete review of biofeedback techniques for the treatment of pain is beyond the scope of this chapter. However, relaxation-assisted biofeedback training appears a promising intervention for patients suffering from headaches,[9] rheumatoid arthritis,[1,18] and back pain.[25] Pain patients who are often reluctant to follow through on mental health center referrals are more receptive to a stress management program with biofeedback. Learning to reduce muscle tension and anxiety associated with the pain by using devices that actually measure muscle tension is itself a valid intervention to the person suffering from chronic pain.

There are three major rationales for the use of biofeedback with back pain.[6] First, biofeedback assists in relaxation training. This rationale assumes that as general arousal is reduced, central nervous system processing of peripheral sensory inputs is also reduced. If anxiety reduces pain tolerance, then reduction of anxiety or a decrease in arousal should result in decreased pain. This model of biofeedback implies that direct, site-specific training is not required for effective biofeedback training.

A second rationale for biofeedback training is based on the pain-spasm-pain model.[10] Training a person to control specific peripheral areas (spasming paraspinal musculature) should directly alter the hyperactivity and reduce pain by altering peripheral factors. Even the training of focal areas is accompanied by relaxation training suggestions.[6]

Alternatively, biofeedback represents a truly psychosomatic intervention. Patients learn how their bodies respond to stress and how to change this physical response. People who previously reported no control over the pain learn to prevent muscle spasms by lowering muscle tension levels and coping better. Biofeedback is therefore both a physical and a psychological intervention.[6]

COGNITIVE RESTRUCTURING TECHINQUES

A growing body of literature emphasizes the role of cognitive factors in pain and pain management. These conditions include thoughts, self-statements, or evaluations when in pain; beliefs and attributions about the meaning of pain; and cognitive appraisals of the impact of pain on a person's life.[55] In a 1981 study, Lefebvre[35] demonstrated that chronic pain patients, like depressed patients, utilize faulty reasoning to misconstrue themselves, their world, and the future in negative terms. Cognitive distortions for patients with pain were restricted to the meaning of pain and the consequences of pain on their lives.

Cognitive-behavioral intervention techniques require a thorough assessment of maladaptive beliefs so that such beliefs can be disputed. Alternative thoughts and beliefs are then shaped. For example, S. R. believed that without surgery, she was too sick to return to school. In an effort to "stabilize" her back, physically bracing actually created impaired posture and hyperactive musculature. She was taught via biofeedback, relaxation training, and body mechanics to reduce paraspinal EMG activity. Her activities were actually severely limited because of deconditioning and low sitting tolerance. A sitting protocol was developed with realistic goals and incremental increases until she was able to tolerate 2 hours of continuous sitting. She then returned to school.

Assessment of attitudes and beliefs should be a part of each assessment with patients. During the acute phase, a clear understanding of the patient's attribution of pain and the meaning of the pain is needed so that appropriate education can occur, as well as early attempts to facilitate compliance with the treatment program. Psychotherapy with a cognitive-behavioral approach requires a skilled, trained professional in behavioral change techniques. Cognitive-behavioral treatment interventions have been demonstrated as useful with headache pain,[30] arthritis,[44] and low back pain.[56] The 1983[54] text by Turk, et al., *Pain and Behavior Medicine: A Cognitive-Behavioral Perspective*, provides a detailed explanation for this approach to pain management.

Because pain typically affects the whole family, interventions should include the family. By role-playing difficult situations, family members can learn to redirect their concern by encouraging and reinforcing activity instead of pain behaviors. Seeing the acute pain patient and spouse together to discuss the diagnosis and treatment can prevent cognitive distortions and reframe the pain before family dysfunction develops.

One complication of bed rest, a commonly prescribed treatment for acute episodes of back pain, is work absenteeism and reduction of other activities. Despite limited evidence of the efficacy of bed rest interventions, 1 to 2 weeks are commonly prescribed. In a well-controlled study, Deyo, Diehl, and Rosenthal[22] randomly assigned 200 patients presenting with mechanical, acute back pain at a primary care center to one of two treatment conditions: 2 days bed rest or seven days bed rest. Compliance with these instructions was variable. The modal days of bed rest was 2.3 days for group 1 and 3.9 days for group 2. Outcome data including functional outcome, physical findings, patient and physician ratings, and the use of medical services did not differ for the two groups; only absenteeism from work varied with group assignment. For mechanical back pain (no evidence of neurologic symptoms), 2 days of bed rest is optimal. Limited bed rest reduces the likelihood of the development of deconditioning often associated with chronic pain.

SUMMARY

Chronic intractable pain is difficult to treat and is best managed by prevention. Efforts should be aimed at preventing dysfunction associated with chronic pain and reducing the incidence of chronic pain through early intervention with high risk groups and through primary prevention efforts.

REFERENCES

1. Achterberg J, McGraw P, Lawlis GF: Rheumatoid arthritis: A study of relaxation of patients and temperature biofeedback as adjunctive therapy. *Biofeedback Self Regul* 1981; 6:207.
2. Addison RG: Chronic pain syndrome. *Am J Med* 1984; 77:54.
3. Andersson GBJ: Epidemiologic aspects on low back pain in industry. *Spine* 1981; 6:53.
4. Aronoff GM, Evans WO, Enders PL: A review of follow-up studies of multidisciplinary pain units. *Pain* 1983; 16:1.
5. Barsky AJ, Klerman GL: Overview: Hypochondriasis, bodily complaints, and somatic styles. *Am J Psychiat* 1983; 140:273.
6. Belar CD, Kibrick SA: Biofeedback in the treatment of chronic back pain. In Holzman AD, Turk DC (editors): *Pain Management: A handbook of psychological treatment approaches*. New York, Pergamon Press, 1986, pp 131–140.
7. Benson H: *Relaxation response*. New York, Avon, 1975, pp 158–163.
8. Biering-Sorensen F: Physical measurements as risk indicators of low-back trouble

over a one-year period. *Spine* 1984; 9:106.

9. Blanchard EB, Andrasik F: Psychological assessment and treatment of headache: Recent developments and emerging issues. *J of Consult Clin Psychol* 1982; 50:859.

10. Bonica JJ: Management of myofascial pain syndromes in general practice. *J Am Med Assoc* 1957; 164:732.

11. Bonica JJ: Pain research and therapy: Past and current status and future needs. In Ng LKY, Bonica JJ (editors): *Pain, discomfort and humanitarian care*. New York, Elsevier/New Holland, 1980, pp 1–46.

12. Cady LD, Bischoff DP, O'Connell ER, Thomas PC, Allan JH: Strength and fitness and subsequent back injuries in firefighters. *J Occup Med* 1979; 21:269.

13. Cannon WB: The James-Lange theory of emotions: A critical examination and an alternative. *Am J Psych* 1927; 39:106.

14. Caplan G: *Principles of preventive psychiatry*. New York, Basic Books, 1964.

15. Carron H, DeGood DE, Tait R: A comparison of low back pain patients in the United States and New Zealand: Psychosocial and economic factors affecting severity of disability. *Pain* 1985; 21:77.

16. Chapman RC: New directions in the understanding and management of pain. *Soc Sci Med* 1984; 19:1261.

17. DeGood DE: Reducing medical patients' reluctance to participate in psychological therapies: The initial session. *Prof Psychol Res Pract* 1983; 14:570.

18. Denver DR, Laveault D, Girard F, et al: Behavioral medicine: Behavioral effects of short-term thermal biofeedback and relaxation in rheumatoid arthritis patients. *Biofeedback Self Regul* 1979; 4:245.

19. DeRogatis LR: *SCL-90-R Manual II: Administration, scoring and procedures*. Baltimore, Clinical Psychometric Research, 1983.

20. Descartes R: L'Homme (1664, trans. M. Foster). In *Lectures on the History of Physiology During the 16th, 17th, and 18th Centuries*. Cambridge, Cambridge University Press, 1901.

21. Deyo RA: Conservative therapy for low back pain. *JAMA* 1983; 250:1057.

22. Deyo RA, Diehl AK, Rosenthal M: How many days of bed rest for acute low back pain? A randomized clinical trial. *New Engl J Med* 1986; 315:1064.

23. Dixon AStJ: Progress and problems in back pain research. *Rheumatol Rehabil* 1973; 12:165.

24. Fisk JR, DiMonte P, Courington SM: Back schools: Past, present and future. *Clin Orthop Rel Res* 1983; 179:18.

25. Flor H, Haag G, Turk DC, Koehler G: Efficacy of EMG biofeedback, pseudotherapy, and conventional medical treatments for chronic rheumatic back pain. *Pain* 1983; 17:21.

26. Flor H, Turk DC, Birbaumer N: Assessment of stress-related psychophysiological reactions in chronic back pain patients. *JCCP* 1985; 53:354.

27. Flor H, Turk DC, Rudy TE: Pain and families II. Assessment and treatment. *Pain* 1987; 30:29.

28. France RD, Houpt JL: Chronic pain: Update from Duke Medical Center. *Gen Hosp Psychiatry* 1984; 6:37.

29. Green JA, Shellenberger RD: Biofeedback research and the ghost in the box: A reply to Roberts. *Am Psychol* 1986; 41:1003.

30. Holroyd KA, Andrasik F, Westbrook T: Cognitive control of tension headache. *Cogn Ther Res* 1977; 1:121.

31. International Association for the Study of Pain Subcommittee on Taxonomy. *Pain* 1986; Suppl.

32. Jacobson E: *Progressive relaxation*. Chicago, University of Chicago Press, 1938.

33. Kelsey JL, Hochberg MC. Epidemiology and prevention of musculoskeletal disorders. *Monogr Epidemiol Biostat* 1982; vol 3.

34. Lankhurst GJ, Van de Stadt RJ, Vogelaar TW, Van de Karst JK, Prevo AJH: The effect of the Swedish back school of chronic idiopathic low back pain. *Scand J Rehabil Med* 1983; 15:141.

35. Lefebvre MF: Cognitive distortion and cognitive errors in depressed psychiatric and low back pain patients. *JCCP* 1981; 49:517.

36. Livingston WK: *Pain mechanisms: A Physiologic Interpretation of Causalgia and its*

Related Stress. New York, Macmillan Co., 1943.

37. Mayer TG, Gatchel RJ, Kishino N, et al: A prospective short-term study of chronic low back pain patients utilizing novel objective functional measurement. *Pain* 1986; 25:53.

38. McGovern H: Comment on Roberts' criticism of biofeedback. *Am Psychologist* 1986; 41:1007.

39. Melzack R: *The puzzle of pain*. New York, Basic Books, Inc., 1973.

40. Melzack R, Wall PD: Pain mechanisms: A new theory. *Science* 1965; 150:971.

41. Melzack R, Wall PD: *The challenge of pain*. New York, Basic Books, Inc., 1982.

42. Merskey H, IASP Subcommittee on Taxonomy: Pain terms, a list with definitions and notes on usage. *Pain* 1979; 6:249.

43. Norris PA: On the status of biofeedback and clinical practice. *Am Psychologist* 1986; 41:1009.

44. Randich SR: Evaluation of a pain management program for rheumatoid arthritis patients. *Arth Rheu* 1982; 25:11.

45. Roberts AH: Biofeedback: Research, training, and clinical roles. *Am Psychologist* 1985; 40:938.

46. Roberts AH: Biofeedback, science, and training. *Am Psychologist* 1986; 41:1010.

47. Russo DC, Bird PO, Masek BJ: Assessment issues in behavioral medicine. *Behavior Assess* 1980; 2:1.

48. Schultz JH, Luthe W: *Autogenic therapy*, Vol 106. New York, Grune & Stratton, 1969.

49. Schwartz DP, DeGood DE: An approach to the psychosocial assessment of the chronic pain patient. *Curr Con Pain* 1983; 1:3.

50. Smith JC: Meditation, biofeedback, and the relaxation controversy: A cognitive-behavioral perspective. *Am Psychologist* 1986; 41:1007.

51. Sternbach RA: Psychological aspects of chronic pain. *Clin Orthop Rel Res* 1977; 129:150.

52. Turk D, Flor H: Etiological theories and treatments for chronic back pain. II. Psychological models and interventions. *Pain* 1984; 19:209.

53. Turk DC, Flor H, Rudy TE: Pain and families. I. Etiology, maintenance, and psychosocial impact. *Pain* 1987; 30:3.

54. Turk DC, Meichenbaum D, Genest M: *Pain and behavioral medicine. A cognitive-behavioral perspective*. New York, Guilford Press, 1983.

55. Turk DC, Rudy TE: Assessment of cognitive factors in chronic pain: A worthwhile surprise? *JCCP* 1986; 54:760.

56. Turner JA: Comparison of group progressive relaxation and cognitive-behavioral group therapy for chronic low back pain. *JCCP* 1982; 50:757.

57. White JC, Sweet WH: *Pain and the neurosurgeon*. Springfield, Ill., Charles C. Thomas, 1969.

58. White L, Tursky B: commentary on Roberts. *Am Psychologist* 1986; 41:1005.

12

Electrodiagnosis in Musculoskeletal Medicine

Richard T. Katz and
Christina A. Marciniak

This chapter is a guide to the electrodiagnosis of neuromuscular problems in patients likely to be seen in a physical medicine clinic for musculoskeletal disease. This discussion is not a comprehensive review of electrodiagnosis but contains enough background to give the clinician a thorough understanding of the neuromuscular problems involved and the diagnostic information that clinical neurophysiology can offer. Electrodiagnostic studies are discussed in three contexts: (1) root entrapment and radiculopathy, (2) focal neuropathies and entrapment syndromes, and (3) problems common to patients with inflammatory arthritis, for which rheumatoid arthritis will serve as the model.

Several excellent reviews and texts[7,52,70,77,155] discuss basic aspects of electrodiagnosis, so only a brief overview is presented here. The three components of electrodiagnosis useful in evaluation of the peripheral nervous system and spinal cord include electromyography, electroneurography (nerve conduction studies), and somatosensory evoked potentials.

ELECTROMYOGRAPHY[8,91]

The electromyographic (EMG) examination involves the introduction of a special recording needle into a muscle belly. Two types of needle are used by nearly all electromyographers: monopolar, composed of a steel wire insulated with Teflon, and concentric, a hollow needle through which runs a fine wire. Monopolar needles may be preferred because they cause less discomfort to the patient, but they require a surface reference electrode. Concentric needles have a more uniform exposed surface and so may be more useful for quantification and comparison of motor unit potentials.

Electrical potentials located within a few millimeters of the needle radius are picked up by the electrode then transmitted from the muscle to a preamplifier placed near the patient. These signals are further amplified and filtered and displayed visually for examination by the electromyographer. The potentials are also transmitted over a loudspeaker, since auditory monitoring is a valuable adjunct to the visual inspection of the waveforms. The EMG examination consists of the analysis of waveforms of various muscles during relaxation, mild contraction, and vigorous contraction. The choice of muscles for study is dictated by the clinical problem and the history and physical examination.

Normally, when a needle is inserted into resting muscle, a brief spurt of electrical potentials (insertional activity) is produced, followed by silence. The electromyographer then moves the needle in several directions in search of spontaneous activity, that is, electrical potentials that occur while the muscle is at rest. Although spontaneous activity may normally occur when the needle is in the vicinity of the neuromuscular junction, in other locations spontaneous activity is generally a pathologic finding. Three particular types of spontaneous activity are of the greatest relevance to this discussion: positive sharp waves, fibrillation potentials, and fasciculations.

Positive sharp waves and fibrillations are small discharges that are the result of spontaneous depolarization of single muscle fiber membranes. Most often these potentials signify denervation. They are not synonymous with axonal loss, however, since they may be seen in a variety of disorders including certain myopathies, periodic paralysis, botulism, and severe upper motor neuron injury. Fasciculations are the involuntary twitching of a bundle of muscle fibers, which are often obvious to the observer on visual inspection. Fasciculation potentials are noted as irregular firing of a motor unit on the EMG screen. They may or may not be a pathologic finding, depending on the clinical and EMG scenario.

When the patient performs a mild muscle contraction, the electromyographyer is able to analyze the motor unit potentials from various territories within the muscle. The motor unit—the anterior horn cell and nerve axons, and the muscle fibers they innervate—is the "final pathway" of motoric activity within the nervous system. Normally, motor units fire in characteristic recruitment patterns. Small motor neurons supplying fatigue-resistant (slow) muscle fibers are recruited first, and larger neurons supplying fatiguable high-tension (fast) muscle fibers are recruited later. The electromyographer can thereby study not only the characteristics (shape, amplitude, duration) of the first few recruited motor units but also their firing rate and pattern. Finally, when the patient performs a vigorous contraction of the muscle, the screen normally fills with motor unit potentials, the interference pattern.

When studying the motor unit potentials during the EMG examination, several important characteristics are evaluated in distinguishing whether a neuropathic or myopathic process is present. Normally motor units range in size from 300–5000 μV and vary greatly (1–16 m/sec) in duration. Motor units generally have two to four phases, while polyphasic motor unit potentials (greater than four phases) are noted no more than 15% of the time.

When a lower motor neuron insult has occurred, the inspection of the motor units may show waveforms of increased amplitude, phase, and duration. These changes are secondary to reinnervation of denervated muscle fibers by surviving neighboring axons. Also, as motor units are lost to the pathologic process, those remaining need to compensate by firing faster than normal. On full recruitment, a drop-out of motor units is seen on the oscilloscope interference pattern while low-pitched motor units are heard over the loudspeaker.

In a myopathic process, motor units tend to be smaller and of shorter duration, as loss of muscle fibers results in a smaller electrical potential. These motor units may also be polyphasic. As weakened muscles produce less desired force, an early recruitment of additional motor units is noted on mild muscular contraction. The interference pattern may be full until late in the disease, and characteristically high-pitched motor units are noted over the loudspeaker.[15]

NERVE CONDUCTION STUDIES[32,94]

EMG can help assess the status of nerve fibers indirectly, but the integrity of large myelinated sensory and motor neurons can be evaluated directly by nerve conduction studies, also known as electroneurography. Nerve conduction studies involve the introduction of an electrical stimulus, by surface electrode or needle, and recording an evoked response, again using either a surface disc electrode or a needle electrode inserted through the skin. Nerve conduction studies can assess motor neurons, sensory neurons, or mixed nerve trunks depending on the strategy employed. Motor conduction studies are performed by applying an electrical shock via the stimulating electrode to a mixed nerve trunk, for example, the median nerve at the wrist (Fig. 12-1). The nerve is usually stimulated at a fixed distance, often 8 or 10 cm proximal to the recording electrodes. Recording electrodes, usually of the disc type, are taped to the skin over a muscle belly innervated by that mixed nerve, in this example the thenar eminence. Upon increasing the stimulus intensity just beyond the point at which the evoked muscle response (compound muscle action potential, CMAP) is maximal, several useful parameters can be measured. The distal latency is a measure of the time from the onset of the stimulus until it reaches the muscle belly. Similarly, the amplitude of the CMAP is recorded and sometimes its duration and shape. The nerve is then stimulated at a more proximal site and identical measurements are made. Using the value obtained by stimulating the nerve at two separate sites then measuring the distance between stimulating sites and subtracting the difference in latencies allows the calculation of the motor conduction velocity (see Fig. 12-1). Although each electrodiagnostic laboratory may develop its own normal values for nerve conduction velocity and amplitude and duration of the CMAP, standardized values are readily obtained in the medical literature. An easy rule of thumb is that upper and lower extremity motor fibers generally conduct at or greater than 50 and 40 m/sec, respectively. Because of their anatomic availability, the most commonly performed motor conduction studies are of the median, ulnar, radial, tibial, and peroneal nerves. Studies of the facial nerve are occasionally useful when cranial nerve function is to be assessed.

Sensory nerve fibers can be studied orthodromically or antidromically. Orthodromic and antidromic conduction mean that the electrical stimulus introduced by the examiner travels in the same direction (ortho-) as the naturally-occurring conduction of that nerve fiber or in the opposite direction (anti-). Stimulating purely sensory fibers and recording proximally over the mixed nerve trunk is orthodromic conduction; for example, placing

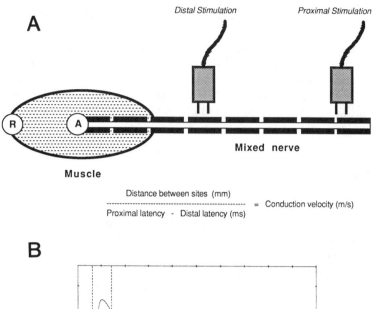

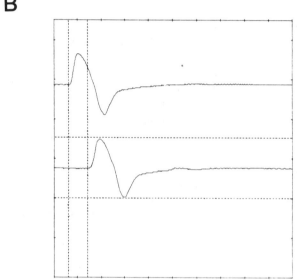

Figure 12-1. Motor conduction study. *(A)* The mixed nerve is stimulated at a proximal and distal site. The distal site is often stimulated 8 cm from the active (A) electrode. The reference electrode (R) is placed more distally over the muscle tendon. *(B)* The top tracing was obtained upon stimulation of the median nerve at the wrist, and recording over the thenar eminence. Latency between stimulation and the onset of the compound muscle action potential is 3.12 m/sec. The bottom tracing was obtained upon stimulation of the median nerve at the elbow and recording over the thenar eminence. Latency is 7.07 m/sec. Both responses are approximately 38,000 mV. The motor nerve conduction velocity from the proximal to distal site can be obtained according to the formula in this figure. The distance between sites is 245 mm, and the difference in the latencies is 3.95, resulting in a velocity of 62 m/sec.

ring electrodes around the index finger to stimulate the median digital nerves and recording with surface electrodes proximally. Antidromic conduction can be achieved by stimulating the mixed nerve at the wrist and recording from purely sensory fibers distally; using the median nerve as an example again, by stimulating at the wrist and recording from the index finger (Fig. 12-2). Fourteen centimeters is a commonly-used distance for distal sensory latencies, but 10 and 12 cm distances are often used as well. Sensory nerve conduction velocity can be studied in a manner analogous to motor conduction velocity. Sensory fibers can be directly stimulated (e.g., at the index finger), and the nerve evoked response can be measured at the wrist and elbow. The distance between recording sites divided by the difference between the latencies is the sensory conduction velocity. Nerves routinely used for sensory conduction studies are the median, ulnar, radial, sural, and peroneal.

Late responses[17] are the third type of nerve conduction study frequently used. Late responses include F waves and the H reflex. The F wave is a small potential occurring after the CMAP during most motor conduction studies. Upon stimulating the mixed nerve trunk, the electrical volley not only travels distally toward the muscle belly but also antidromically toward the spinal cord. The F wave is believed to originate from electrical volleys that "bounce back" from the anterior horn cell and then travel back down the nerve, orthodromically, to the muscle belly. As the F wave has a substantially longer path than the orthodromically travelling motor latency, it occurs later, hence its descriptor as a "late" response.

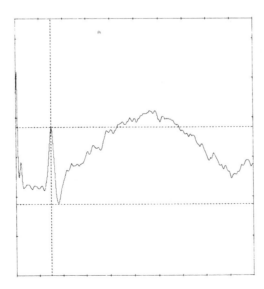

Figure 12-2. Sensory nerve action potential. The median nerve was stimulated 14 cm proximal to the metacarpal-phalangeal joint of the index finger. The latency to the peak of the sensory nerve action potential is 3.02 m/sec with an amplitude of 40 mV.

The H reflex has a similar latency to the F wave but traverses different pathways. Upon stimulating the tibial nerve at the popliteal fossa (since this is the only site where the H reflex is readily elicited in normal subjects), the electrical volley travels orthodromically along A1 sensory afferent fibers, makes a primarily monosynaptic connection within the spinal cord, and continues orthodromically down motor fibers toward the muscle belly. Thus, the H reflex is similar to a muscle stretch reflex, except that it bypasses the mechanically induced stretch of the muscle belly. The H reflex can be reliably distinguished from the F wave by its shape and the stimulus intensity used. The H reflex is obtained with a submaximal stimulus (insufficient to generate a CMAP of maximal amplitude), while the F wave is elicited best by one that is supramaximal.

The F wave can be an extremely valuable adjunct to more distal nerve conduction studies, since it can be used to evaluate the more proximal portions of the nerves. These sites are not as easily studied by usual nerve conductions because they require quite uncomfortable stimulus intensities to reach the nerves positioned more deeply within the limb. In patients with early stages of peripheral neuropathy, F waves may show abnormalities when conventional motor and sensory studies are unremarkable.[84] H reflexes are routinely used primarily for the evaluation of S1 radiculopathy.

NEUROMUSCULAR JUNCTION STUDIES[64]

Two techniques are especially useful for assessing disorders of the neuromuscular junction: repetitive stimulation and single fiber EMG. Repetitive stimulation studies are begun in the manner of a nerve conduction study (described above) except that the nerve is stimulated not once but many times. The examiner stimulates the mixed nerve and records the CMAP from a muscle innervated by that nerve. Before repetitive stimulation is attempted, the stimulation intensity is increased until a maximum CMAP is achieved. At that point the nerve is stimulated at a rate of two to three times per second until a total of ten CMAPs are recorded. These potentials are then analyzed to determine if a decrement in the amplitude of the CMAP has occurred. A loss in amplitude of greater than 8%–10% is suggestive of a postsynaptic neuromuscular junction disorder, such as myasthenia gravis. Presynaptic junction disorders may be diagnosed by stimulating the nerve at very high frequencies, for example 40 Hz, and noting a dramatic increase in the size of the CMAP. Such an incremental response is seen in Eaton-Lambert syndrome.

Single fiber EMG is an even more sensitive (but less specific) method of diagnosing disorders of the neuromuscular junction. It involves the use of a special needle to pick up electrical potentials from a very small area of muscle tissue—generally only one or two muscle fibers. Using this method one is able to assess single or paired potentials generated from one or two muscle fibers rather than the three to five muscle fibers recorded using a conventional EMG needle. When two single fiber potentials are recorded at the same site, the first potential is used to activate a type of trigger, which activates the sweep on the oscilloscope screen. Measuring the time from the activation of the recording until the beginning of the second potential gives the interpotential interval. Two single muscle fibers of the same motor unit do not necessarily fire at the same time but with a small disparity. The interpotential interval between two single fiber potentials is normally quite stable, varying only slightly, on the order of 35 μsec. The variability of the interpotential interval, known as jitter, is due to fluctuating conduction time through terminal axons, the myoneural junction, and muscle fiber membranes. Jitter is characteristically increased in neuromuscular junction disorders as well as some neuropathies and myopathies. The discovery of abnormally increased jitter is the most sensitive method of electrodiagnostically demonstrating a neuromuscular junction disorder.[81,135]

SOMATOSENSORY EVOKED POTENTIALS[18,40]

Somatosensory evoked potentials (SSEP) are occasionally useful as an adjunct to EMG and nerve conduction studies in the diagnosis of neuromuscular difficulties in patients with musculoskeletal disease. SSEP are most commonly obtained by stimulating a peripheral mixed nerve at a frequency of approximately 2–5 Hz. The afferent volleys travel primarily via A1 afferents to the spinal cord and ascend the dorsal columns to synapse in the nucleus cuneatus and nucleus gracilis. The volley crosses to the contralateral side as it is traverses the medial lemniscus to the thalamus, then finally ascends to the somatosensory cortex.

With stimulation using a surface cup or subcutaneous electroencephalographic needle electrodes, a tiny evoked response is recorded over the parietal cortex contralateral to the site of stimulation. Evoked responses can also be measured from the brachial plexus and cervicomedullary junction upon upper extremity nerve trunk stimulation as well as the lumbosacral plexus upon lower extremity stimulation. These tiny potentials can be examined only with the use of an electronic averaging system, in which the time-locked evoked response emerges from background electrical noise. SSEP are most often elicited upon stimulating the median, ulnar, tibial, and peroneal nerves.

RADICULOPATHY

Electrodiagnosis is an extremely valuable technique in assessing lesions of the spinal root and serves as an integral part of the diagnostic work up. Whereas myelography and computerized tomography (CT) each have their advocates as the primary modality for assessing the anatomy of root impingement,[75] electrodiagnosis addresses the neurophysiologic status of the nerve fibers. This is particularly important in light of the large number of abnormal radiologic findings in asymptomatic individuals.[161] EMG has been shown to correlate better with clinical findings than do plain radiographs in the evaluation of cervical radiculopathy.[61] Electrodiagnostic studies are of similar use in the evaluation of lumbosacral radiculopathy and serve as a useful adjunct to myelography.[148] EMG correlates better than CT with the demonstrated course of radiculopathy.[74] CT data should be correlated with EMG findings when considering patient management.[136]

Radiculopathy is generally due to degenerative disease of the disc, with the C7 root most commonly involved in the cervical region and the L5 and S1 roots most commonly involved in the lumbosacral region. The specific root involved in any given case is determined by correlation of clinical symptoms and neurologic examination with the EMG needle examination. The specific syndromes associated with each root level may present as weakness in muscles innervated by that root (Tables 12-1, 12-2), sensory symptoms in characteristic distributions (Table 12-3), and loss of diminuation of muscle stretch reflexes (Table 12-4). The muscles frequently examined during EMG and their corresponding root levels are included in Tables 12-1 and 12-2.

EMG is the principal electrophysiologic modality used in the assessment of root impingement, although rarely late responses and SSEP techniques may provide additional information. The most dramatic evidence of radiculopathy on EMG examination is the presence of positive sharp waves and fibrillation potentials, indicating that denervation has occurred. These changes may occur as early as 1–2 weeks following nerve injury, depending on the proximity of the muscles to the site of compromise. Paraspinal muscles develop changes before proximal limb muscles, while distal muscles may take several weeks or more to develop electrical evidence of denervation. Occasionally other types of spontaneous activity, such as fasciculations and complex repetitive discharges, may also be found in a radiculopathy.

Since electrophysiologic signs of denervation may not develop for several weeks, more subtle abnormalities must be sought shortly after the acute onset of radiculopathy. The experienced electromyographer may note changes in the recruitment pattern of individual muscles soon after the injury. These changes include an increased firing rate of the remaining functional motor units and a reduction in motor units participating in the full interference pattern. In chronic radiculopathy, small amplitude polyphasic potentials may herald reinnervation. The specific nerve root involved is identified by noting these

TABLE 12-1. SEGMENTAL INNERVATION OF THE UPPER LIMB

Muscles	Roots	Nerves
Rhomboids	C5	Dorsal scapular
Supraspinatus	C5,6	Suprascapular
Infraspinatus	C5,6	Suprascapular
Deltoid	C5,6	Axillary
Biceps	C5,6	Musculocutaneous
Brachioradialis	C5,6	Radial
Serratus anterior	C5,6,7	Long thoracic
Pectoralis major (clavicular head)	C5,6,7	Lateral pectoral
Pronator teres	C6,7	Median
Extensor carpi radialis	C6,7	Radial
Flexor carpi radialis	C6,7,8	Median
Latissimus dorsi	C6,7,8	Thoracocorsal
Triceps (lateral head)	C6,7,8	Radial
Flexor carpi ulnaris	C7,8	Ulnar
Anconeus	C7,8	Radial
Triceps (long head)	C7,8	Radial
Extensor carpi ulnaris	C7,8	Posterior interosseous
Extensor indicis propius	C7,8	Posterior interosseous
Pronator quadratus	C7,8 T1	Anterior interosseous
Pectoralis major (sternal head)	C7,8 T1	Medial pectoral
Intrinsics	C8 T1	Median, ulnar
Paraspinal muscles	C5 T1	Dorsal rami

TABLE 12-2. SEGMENTAL INNERVATION OF THE LOWER LIMB

Muscles	Roots	Nerves
Adductor longus	L2,3,4	Obturator
Vastus medialis	L2,3,4	Femoral
Vastus lateralis	L2,3,4	Femoral
Rectus femoris	L2,3,4	Femoral
Tibialis anterior	L4,5	Deep peroneal
Extensor hallucis longus	L4,5 S1	Deep peroneal
Medial hamstring	L4,5 S1	Sciatic (tibial division)
Gluteus medius	L4,5 S1	Superior gluteal
Tensor fascia latae	L4,5 S1	Superior gluteal
Flexor digitorum longus	L5 S1	Tibial
Tibialis posterior	L5 S1	Tibial
Extensor digitorum brevis	L5 S1	Deep peroneal
Biceps femoris (short head)	L5 S1	Sciatic (peroneal division)
Peroneus longus	L5 S1	Superficial peroneal
Peroneus brevis	L5 S1	Superficial peroneal
Biceps femoris (long head)	L5 S1	Sciatic (tibial division)
Gluteus maximus	L5 S1,2	Inf gluteal
Lateral gastrocnemius	L5 S1,2	Tibial
Medial gastrocnemius	S1,2	Tibial
First dorsal interosseous	S1,2	Lateral plantar
Paraspinal muscles	L2-S1	Dorsal rami

TABLE 12-3. SENSORY SYNDROMES ASSOCIATED WITH COMMON RADICULOPATHIES

C5	Intrascapular pain radiating to the lateral aspect of the arm only as far as the elbow
C6	Pain radiating down the lateral arm & forearm to the thumb and index fingers
C7	Pain radiating down the entire arm to the middle finger
L4	Pain radiating down from the knee to the medial malleolus
L5	Pain radiating from the buttock along the posterior thigh to the lateral aspect of the leg and into the medial side of the foot and toes
S1	Pain similar to L5, but extending to the lateral aspect of the foot

TABLE 12-4. MUSCLE STRETCH REFLEXES IN THE DIAGNOSIS OF RADICULOPATHIES

Biceps reflex	C5,6
Brachioradialis	C5,6
Pronator teres	C6,7
Triceps	C6,7,8
Quadriceps	L2,3,4
Medial Hamstring	L5 S1
Achilles tendon	S1,2

electrophysiologic changes in certain muscles (as listed in Tables 12-1 and 12-2) and determining which common root or roots innervate those muscles that show pathologic findings.

The needle examination of the paraspinal musculature is especially important for two reasons. Although overlapping innervation between nerve root levels precludes precise localization on paraspinal EMG alone, findings in these muscles assure that the nerve lesion is proximal to the plexus, that is, at the root level. Second, the proximity of these muscles to the nerve injury often means that pathologic findings will appear in these muscles before they do in the limb. The pronator teres and the medial head of the triceps are especially useful for identifying radiculopathy in the upper extremity, while the gluteus maximus and gastrocnemius are the muscles in the lower extremity in which abnormal findings are often seen.

Of the late responses, only the H reflex has some use in the diagnosis of radiculopathy. Unfortunately, in the absence of an upper neuron lesion, the H reflex is easily elicited only from the triceps surae, so its use is limited to the evaluation of an S1 radiculopathy. Side to side comparison of the H reflex may demonstrate a significant delay in S1 radiculopathy patients who do not demonstrate EMG changes. F wave studies are of little use.[39]

SSEP also have not been shown to be of help in the electrophysiologic study of radiculopathy, probably because of the multiple nerve roots that supply the nerves for study. For example, the tibial nerve receives contributions from the L5, S1, and S2 nerve roots. In an effort to improve this, investigators have attempted to use dermatomal stimulation techniques to improve the yield of evoked potentials in radiculopathy. Dermatomal techniques involve the stimulation of particular patches of skin innervated by single dermatomal levels or of cutaneous nerves with largely one contributing root. Unfortunately, these tests are extremely time-consuming and have added only marginally to diagnosis.[6]

Although conservative therapy of radiculopathy is being more widely accepted as the predominant treatment of choice, some patients will inevitably undergo surgical treatment. Laminectomies produce EMG changes that may confuse subsequent diagnostic attempts. Paraspinal muscles continue to show evidence of denervation adjacent to the laminectomy site for many years. Sometimes, however, the astute electromyographer may note a pocket of intense fibrillations and sharp waves among a group of paraspinal muscles where only occasional findings are noted, and this may give a clue that a further nerve impingement has occurred.

In summary, EMG is a valuable and accurate method of electrophysiologically assessing the status of motor nerve involvement secondary to radiculopathy. The process is complementary to anatomic studies (myelography, CT, MRI) in the precise localization of and prognostication for radiculopathic disease. However, 2 or more weeks should pass before definitive EMG changes occur. Fortunately, this fits neatly into the widespread strategy of 3–5 days of bed rest followed by slow mobilization.

FOCAL NEUROPATHY[30]

Entrapment neuropathy is a generic term that refers to any localized lesion of a particular nerve. Since nerve insult may result from various etiologies—entrapment, trauma, laceration, vasculitis—perhaps a better descriptor is focal neuropathy or mononeuropathy. Focal neuropathies are relatively common, especially within the carpal tunnel. Focal weakness and atrophy in muscles innervated distal to the lesion are most useful in the clinical diagnosis. Sensory findings may appear earlier but are less reliable because of the overlap of dermatomes.

Focal pathology falls into two principal categories, local demyelination and axonal damage. Nerve entrapment often results in local demyelination of nerve fibers in the vicinity of the insult, and these changes are most notable in large myelinated sensory and motor fibers. If the myelin loss is severe enough to cause conduction block within nerve fibers, a neuropraxia has occurred. If the focal injury is more severe, axonal disruption or axonotmesis may occur, and the nerve fiber degenerates distal to the site over the ensuing days to weeks. Within a nerve trunk, a combination of neurapraxia and axonotmesis may occur. Neurapraxia and axonotmesis may be differentiated electrodiagnostically. Nerve conduction studies demonstrate that fibers are electrically excitable distal to the lesion in neurapraxia, while degenerated fibers are not excitable in an axonotmetic lesion. On needle examination, axonotmetic lesions produce fibrillations and positive sharp waves in muscles innervated by the involved nerve. These potentials are absent in neurapraxic injury. Severe focal neuropathies often are a combination of neurapraxic and axonotmetic injury in various fibers. Milder injuries may be largely neurapraxic. In these cases, the most sensitive method of finding such an entrapment is noting a loss of amplitude proximal to the entrapment site and a slowing of conduction across the site of compromise.[48,58] Recent investigation has shown that the integrated area of the CMAP is more sensitive than amplitude in diagnosing and quantifying compression block.[108] (See Table 12-5.)

Cranial Neuropathies

Both the facial and accessory nerves can be assessed in the electrodiagnostic laboratory. In Bell's palsy, involvement of the interosseous portion of the facial nerve may result in weakness in the lower two thirds of the face and alteration in taste in the anterior part of the tongue. EMG and nerve conduction studies may both be useful in prognosticating this condition.[54] The facial nerve may also be involved in peripheral neuropathies, multiple sclerosis, tumors, and trauma to the skull.

TABLE 12-5. ELECTRODIAGNOSTIC FINDINGS IN FOCAL NEUROPATHY

	Neurapraxia	Axonotmesis
Electromyogram		
Muscle at rest	Normal Fibrillation potentials	Positive sharp waves
Recruitment pattern	Decreased	Decreased
Interference pattern	Decreased	Decreased
Nerve Conduction Studies		
Stimulation proximal to the lesion	No CMAP elicited	No CMAP elicited
Stimulation distal to the lesion	CMAP elicited	No CMAP elicited

The spinal accessory nerve may be compromised by trauma or tumor, resulting in weakness of the trapezius and sternocleidomastoid.[13] Winging secondary to trapezius weakness may be differentiated from long thoracic nerve palsy by noting maximal winging on arm abduction. The accessory nerve may be assessed by nerve conduction and needle examination of the trapezius and sternocleidomastoid.[113]

Proximal Nerves of the Brachial Plexus

The long thoracic nerve may be injured in isolation or as part of a more diffuse traumatic brachial plexopathy. It is particularly vulnerable to traction because of its long and straight course. It may also be damaged by piercing injuries, shoulder bags, or shoulder bracing. Weakness of the serratus anterior results in winging of scapula, which is most prominent when pushing against a wall with outstretched arms. The diagnosis is made by the clinical findings and EMG examination.

The differential diagnosis of shoulder pain should include examination of the suprascapular nerve. The supraspinatus and infraspinatus muscles may be weakened by a mononeuropathy of the suprascapular nerve, most commonly occurring at the suprascapular notch. The suprascapular nerve has no cutaneous innervation, but aching shoulder pain may be noted because of innervation of part of the shoulder capsule. Suprascapular nerve palsy can result from improper crutch use, as can axillary and radial neuropathies.[129] Rarely, the suprascapular nerve can become entrapped more distally at the spinoglenoid notch.[1] Electrodiagnostic findings may include denervation in the appropriate muscles and a delay in the motor latency upon stimulation at Erb's point or upon nerve root stimulation (see below under Thoracic Outlet Syndrome).

Weakness of the rhomboid muscles and levator scapulae results from entrapment of the dorsal scapular nerve, resulting in winging of the scapula on wide abduction of the arm.

Musculocutaneous Nerve

The musculocutaneous nerve is uncommonly injured without concommitant brachial plexus injury.[76] It may be involved after humeral fractures or dislocations,[88] penetrating wounds, and heavy exercise. Weakness may be noted in the coracobrachialis, biceps, and brachialis muscles. The sensory branch of the musculocutaneous nerve and the lateral cutaneous nerve of the forearm may be compressed in the proximal forearm.[47] This syndrome includes proximal forearm pain and hypesthesias or paresthesias in the lateral forearm. Conduction techniques are available for both the musculocutaneous nerve and its sensory branch to help confirm the clinical diagnosis.

Axillary Nerve

The axillary nerve is most often injured by trauma. Injury produces weakness of the deltoid and teres minor and hypesthesia in a patch of skin over the lateral brachium.[12] The axillary nerve may be injured by anterior dislocation of the shoulder[88] or by incorrect crutch use.

Median Neuropathies[157]

The carpal tunnel syndrome, or entrapment of the median nerve within the carpal tunnel under the flexor retinaculum, is probably the most common type of focal neuropathy.

The patient usually complains of pain, often worse at night, which may continue proximally through the forearm to the elbow and even to the shoulder. Numbness may present in the classic median distribution (radial 3 1/2 digits) or may include all fingers. On physical examination, sensory loss may be found in this distribution. Tapping over the nerve (Tinel's sign) at the wrist and unforced flexion of the wrists (Phalen's sign) may exacerbate the symptoms. Motor weakness in the thenar intrinsic muscles, especially the abductor pollicis brevis, or thenar atrophy can occur in more severe cases. A small cross-sectional area within the carpal tunnel may predispose patients to this syndrome. It also may be associated with rheumatoid flexor tenosyovitis (see below), trauma, space-occupying lesions (tumors, granulomas, vascular abnormalities), hereditary neuropathies, pregnancy, amyloidosis, endocrine disorder (myxedema, acromegaly), and vitamin deficiency (pyridoxine deficiency).

Since the diagnosis of carpal tunnel syndrome is so frequently made by the electromyographer, several specialized techniques of diagnosing this disorder appear in the electrophysiologic literature. A few principles of use in the diagnosis are: (1) Carpal tunnel syndrome usually involves both large sensory and motor fibers, but occasionally only one type of nerve fiber is involved;[68] this would imply that sensory and motor studies are almost always necessary. (2) Nerve conduction studies of nerves other than the median should always be carried out to assure that the median nerve slowing is caused by a focal entrapment and not a peripheral neuropathy. (3) Although standard conduction latencies have been calculated for each nerve at fixed distances in normal adults, the sensitivity of the test can be increased by comparing the median study to that of an adjacent nerve[103]—ulnar or superficial radial—or the corresponding nerve in the contralateral limb. (4) Palmar stimulation techniques may increase the diagnostic accuracy.[35,102] (5) It is frequently useful to perform an EMG of the entire upper extremity and corresponding paraspinal muscles in addition to that of the thenar eminence; some believe that the presence of radiculopathy may predispose the patient to carpal tunnel syndrome (the "double-crush" syndrome).[63,150] Electrodiagnosis of carpal tunnel syndrome has been recently reviewed.[138] Although nerve conduction studies do not correlate well with the severity of the patient's complaints, they are nonetheless an extremely sensitive and specific test for carpal tunnel syndrome. Electrophysiology can also be helpful in planning therapy since milder cases may respond to conservative therapy.

Traumatic compression of the median nerve can occur more proximally at the shoulder or humerus secondary to anterior shoulder dislocation, humeral fracture, axillary crutches, or a tight gunstrap. The patient may exhibit weakness of all median arm and hand musculature and sensory loss in the radial 3 1/2 fingers of the hand.

The median nerve may be compressed above the elbow by a rare anomaly, the ligament of Struthers.[141] This ligament stretches from a spur of bone on the distal anteromedial surface of the humerus to the medial epicondyle and may entrap both the median nerve and brachial artery. The patient complains of elbow pain and tenderness, but the median nerve complaints are often nonspecific since the entrapment is mild. Diagnosis is aided by discovery of the supercondylar spur on radiography.

The median nerve can be entrapped within a tendinous band of the pronator teres muscle, a condition called the pronator syndrome.[59] This causes aching in the proximal forearm, often after repetitive elbow motions. Motor and sensory symptoms are poorly defined. Tenderness may be present over the muscle, but objective motor and sensory findings may be lacking. EMG can help define the location of the pathologic condition, aided by specific nerve conduction techniques.[25]

The anterior interosseous nerve, a major branch of the median nerve, can be compressed by tendinous or muscular structures, aberrant vessels, or trauma.[62,127] Symptoms include pain in the proximal forearm and weakness in the flexor pollicis longus, flexor digitorum longus (II/III), and pronator quadratus, without sensory symptoms. Upon pinching the thumb and index finger, a characteristically abnormal diamond-shaped posture is assumed because of weakness of the thumb and finger flexors. The diagnosis is confirmed by EMG and prolongation of the distal motor latency to the pronator quadratus muscle.

Ulnar Nerve

The ulnar nerve is most commonly compromised at the elbow. With entrapment at this level, the patient complains of numbness over the ulnar side of the hand.[122] On physical examination, numbness may be detected in the ulnar 1 1/2 fingers, not extending proximal to the distal wrist crease. Weakness may be noted in the flexor carpi ulnaris, the ulnar two tendons of the flexor digitorum profundus, hypothenar muscles, ulnar lumbricals, dorsal and palmar interossei, and the adductor pollicis. Weakness can be easily detected in the adductor pollicis by having the patient pinch a piece of paper then pulling it away and by isolated manual muscle testing of the first dorsal interosseous muscle.

Entrapments in the elbow region most commonly occur at the cubital tunnel because of constriction by the aponeurosis of the flexor carpi ulnaris located approximately 2–3 cm below the medial epicondyle. This is known as the cubital tunnel syndrome.[101] Other etiologies include trauma to the nerve within the cubital tunnel, chronic subluxation of the nerve, compression associated with bony misalignment of the cubital tunnel, and scar impingement at the elbow. Nerve conduction studies may show slowing across the elbow nerve segment. The inching technique, stimulation of the ulnar nerve in sequential steps approximately 1 in. apart, may help precisely localize the site of entrapment. The sensory nerve action potential (SNAP) can be diminished in size if sensory fibers are involved. EMG may also help localize the site of entrapment.[41,78]

More distal entrapments of the ulnar nerve may result from trauma, extrinsic compression, carpal fractures, repetitive occupations, and space-occupying lesions. Lesions in Guyon's canal (the ulnar tunnel bounded by the volar carpal ligament, transverse carpal ligaments, and carpal bones) can cause sensory loss on the volar ulnar 1 1/2 digits and ulnar intrinsic weakness.[55] The dorsal aspect of these digits, supplied by the dorsal sensory branch of the ulnar nerve, are spared because this branch arises 6–8 cm proximal to the wrist. Occasionally only sensory symptoms are noted. The deep palmar motor branch alone may be affected when the nerve lesion occurs distal to Guyon's canal.[163] This entrapment results in weakness of the hand intrinsics but often spares the hypothenar muscles. Lesions proximal to Guyon's canal may or may not involve the dorsal sensory branch. Isolated dorsal sensory branch compression neuropathy rarely occurs, resulting in loss of sensation over the dorsal aspect of the ulnar 1 1/2 fingers. It is usually associated with blunt trauma or laceration. Motor studies across the ulnar canal and the deep motor branch, SNAP to the little finger and dorsal branch of the ulnar nerve,[65] and EMG needle examination may all aid in localization of distal ulnar entrapments.

Radial Nerve

Compression of the radial nerve may occur in the axillary region or distally within the spiral groove. High axillary lesions are sometimes distinguished from those more distal by the involvement of the triceps muscle. Additional muscular weakness may be noted in the anconeus, brachioradialis, brachialis, extensor carpi radialis longus and brevis, supinator, extensor digitorum, extensor digiti quinti, extensor carpi ulnaris, abductor pollcis longus, extensor pollicis longus and brevis, and most distally, the extensor indicis. Numbness may be noted over the radial dorsum of the hand extending variably over the thumb and part of the index finger. Trauma produces the majority of these palsies. Among the etiologies are humeral fractures, crutch injury, or an arm left hanging over a bench ("Saturday night palsy"). Localization of the nerve entrapment or insult can be made by EMG examination and motor conduction studies along the radial nerve. Because of the position of the nerve within the arm, needle stimulation and pick-up electrodes are often used. SNAPs can be used to assess viability of sensory axons within the superficial radial nerve.

The posterior interosseous nerve may be entrapped just distal to its bifurcation from the superficial sensory branch, where it plunges into the supinator muscle under the arcade of Frohse—the supinator syndrome.[16] The extensor carpi radialis brevis and longus, supinator, and brachoradialis are generally spared since they are innervated more proximally. More distal muscles (listed above) are weakened. The posterior interosseous syndrome may be caused by tumors, such as lipomas;[117] cysts, or ganglia: trauma (elbow dislocation, radial/ulnar fractures); or rheumatoid synovitis (see below). Occasionally the nerve may be injured by a sharp tendinous band at the arcade of Frohse without previous trauma. Posterior interosseous nerve entrapment may cause marked lateral epicondylar pain, making it difficult to discern from lateral epicondylitis (tennis elbow). Finally, the radial nerve may be injured in its superficial radial branch[34] because of lacerations at the wrist, injections,[26] or tight wrist bands (e.g., handcuff neuropathy). Handcuff neuropathy may also include the median nerve.[87,121] Diagnosis can be verified by examination of the superficial radial SNAP.

Digital Neuropathy

Although far less common than more proximal compressions, pain, hypesthesias, paresthesias, and dysfunction of the hand may occur because of compression neuropathies of the palm and digits. These may be due to chronic trauma (gripping a pen too tightly or throwing a bowling ball) or space-occupying lesions (cysts, osteophytes, tumors, Dupuytren's contracture).

Thoracic Outlet Syndrome

Constriction of the brachial plexus can occur at various sites between the neck and shoulder, resulting in poorly defined upper extremity pain and tingling. Most often the compression is due to an anomalous fibrous band, which may or may not be attached to a rudimentary cervical rib. Such syndromes have been lumped under the broad heading of thoracic outlet syndrome. While the medical literature associated with this syndrome is extensive, it is confusing. Clarity may be gained by attempting to divide thoracic outlet syndrome into neurogenic and vascular causes.[56]

Vascular compression may produce symptoms of coldness or aching pain on exercise of the arm. The hand may lose strength and turn pale. Rarely, arterial emboli or digital gangrene may occur. Physical signs such as Adson's maneuver have little demonstrable value, since they may be "abnormal" in a large population without symptoms. Evidence of vascular insufficiency can be documented by angiography, and true compressive lesions may be surgically repaired. Neurogenic symptoms include paresthesias and pain in the arm and hand, more prominent on the ulnar side of the arm and forearm. Weakness and atrophy of the hand intrinsics, sometimes selectively affecting the thenar eminence, may occur in severe cases.[30]

Electrophysiologic diagnosis is generally unremarkable in patients believed to have thoracic outlet syndrome. Probably the most useful role of electrodiagnosis in these patients is to rule out other causes of the symptoms such as carpal tunnel syndrome, ulnar neuropathy, and cervical radiculopathy. In the rare case of "true neurogenic" thoracic outlet syndrome, findings may include a diminution in the amplitude of the SNAP, delay in F wave studies secondary to slowing across the brachial plexus, and, less commonly, fibrillations and positive sharp waves in hand intrinsic or forearm muscles. Motor conduction studies upon stimulation at Erb's point (anterior to the sternocleidomastoid muscle in the supraclavicular fossa) are not valuable, because conduction is done distal to the site of supposed compression.[158] If motor conduction is to be assessed, stimulation of the C8-T1 roots can be achieved by inserting needle electrodes deep into the cervical paraspinal muscles.[93] Delay in SSEP from ulnar nerve stimulation has been suggested as a diagnostic aid,[27,50,67,166] but its true value remains uncertain.

Brachial Plexopathy

Brachial plexopathy may be traumatic or nontraumatic. Traumatic causes include firearm injuries, difficult births, clavicular fractures, humeral fractures or dislocations, and arterial catheterizations.[159] Nontraumatic bracial plexopathy is most often idiopathic, the neuralgic amyotrophy known as Parsonage-Turner syndrome. Brachial neuritis most commonly presents as weakness and aching in the shoulder and arm region. The C5 and C6 dermatomes are usually involved, but it may involve also a variety of roots, nerves, and portions of the plexus.[42,154] A hereditary component has been suggested in some cases.[36,115] Neoplastic invasion and postmastectomy radiation effects are other notable causes of nontraumatic brachial plexopathy.

The electrophysiologic diagnosis of brachial plexopathy is an interesting lesson in applied anatomy, which can be reviewed in standard texts.[60] The pattern of EMG abnormalities is the most valuable electrodiagnostic method in assessing which roots, trunks, or cords are involved. Slowing of conduction and loss of CMAP amplitude across the lesion are often present, but usually do not add greatly to the clinical impression. SNAPs are a vital part of the the evaluation. The presence of SNAP distal to the lesion implies integrity of sensory fibers between the periphery and the dorsal root ganglion. By studying a variety of sensory nerves—lateral antebrachial cutaneous, superficial radial, median, and ulnar—one can gain insight into the remaining nerve fibers across the plexus. Unfortunately, in cases of nerve root lesions, SNAP may be intact, and yet the patient may complain of complete anesthesia in the sensory nerve territory. Thus complete anesthesia despite good SNAP is a poor prognostic sign, since it implies nerve root avulsion. Cortical evoked potentials upon stimulation of these same sensory nerves may be useful in such a situation since they can help confirm the integrity of sensory fibers proximal to the dorsal root ganglion.[5,80,144,145,165]

Electrodiagnosis may be helpful in the evaluation of arm weakness in women who have undergone postmastectomy radiation subsequent to breast malignancy. Late weakness and pain can result from tumor recurrence or postradiation neuropathy. Patients with neoplastic plexopathy more often have pain, lower trunk involvement, and Horner's syndrome, while postradiation plexopathy was associated with paresthesias, upper trunk involvement, and rarely Horner's syndrome. Radiation plexopathy is more frequently associated with abnormal conduction studies as well as fasciculations or myokymia (a rare form of spontaneous activity often seen in radiation-induced nerve injury) on the needle examination.[79,86]

Proximal Nerves of the Pelvic Girdle

Ilioinguinal nerve entrapment may occur slightly medial to the anterior iliac spine, resulting in burning pain the lower abdomen, inner thigh, and genitalia. The nerve may be injured by trauma or surgery to the abdominal wall. Electrodiagnostic studies are not of value.

The genitofemoral nerve may be injured by trauma in the groin or secondary to surgery. Pain is noted in the inguinal region and loss of sensation occurs over the femoral triangle.

The superior and inferior gluteal nerves may be injured by intramuscular injection, piriformis compression, or upper femoral fracture.[120] Weakness and EMG findings may be noted in the gluteus medius and minimums and the gluteus maximus, respectively.

Obturator Nerve

The obturator nerve may be injured as a consequence of labor, pelvic fracture, hip prosthetic surgery,[130] or obturator hernial repair. Besides innervating the majority of adductor musculature, the obturator nerve supplies sensation to the hip and knee joints and medial thigh. Dysfunction of the knee adductors can profoundly disturb gait function.

Femoral Nerve

The femoral nerve may be injured in the retroperitoneal space, within the pelvis, or under the inguinal ligament.[132] Causes include retroperitoneal or intrapelvic hematoma,[123,131,142] tumors and space-occupying lesions, and extreme hip flexion or extension.[102] The nerve may be directly injured by femoral fractures or cardiac catheterization, and is a frequent site of diabetic mononeuropathy. The patient may demonstrate weakness in the iliopsoas and quadriceps muscles and sensory loss over the anterior thigh and medial calf, depending on the site of compromise. The terminal sensory branch of the femoral nerve is the saphenous nerve, and this may be injured at its exit from Hunter's canal. Electrodiagnostic evaluation includes needle examination, femoral nerve conduction studies, saphenous nerve sensory conduction studies, and saphenous nerve evoked potentials.

The lateral femoral cutaneous nerve of the thigh, a purely sensory nerve, may become entrapped under the inguinal ligament. The resulting paresthesia and hypesthesia of the upper lateral aspect of the thigh has been called meralgia paresthetica. Nerve conduction studies are possible but are often difficult. Comparing side to side cortical sensory evoked potentials upon stimulation of a homologous patch of skin may help establish the diagnosis.[38]

Sciatic Neuropathies

The sciatic nerve may be entrapped in the piriformis muscle (piriformis syndrome), although this syndrome is not universally acknowledged. Cases have been reported in which the sciatic nerve was entrapped by a myofascial band secondary to hip prosthetic surgery,[98] caused by fibrosis induced by pentazocine, or secondary to retroperitoneal bleeding after anticoagulant therapy. Electrophysiologic evaluation would include EMG needle examination, while lumbosacral root stimulation techniques may demonstrate focal motor slowing.[92] The tibial nerve may be entrapped in the popliteal region by a Baker's cyst.[72]

Tarsal Tunnel Syndrome

Analogous to the median nerve at the wrist, the tibial nerve passes under a flexor retinaculum posterior to the medial malleolus, where it may be entrapped. Tarsal tunnel syndrome should be considered in the differential diagnosis in a patient complaining of burning pain in the sole of the foot. This pain may be exacerbated by tapping the nerve at the entrapment site and noting local pain tingling distally (Tinel's sign). Weakness and sensory loss are seldom seen. Electrodiagnosic changes include a decrease in the amplitude of the CMAP[71] and prolongation of motor latencies upon stimulation proximal to the entrapment. Delay of the orthodromic SNAP upon stimulating the toes and recording proximal to the medial malleolus is also helpful in evaluation.[33] Sensory studies may be difficult unless extensive averaging or near-nerve techniques (placing a needle recording electrode just adjacent to the nerve trunk) are used.[107]

Digital Neuropathy

Morton's neuralgia, a painful entrapment of the interdigital nerves of the foot, usually causes pain between the third and fourth digits. The pain is often localized over the fourth metatarsal bone and may be confused with metatarsalgia. Although sensory conduction studies have been proposed for diagnosis of this condition,[44,106] these have not been successful in all hands.

Peroneal Neuropathies

The peroneal nerve is most commonly injured at the head of the fibula because of fractures, immobilization casts, tight stockings, improper positioning during surgery, mass lesions, and several miscellaneous conditions.[160] Weakness may occur in the anterior and lateral compartments of the leg, depending on whether the deep or superficial peroneal fibers are involved. The patient most often complains of foot-drop during ambulation. Involvement of the superficial peroneal may cause sensory loss over the anterolateral calf and dorsolateral foot, while deep peroneal fibers innervate only a patch of skin in the web space between the great and second toe. Motor conduction studies, concentrating on segments of nerve about the fibular head, may help localize the lesion.[116] Inching techniques may be especially valuable in this setting (described above). EMG can help assess the site of injury according to the distribution of abnormalities.

The deep peroneal nerve may become entrapped under the dense superficial fascia of the ankle, the so-called anterior tarsal tunnel syndrome. Symptoms generally involve pain and tingling over the dorsum of the foot and loss of sensation in the first web space. A delay in the distal motor latency of the deep branch may be noted. An "anterolateral tarsal tunnel syndrome" due to compression of terminal fibers of the superficial peroneal nerve has also been described.[134]

Sural Nerve

The sural nerve supplies sensation beneath the lateral malleolus and on the lateral side of the foot. It is probably most often damaged intentionally for nerve biopsy. Sural nerve studies are easily and routinely performed.

Lumbosacral Plexopathy

Lumbosacral plexopathy is caused by trauma, neoplasms, hematomas, or idiopathic neuritis.[159] Neoplastic lumbosacral plexopathy is associated with pain, weakness, edema, a rectal mass, and hydronephrosis.[66] Unlike the brachial plexus, lumbosacral plexopathy caused by radiation after the treatment of malignancies is rare.[147] Individual nerves or parts of the plexus may be involved in lumbosacral plexus neuritis.[43,128] Diagnosis is principally based on the needle examination, but F wave and H reflex studies may be of some ancillary value.

Polyneuropathy and Mononeuritis Multiplex

Polyneuropathies are generalized neuropathic processes of the lower motor neuron. Electrodiagnostic studies are extremely useful in determining whether (1) a peripheral neuropathy is present, (2) the neuropathy affects predominantly motor or sensory fibers, and (3) the pathologic process is demyelinating or axonopathic. The development of nerve conduction studies and their application to peripheral neuropathy have been recently reviewed.[49]

Axonal neuropathies may display little slowing on conduction studies, but more characteristically, the CMAP may be slightly diminished in amplitude. Needle examination at rest often shows spontaneous activity in distal muscles (suggestive of denervation), and a decrease in recruited motor units is seen upon voluntary muscle contraction. These motor units may be increased in size if significant reinnervation has occurred. Demyelinating neuropathies characteristically show little denervation but often display significant slowing of the motor and sensory nerve latency and velocity.

Mononeuritis multiplex may manifest as weakness or sensory loss in the distribution of various peripheral nerves rather than the symmetric stocking-glove distribution often accompanying peripheral neuropathy. Mononeuritis multiplex results from infarction of the nerve axons[1] vessels that occurs when vasa nervorum are occluded because of a vasculitic process.[110]

Differential Diagnosis of Peripheral Neuropathy

When evaluating a patient who may have a focal neuropathy, the clinician should carefully consider other neurologic conditions. Radiculopathy should be considered when weakness or sensory symptoms are noted. For example, C6 radiculopathy and carpal tunnel syndrome can both cause paresthesias in the thumb and index finger, and L2 radiculopathy and femoral neuropathy can both cause knee extensor weakness. Polyneuropathy and mononeuritis multiplex may sometimes affect certain nerves preferentially. These are discussed briefly in the section below on the rheumatoid arthritis patient. Brachial plexopathy, an idiopathic axonal disruption of the brachial or lumbosacral plexus, should also be considered when weakness is noted. Motor neuron disease, or amytrophic lateral sclerosis, may occasionally present as monomelic weakness and can be confused with mononeuropathy.[30]

NEUROMUSCULAR DISEASE IN THE RHEUMATOID ARTHRITIS PATIENT

Peripheral nerve or muscular involvement is a common problem in the patient with rheumatoid arthritis (Table 12-6). Clinical neurophysiology can be helpful in the evaluation of complications of this disease and its treatment.[149]

Various distinct syndromes of peripheral neuropathy have been noted in patients with rheumatoid arthritis.[19,37,51,82,85,97,105,109,112,151,156] Chamberlain and Bruckner[19] described two forms of peripheral nerve involvement in rheumatoid arthritis. The first group of patients had relatively benign rheumatoid arthritis with a patchy stocking-glove, predominantly sensory peripheral neuropathy. Electrodiagnostic studies in this group showed marked slowing of motor conduction accompanied by a diminution or loss of SNAP. The second group had more malignant rheumatoid arthritis and demonstrated more lower than upper extremity motor weakness. This group displayed severe denerva-

---

**TABLE 12-6. NEUROMUSCULAR DISORDERS IN
THE RHEUMATOID ARTHRITIS PATIENT**

Neuropathy
  Peripheral Neuropathy
    Predominantly sensory, stocking-glove distribution
    Sensorimotor, lower greater than upper extremity
      weakness
    Secondary to gold therapy
    Secondary to penicillamine therapy
  Mononeuritis Multiplex
  Focal Neuropathy
    Median (carpal tunnel)
    Posterior interosseous nerve (elbow)
    Ulnar nerve (elbow, wrist)
    Tibial (popliteal, tarsal tunnel)
    Peroneal (popliteal)
Myopathy
  Myositis due to vasculitis
  Steroid myopathy
  Myopathy secondary to penicillamine therapy
Neuromuscular Junction Disorders
  Penicillamine-induced myasthenia gravis
Cervical Myelopathy

---

tion on needle examination.[19] Pallis and Scott described similar patterns of peripheral neuropathy but did not utilize electrodiagnostic studies.[109] Autonomic nervous system involvement[37,151] and restless leg syndrome[124] have also been described in patients with rheumatoid arthritis.

Pathologic studies of sural nerve biopsies have recently helped to define the peripheral neuropathy of rheumatoid arthritis.[28,114,152] Conn, et al.[28] studied five patients with rheumatoid arthritis and sensorimotor neuropathy documented by electrodiagnostic studies. All biopsies demonstrated axonal degeneration. This was thought to be due to immune-complex deposition within vessel walls causing an acute necrotizing arteritis, and these vasculitic vessels resulted in infarction of nerve axons. Van Lis and Jennekens[152] describe a single sural nerve biopsy from a rheumatoid arthritis patient with neuropathy in which epineurial vessels were heavily infiltrated by mononuclear cells. In addition, necrotic and fibrinoid material was found in the lumen and intima while immunoglobulins and complement were seen on immunofluorescent studies. Arterioles without necrotic and fibrinoid material showed fibrocellular proliferation and mild infiltration by inflammatory cells.

It has been suspected for many years that the peripheral neuropathy of rheumatoid arthritis is a form of mononeuropathy multiplex.[109] Sabin[126] hypothesized that random infarcts of nerve axons due to vasculitic occlusion of the vasa nervorum might result in a neuropathy with a stocking-glove distribution. The statistical likelihood of an axonal insult occurring in longer fibers is greater than in shorter fibers (if infarcts do occur randomly). Thus, the chance that distal symptoms arise because of infarction of longer axons is statistically greater.

Pharmacologic treatment of rheumatoid arthritis may in itself be a cause of neuropathologic conditions of the peripheral nervous system. Fam, et al. described three patients with rheumatoid arthritis who developed cranial neuropathies (nerves IV, V, VI, VII) as well as a peripheral neuropathy after treatment with gold.[45] Pool, et al. reported a case of profound sensorimotor neuropathy responding to pyridoxine in a patient treated with D-penicillamine, a known pyridoxine antagonist.[119]

The vasculitis of rheumatoid arthritis often manifests as a mononeuritis multiplex, as discussed above. Clinical characteristics depend on the nerve trunks involved. Onset is generally acute or subacute and may be associated with poorly localized aching pain. Proximal neuropathy predominantly affects motor nerves, while distal nerve trunk involvement gives a mixed involvement of the sensory and motor systems. The electrodiagnostic findings are principally those of axonal damage—namely, positive sharp waves and fibrillations on needle examination. Large-amplitude polyphasic motor units may occur later in the disorder as reinnervation occurs.[109] Mononeuritis multiplex characteristically occurs within 1 year of the onset of systemic vasculitis. Provided that no recurrences of the systemic vasculitis occur within 18 months of the initial neurologic insult, the prognosis for recovery from nerve damage is good.[22]

Focal Neuropathies

Several studies document the occurrence of carpal tunnel syndrome in rheumatoid arthritis patients.[10,14,20,153] The entrapment may be caused by swelling (synovitis) on the volar aspect of the wrist.[14,82] In one study 20 patients with documented rheumatoid arthritis were examined for carpal tunnel syndrome. Of the 40 wrists studied 8 demonstrated findings consistent with carpal tunnel syndrome, and all improved with anti-inflammatory medication and splinting.[153] An older series of 45 patients with classic or definite rheumatoid arthritis demonstrated electrodiagnostic abnormalities consistent with carpal tunnel syndrome in 49% of the patients; 33% also had clinical evidence of the disorder.[10]

Tarsal tunnel syndrome, entrapment of the tibial nerve under the flexor retinaculum (laciniate ligament) at the ankle, has been documented in 5%–25% of rheumatoid arthritics[11,53] who did not have findings of peripheral neuropathy. The posterior interosseous nerve may be entrapped against the supinator muscle by swollen synovium at the elbow,[21,96,100] the ulnar nerve at the elbow and wrist[9,146] and the peroneal nerve and tibial nerves can be compressed in the popliteal region.[69,82] Digital neuropathy of the upper extremities has also been described.[97] Winging of the scapula in rheumatoid patients may be due to disruption of the serratus anterior and should not be confused with a palsy of the long thoracic nerve.[99]

Myopathy[15]

Myopathy is perhaps less appreciated in rheumatoid arthritis than are peripheral or focal neuropathies, yet it has been frequently reported in the arthritis literature.[56,118,137,140,162,164] EMG examination of rheumatoid patients consistently showed changes suggestive of myopathy—most commonly a myopathic recruitment pattern and occasionally fibrillations and positive sharp waves. EMG changes do not necessarily correlate with findings on muscle biopsy,[140] which characteristically shows diffuse inflammatory cell infiltrates and fiber necrosis.[57,118] EMG findings are most marked in patients with elevated creatine phosphokinase.[57]

A confusing picture may arise in patients treated with steroids, which itself may result in myopathy.[164] Steroids result in atrophy of skeletal muscle due to inhibition of protein synthesis, an effect most notable in fast-twitch fibers.[73] The EMG may be extremely useful here—fibrillation potentials and positive sharp waves rapidly disappear in vasculitic myopathy treated with steroids. Recurrence of myositic weakness can be distinguished from steroid myopathy by the presence or absence of these potentials.[77]

Polymyositis and dermatomyositis have been described as ill-effects of penicillamine in two cases. One rheumatoid patient developed polymyositis 1 month after beginning the drug. Notable was the association of myocarditis, proteinuria, and hypocomplementemia, which are not typical of idiopathic polymyositis.[29] A second patient developed a severe dermatomyopathy that remitted on discontinuation of the drug. Muscle biopsy was compatible with severe necrotizing myositis, and skin biopsy showed atrophic epidermis and basophilic degeneration of connective tissue.[89]

Neuromuscular Junction Disorders

The administration of D-penicillamine as a remittive agent in rheumatoid arthritis may lead to weakness similar to that in myasthenia gravis.[3] The myasthenia usually resolves 3 to 6 months after discontinuation of the drug.[3,46] D-penicillamine predisposes certain patients to myasthenia gravis but does not directly interfere with neuromuscular transmission.[2,143] The myasthenia may be due to penicillamine-induced synthesis of acetylcholine-receptor antibodies,[46,83] whose titers fall dramatically when the drug is discontinued.[143]

One case has been reported of myasthenia gravis due to penicillamine that was reactivated by gold.[104] Steroids and azathioprine both act presynaptically to disturb neuromuscular transmission, but it is not known if their effects are clinically significant.[143]

Cervical Myelopathy in Rheumatoid Arthritis

Rheumatoid arthritis can result in degeneration of connective tissue structures responsible for stability of the upper cervical spine. Atlantoaxial instability may cause anterior compression of the spinal cord. The patient may present with signs of spinal cord compression including pain, sensory symptoms, weakness, spastic hypertonia, and rarely fifth cranial nerve involvement. The diagnosis may be delayed if the neurologic features are wrongly attributed to peripheral nerve changes.[95,139]

The diagnosis of myelopathy in rheumatoid arthritis is not always clear-cut since the degree of atlantoaxial subluxation determined on plain radiographs is not closely correlated to the clinical picture.[111] Computerized myelography may add to the diagnostic impression,[139] and SSEP may be especially useful in this setting. Upon stimulation of the upper extremity, one may find that the brachial plexus potential is of normal amplitude and latency, while the evoked responses measured over the cervicomedullary junction and contralateral somatosensory cortex may show delay and loss of amplitude.[24,133,167] Neurologic progression with corroborating laboratory studies suggests the need for rapid surgical intervention.[95,139]

SUMMARY

The principal electrodiagnostic tools in patients with musculoskeletal disease are electromyography, nerve conduction studies, neuromuscular junction studies, and somatosensory evoked potentials. EMG is useful in detecting disease involving the motor unit and localizing disease to a particular distribution, i.e., muscle, nerve, root, or plexus. EMG may help prognosticate the severity of focal nerve lesions. Nerve conduction studies identify pathologic conditions along the length of a nerve and can help localize a focal nerve lesion. EMG and nerve conduction studies may help differentiate demyelinating from axonal neuropathic processes. Late responses—F waves and H reflexes—may assist in discovering neuropathologic conditions occurring proximally in nerve trunks. Repetitive stimulation and single fiber EMG are used to delineate disorders of the neuromuscular junction. SSEP can be used to verify the presence of myelopathy in the rheumatoid patient with cervical myelopathy.

Electrodiagnosis is an essential part of the evaluation of the patient with radiculopathy. Myelography, CT, and MRI all address anatomy, while electrodiagnostic studies evaluate the neurophysiologic status of the root. C7 is most commonly involved in the cervical region and L5 and S1 in the lumbosacral region. Careful evaluation of muscle weakness, sensory symptoms, and reflex changes are critical in clinical diagnosis. EMG is the most valuable electrophysiologic tool for identifying radiculopathy, although occasionally late responses and sensory evoked potentials may give ancillary information. Signs of denervation may not be evident on EMG for up to 2 weeks after acute root impingement.

Electrodiagnostic studies remain the primary laboratory modality for evaluating focal neuropathies. EMG and nerve conduction studies help not only in diagnosis but also in establishing the severity and prognosis of the lesion. Focal neuropathies may be secondary to local demyelination, causing conduction block (neuropraxia) or axonal disruption (axonotmesis). Electrodiagnostic studies give the clinician a good impression as to which of these two pathologic processes predominates.

The facial and accessory nerves are the most common cranial neuropathies evaluated in the EMG laboratory. Long thoracic nerve palsy manifests as weakness of the serratus anterior with characteristic winging of the scapula. Suprascapular nerve involvement results in shoulder pain and weakness of the supra- and infraspinatus muscles. The musculocutaneous nerve is generally injured as part of a concommitant brachial plexopathy. Isolated axillary nerve palsies can occur after anterior shoulder dislocation and result in shoulder abduction weakness.

The median nerve is most commonly entrapped at the carpal tunnel, resulting in characteristic sensory findings in the radial 3 1/2 digits and resulting in weakness only in more severe cases. Special nerve conduction techniques may greatly increase the diagnostic yield. The median nerve may also become entrapped at the ligament of Struthers above the medial epicondyle or at the proximal humerus. Entrapment of the median nerve within the pronator teres muscle is called the pronator syndrome and may mimic the symptoms of carpal tunnel syndrome. The anterior interosseous nerve is purely motor, and entrapment results in weakness of the long flexors of first three digits.

Ulnar neuropathies most often occur at or just below the elbow in the aponeurosis of the flexor carpi ulnaris (cubital tunnel syndrome). Distal entrapments of the ulnar nerve at the wrist may involve any combination of the deep motor branch or the two distal sensory branches. Radial nerve compression generally occurs as the nerve courses through the radial spiral groove. More distally the nerve may be entrapped at the arcade of Frohse (the supinator syndrome), which can be confused with tennis elbow. The superficial sensory branch of the radial nerve may be injured at the wrist, causing the condition referred to as handcuff neuropathy. Digital nerve entrapments rarely occur.

The thoracic outlet syndrome has received extensive attention in the medical literature, but true neurogenic entrapment at the thoracic outlet is uncommon. Most likely the syndrome is better understood by thinking in terms neurogenic and vascular groups. Electrodiagnosis is probably of more value in ruling out carpal tunnel syndrome, radiculopathy, and ulnar nerve entrapment as part of the differential diagnosis. Electrodiagnostic findings are seen in patients with true neurogenic thoracic outlet syndrome.

Brachial plexopathy is most often due to trauma. Nontraumatic plexopathy is generally idiopathic, resulting in axonal involvement. EMG is the most useful tool in assessing the extent of the lesion, but SNAP and SSEP may add additional information.

Ilioinguinal and genitofemoral nerve mononeuropathies are not often recognized as causes of groin pain. Superior and inferior gluteal nerve neuropathies result in weakness of gluteal musculature. The obturator nerve is occasionally injured in labor, resulting in weakness in the adductors and mild medial thigh sensory loss. Femoral nerve syndromes may cause weakness in the hip flexors and knee extensors and sensory loss in the saphenous nerve territory. The saphenous nerve may be entrapped at its exit from Hunter's canal. Meralgia paresthetica refers to sensory changes on the outside of the thigh due to entrapment of the lateral femoral cutaneous nerve.

The sciatic nerve may be injured by intragluteal needle injections, but entrapment by the piriformis muscle has not been clearly demonstrated. The tibial nerve may be compressed by a Baker's cyst at the popliteal fossa but is more often entrapped in the flexor retinaculum of the ankle—the tarsal tunnel syndrome. Morton's neuroma is a painful entrapment of the interdigital nerves of the foot, usually between the third and fourth digits.

Peroneal neuropathies generally occur at the head of the fibula, although the deep and superficial peroneal nerve fibers may be entrapped at the level of the ankle. The sural nerve is most often insulted by intentional nerve biopsy. Lumbosacral plexopathy may be due to a variety of space-occupying lesions or idiopathic axonal involvement, similar to the brachial plexus. The differential diagnosis of entrapment neuropathy should include radiculopathy, mononeuritis multiplex, polyneuropathy, and motor neuron disease.

Patients with rheumatoid arthritis may suffer from a variety of neuromuscular disorders. Electrodiagnosis may aid greatly in the diagnosis of these problems. Rheumatoid patients may suffer from a predominantly sensory peripheral neuropathy with a stocking-glove distribution or a sensorimotor neuropathy with profound weakness. They may also suffer from an asymmetric nerve involvement, mononeuropathy multiplex, in which few nerves are involved. All of these problems may be due to axonal injury caused by the vasculitis associated with rheumatoid arthritis. Vasculitic involvement of the vasa nervorum may result in an infarction of nerve fibers supplied by the diseased vessels. Inflammatory products have been documented in nerve biopsies of afflicted patients. Peripheral nerve involvement may also be caused by treatment of rheumatoid arthritis by gold compounds and D-penicillamine.

The most common entrapment neuropathy in rheumatoid arthritics is the familiar carpal tunnel syndrome. Diagnostic accuracy of the carpal tunnel syndrome can be improved by comparing median nerve conduction with adjacent upper extremity nerve trunks. Other focal neuropathies may be caused in arthritic patients by tenosynovitis or articular swelling. These include entrapment of the tibial nerve at the knee and ankle, the posterior interosseous nerve, the peroneal nerve, and digital neuropathy of the hand. Nerve conduction studies and EMG can help determine if a focal nerve injury is primarily axonotmetic or neurapraxic. Neurapraxic injuries carry a better prognosis.

Myopathy may be an important source of weakness in the rheumatoid patient and can be diagnosed with the aid of EMG. The EMG can also help distinguish weakness due to inflammatory myopathy from that due to steroid treatment. A myopathic picture has been associated with D-penicillamine treatment. Neuromuscular junction abnormalities are the most common neuromuscular problem associated with D-penicillamine, however. SSEP are a useful adjunct in assessing the patient with long-tract signs and symptoms due to rheumatoid degeneration of the upper cervical spine.

REFERENCES

1. Aiello I, Serra G, Traina GC, Tugnoli V. Entrapment of the suprascapular nerve at the spinoglenoid notch. *Ann Neurol* 1982; 12:314.
2. Albers JW, Beals CA, Levine SP. Neuro-

muscular transmission in rheumatoid arthritis, with and without penicillamine treatment. *Neurology* 1981; 31:1562.
3. Albers JW, Hodach RJ, Kimmel DW, Treacy WL. Penicillamine-associated

myasthenia gravis. *Neurology* 1980; 30:1246.

4. Amick LD. Muscle atrophy in rheumatoid arthritis: An electrodiagnostic study. *Arthr Rheum* 1960; 3:54.

5. Aminoff MJ. Electromyography. In Aminoff MJ (editor): *Electrodiagnosis in Clinical Neurology*, 2nd ed. New York, Churchill Livingstone, 1986.

6. Aminoff MJ. The clinical role of somaosensory evoked potential studies: A critical appraisal. *Muscle Nerve* 1984; 7:345.

7. Aminoff MJ. (editor): *Electrodiagnosis in Clinical Neurology*, 2nd ed. New York, Churchill Livingstone, 1986.

8. Aminoff MJ, Goodin DS, Barbaro NM, Weinstein PR, Rosenblum ML. Dermatomal somatosensory evoked potentials in unilateral lumbosacral radiculopathy. *Ann Neurol* 1985; 17:171.

9. Balagtas-Balmaseda OM, Grabois M, Balmaseda PF, Lidsky MD. Cubital tunnel syndrome in rheumatoid arthritis. *Arch Phys Med Rehabil* 1983; 64:163.

10. Barnes CG, Currey HLF. Carpal tunnel syndrome in rheumatoid arthritis. A clinical and electrodiagnostic survey. *Ann Rheum Dis* 1967; 26:226.

11. Baylan SP, Paik SW, Barnert AL, et al: Prevalence of the tarsal tunnel syndrome in rheumatoid arthritis. *Rheum Rehabil* 1981; 20:148.

12. Berry H, Bril V. Axillary nerve palsy following blunt trauma to the shoulder region: A clinical and electrophysiological review. *J Neurol Neurosurg Psychiat* 1982; 45:1027.

13. Bigliani L, Perez-Sanz JR, Wolfe IN. Treatment of trapezius paralysis. *J Bone Joint Surg* 1985; 67A:871.

14. Brumfield RH. Carpal tunnel syndrome in rheumatoid arthritis. *Orthop Rev* 1983; 12:69.

15. Buchtal F. Electromyography in the evaluation of muscle diseases. *Neurol Clin* 1985; 3:573.

16. Carfi J, Ma DM. Posterior interosseous syndrome revisied. *Muscle Nerve* 1985; 8:499.

17. Cassvan A, Pease WS, MacLean IC, Ma DM, Johnson EW. Electrodiagnosis: central evoked potentials. *Arch Phys Med Rehabil* 1987a; 68:S23.

18. Cassvan A, Pease WS, MacLean IC, Ma DM, Johnson EW. Electrodiagnosis: Late responses. *Arch Phys Med Rehabil* 1987; 68:S19.

19. Chamberlain MA, Bruckner FE. Rheumatoid neuropathy: Clinical and electrophysiological features. *Ann Rheum Dis* 1970; 29:609.

20. Chamberlain MA, Corbett M. Carpal tunnel syndrome in early rheumatoid arthritis. *Ann Rheum Dis* 1970a; 29:149.

21. Chang LW, Gowans JDC, Granger CV, Millender LH. Entrapment neuropathy of the posterior interosseous nerve: A complication of rheumatoid arthritis. *Arthr Rheum* 1972; 15:350.

22. Chang RW, Bell CL, Hallett M. Clinical characteristics and prognosis of vasculitic mononeuropathy multiplex. *Arch Neurol* 1984; 41:608.

23. Chiappa KH. *Evoked Potentials in Clinical Medicine*. New York, Raven Press, 1983.

24. Chiappa KH, Ropper AH. Evoked potentials in clinical medicine. *New Engl J Med* 1982; 306:1205.

25. Cho DS, MacLean IC. Pronator syndrome: Establishment of electrophysiologic parameters (abstract). *Arch Phys Med Rehabil* 1981; 62:531.

26. Chodoroff G, Honet JC. Cheiralgia paresthetica and linear atrophy as a complication of local steroid injection. *Arch Phys Med Rehabil* 1985; 66:637.

27. Chodoroff G, Lee DW, Honet JC. Dynamic approach in the diagnosis of thoracic outlet syndrome using somatosensory evoked responses. *Arch Phys Med Rehabil* 1985a; 66:3.

28. Conn DL, McDuffie FC, Dyck PJ. Immunopathologic study of sural nerves in rheumatoid arthritis. *Arthr Rheum* 1972; 15:135.

29. Cucher BG, Goldman AL. D-penicillamine-induced polymyositis in rheumatoid arthritis. *Ann Int Med* 1976; 85:615.

30. Dawson DM, Hallett M, Millender LH. *Entrapment Neuropathies*. Boston, Little, Brown & Co., 1983.

31. Daube J. Nerve conduction studies. In Aminoff MJ (editor): *Electrodiagnosis in Clinical Neurology*, 2nd ed. New York, Churchill Livingstone, 1986.

32. DeLisa JA, MacKenzie K, Barran M. *Manual of Nerve Conduction Velocities and Somatosensory Evoked Potentials*, 2nd ed. New York, Raven, 1987.

33. DeLisa JA, Saeed MA. The tarsal tunnel syndrome. *Muscle Nerve* 1983; 6:664.

34. Dellon AL, Mackinnon SE. Radial sensory nerve entrapment. *Arch Neurol* 1986; 43:833.

35. Diagnosis of the carpal tunnel syndrome. *Lancet* 1985; April 13; 854.

36. Dillin L, Hoaglund FT, Scheck M. Brachial neuritis. *J Bone Joint Surg* 1985; 67:878.

37. Edmonds ME, Jones TC, Saunders WA, Sturrock RD. Autonomic neuropathy in rheumatoid arthritis. *Br Med J* 1979; 2:173.

38. Eisen A. Electrodiagnosis of radiculopathies. *Neurol Clin* 1985; 3:495.

39. Eisen A. The somatosensory evoked potential. *Can J Neurol Sci* 1982; 9:65.

40. Eisen A, Aminoff MJ. Somatosensory Evoked Potentials. In Aminoff MJ (editor): *Electrodiagnosis in Clinical Neurology*, 2nd ed. New York, Churchill Livingstone, 1986.

41. Electrodiagnosis of ulnar neuropathies. *Lancet* 1987; July 4: 25.

42. England JD, Sumner AJ. Neuralgic amyotropy: An increasingly diverse entity. *Muscle Nerve* 1987; 10:60.

43. Evans BA, Stevens JC, Dyck PJ. Lumbosacral plexus neuropathy. *Neurology* 1981; 31:1327.

44. Falck B, Hurme M, Hakkarainen S, AArnio P. Sensory conduction velocity of plantar digital nerves in Morton's metatarsalgia. *Neurology* 1984; 34:698.

45. Fam AG, Gordon DA, Sarkozi J, et al: Neurologic complications associated with gold therapy for rheumatoid arthritis.

46. Fawcett PRW, McLachlan SM, Nicholson LVB, Zohar A, Mastaglia F. D-Penicillamine-associated myasthenia gravis: Immunological and electrophysiological studies. *Muscle Nerve* 1982; 5:328.

47. Felsenthal G, Mondell DL, Reischer MA, Mack RH. Forearm pain secondary to compression syndrome of the lateral cutaneous nerve of the forearm. *Arch Phys Med Rehabil* 1984; 65:139.

48. Gilliatt RW. Electrophysiology of peripheral neuropathies—an overview. *Muscle Nerve* 1982; 5:S108.

49. Gilliatt RW. Physical injury to peripheral nerves—physiologic and electrodiagnostic aspects. *Mayo Clin Proc* 1981; 56:361.

50. Glover JL, Worth RM, Bendick PJ, Hall PV, Markand OM. Evoked responses in the diagnosis of thoracic outlet syndrome. *Surgery* 1981; 89:86.

51. Good AE, Christopher RP, Koepke GH, Bender LF, Tarter ME. Peripheral neuropathy associated with rheumatoid arthritis. *Ann Intern Med* 1965; 63:87.

52. Goodgold J, Eberstein A. *Electrodiagnosis of Neuromuscular Diseases*, 3rd ed. Baltimore, Williams & Wilkins, 1983.

53. Grabois M, Puents J, Lidsky M. Tarsal tunnel syndrome in rheumatoid arthritis. *Arch Phys Med Rehabil* 1981; 62:401.

54. Granger CV. Toward an earlier forecast of recovery in Bell's Palsy. *Arch Phys Med Rehabil* 1967; 48:273.

55. Gross MS. Gelberman Rh. The anatomy of the distal ulnar tunnel. *Clin Orthop Relat Res* 1985; 196:238.

56. Hall CD. Neurovascular syndromes at the thoracic outlet. In Hadler NM (editor): *Clinical Concepts in Regional Musculoskeletal Illness*. London, Grune & Stratton, Inc., 1987.

57. Halla JT, Koopman WJ, Fallahi S, et al: Rheumatoid Myositis. Clinical and histological features and possible pathogenesis. *Arthr Rheum* 1984; 27:737.

58. Hallett M. Electrophysiologic approaches to the diagnosis of entrapment neuropathies. *Neurol Clin* 1985; 3:531.

59. Hartz CR, Linscheid RL, Gramse RR, Daube JR. Pronator teres syndrome: Compressive neuropathy of the median nerve. *J Bone Joint Surg* 1981; 63A:885.

60. Hollinshead WH, Jenkins DB. *Functional Anatomy of the Limbs and Back*. Philadelphia, W.B. Saunders Co., 1981.

61. Hong CZ, Lee S, Lum P. Cervical radiculopathy—clinical, radiographic and EMG findings. *Orthop Rev* 1986; 15:31.

62. Howard FM, Hill NC. Entrapment of the anterior interosseous nerve. *Contemp Orthop* 1986; 12:43.

63. Hurst LC, Weissberg D, Carroll RE. The relationship of the double crush to carpal tunnel syndrome (an analysis of 1000 cases of carpal tunnel syndrome). *J Hand Surg* 1985; 10B:202.

64. Jablecki CK. Electrodiagnostic evaluation of patients with myasthenia gravis and related disorders. *Neurol Clin* 1985; 3:557.

65. Jabre JF. Ulnar nerve lesions at the wrist: New technique for recording from the sensory dorsal branch of the ulnar nerve. *Neurology* 1980; 30:873.

66. Jaeckle KA, Young DF, Foley KM. The natural history of lumbosacral plexopathy in cancer. *Neurology* 1985; 35:8.

67. Jerrett SA, Cuzzone LJ, Pasternak BM. Thoracic outlet syndrome—electrophysiologic reappraisal. *Arch Neurol* 1984; 41:960.

68. Jhee WH, Oryshkevich RS, Wilcox R. Severe carpal tunnel syndrome with sparing of sensory fibers. *Orthop Rev* 1986; 15:93.

69. Jimenea J, Uddin J. Nerve conduction studies in neuropathy associated with rheumatoid arthritis. *Am J Phys Med* 1971; 50:161.

70. Johnson EW. *Practical Electromyography*. Baltimore, Williams & Wilkins, 1980.

71. Kaplan PE, Kernahan WT. Tarsal tunnel syndrome. *J Bone Joint Surg* 1981; 63A:96.

72. Kashani SR, Moon AH, Gaunt WD. Tibial nerve entrapment by a Baker cyst: Case report. *Arch Phys Med Rehabil* 1985; 66:49.

73. Kelly FJ, McGrath JA, Goldspink DF, Cullen MJ. A morphological/biochemical study on the actions of corticosteroids on rat skeletal muscle. *Muscle Nerve* 1986; 9:1.

74. Khatri BO, Baruah J, McQuillen MP. Correlation of electromyography with computed tomography in evaluation of lower back pain. *Arch Neurol* 1984; 41:594.

75. Kieffer SA, Cacayorin ED, Sherry RG. The radiological diagnosis of herniated lumbar intervertebral disk—a current controversy. *J Am Med Assoc* 1984; 251:1192.

76. Kim SM, Goodrich A. Isolated proximal musculocutaneous nerve palsy. *Arch Phys Med Rehab* 1984; 65:735.

77. Kimura J. *Electrodiagnosis in Diseases of Nerve and Muscle*. Philadelphia, F.A. Davis, 1983.

78. Kincaid JC, Phillips LH, Daube JR. The evaluation of suspected ulnar neuropathy at the elbow. *Arch Neurol* 1986; 43:44.

79. Kori SH, Foley KM, Posner JB. Brachial plexus lesions in patients with cancer: 100 cases. *Neurology* 1981; 31:45.

80. Kline DG, Hackett ER, Happel LH. Surgery for lesions of the brachial plexus. *Arch Neurol* 1986; 43:170.

81. Kramer LD, Ruth RA, Johns ME, Sanders DB. A comparison of stapedial reflex fatigue with repetitive stimulation and single-fiber EMG in myasthenia gravis. *Ann Neurol* 1981; 9:531.

82. Krane SM, Simon LS. Rheumatoid arthritis: Clinical features and pathogenetic mechanisms. *Adv Rheum* 1986; 70:263.

83. Kunci RW, Pestronk A, Drachman DB, Rechthand E. The pathophysiology of penicillamine-induced myasthenia gravis. *Ann Neurol* 1986; 20:740.

84. Lachman T, Shahani BT, Young RR. Late responses as aids to diagnosis in peripheral neuropathy. *J Neurol Neurosurg Psychiatr* 1980; 43:156.

85. Lang AH, Kalliomaki JL, Puusa A, Halonen JP. Sensory neuropathy in rheumatoid arthritis. *Scand J Rheum* 1981; 10:81.

86. Lederman RJ, Wilbourn AJ. Brachial plexopathy: Recurrent cancer or radiation? *Neurology* 1984; 34:1331.

87. Levin RA, Felsenthal G. Handcuff neuropathy: Two unusual cases. *Arch Phys Med Rehabil* 1984; 65:41.

88. Liveson JA. Nerve lesions associated with shoulder dislocation: An electrodiagnostic study of eleven cases. *J Neurol Neurosurg Psychiat* 1984; 47:742.

89. Lund HI, Nielsen M. Penicillamine-induced dermatomyositis. *Scand J Rheum* 1983; 12:350.

90. Ma DM, Liveson JA. *Nerve Conduction Handbook*. Philadelphia, F.A. Davis, 1983.

91. Ma MD, Pease WS, MacLean IC, Cassvan A, Johnson EW. Electrodiagnosis: Needle examination. *Arch Phys Med Rehabil* 1987; 68:S7.

92. MacLean IC. Nerve root stimulation to evaluate conduction across the brachial and lumbosacral plexuses. Third Annual Continuing Education Course, American Association of Electromyography and Electrodiagnosis, September 25, 1980, Philadelphia.

93. MacLean IC, Pease W, Ma DM, Cassvan A, Johnson EW. Electrodiagnosis: Peripheral evoked potentials. *Arch Phys Med Rehabil* 1987; 68:S13.

94. MacLean IC, Taylor RS. Nerve root stimulation to evaluate brachial plexus conduction. Abstracts of Communications of the Fifth International Congress of Electromyography, Rochester, Minnesota, 1975, p. 47.

95. Marks JS. Rheumatoid neck. *Br J Hosp Med* 1985; 33:96.

96. Marmor L, Lawrence JF, Dubois E. Posterior interosseous nerve paralysis due to rheumatoid arthritis. *J Bone Joint Surg* 1967; 49A:381.

97. Massey EW. Digitalgia paresthetica with digital neuropathy in rheumatoid arthritis. *South Med J* 1983; 76:923.

98. McLean M. Total hip replacement and sciatic nerve trauma. *Orthopedics* 1986; 9:1121.

99. Meythaler JM, Reddy NM, Mitz M. Serratus anterior disruption: A complication of rheumatoid arthritis. *Arch Phys Med Rehabil* 1986; 67:770.

100. Millender LH, Nalebuff EA, Holdsworth DE. Posterior interosseous nerve syndrome secondary to rheumatoid synovitis. *J Bone Joint Surg* 1973; 55A:375.

101. Miller EH, Benedict FE. Stretch of the femoral nerve in a dancer. *J Bone Joint Surg* 1985; 67A:315.

102. Miller RG. The cubital tunnel syndrome. Precise localization and diagnosis. *Ann Neurol* 1979; 6:56.

103. Monga TN, Shanks GL, Poole BJ. Sensory palmar stimulation in the diagnosis of carpal tunnel syndrome. *Arch Phys Med Rehabil* 1985; 66:598.

104. Moore AP, Williams AC, Hillenbrand P. Penicillamine induced myasthenia gravis reactivated by gold. *Br Med J* 1984; 288:192.

105. Moritz U. Studies on motor nerve conduction in rheumatoid arthritis. *Acta Rheum Scan* 1984; 10:99.

106. Oh SHJ, Kim HS, Ahmad BK. Electrophysiological diagnosis of interdigital neuropathy of the foot. *Muscle Nerve* 1984; 7:218.

107. Oh SHJ, Kim HS, Ahmad BK. The near-nerve sensory nerve conduction in tarsal tunnel syndrome. *J Neurol Neurosurg Psychiat* 1985; 48:999.

108. Olney R, Miller RG. Copnduction block in compression neuropathy: Recognition and quantification. *Muscle Nerve* 1984; 7:662.

109. Pallis CA, Scott JT. Peripheral neuropathy in rheumatic arthritis. *Br Med J* 1965; 1:1141.

110. Parry GJG. Mononeuropathy multiplex. *Muscle Nerve* 1985; 8:493.

111. Pellicci PM, Ranawat CS, Tsairis P, Bryan WJ. A prospective study of the progression of rheumatoid arthritis of the cervical spine. *J Bone Joint Surg* 1981; 63A:342.

112. Petersen I. An electromyographic study of the atrophied first dorsal interosseous muscle in rheumatoid arthritis. *Acta Rheum Scand* 1955; 1:67.

113. Petrera JE, Trojaborg W. Conduction

studies along the accessory nerve and follow-up of patients with trapezius palsy. *J Neurol Neurosurg Psychiat* 1984; 47:630.

114. Peyronnard JM, Charron L, Beaudet F, Couture F. Vasculitic neuropathy in rheumatoid disease and Sjogren syndrome. *Neurology* 1982; 32:839.

115. Phillips LH. Familial long thoracic nerve palsy: A manifestation of brachial plexus neuropathy. *Neurology* 1986; 36:1251.

116. Pickett JB. Localizing peroneal nerve lesions to the knee by motor conduction studies. *Arch Neurol* 1984; 41:192.

117. Pidgeon KJ, Abadee P, Kanakamedala R, Uchizono M. Posterior interosseous nerve syndrome caused by an intermuscular lipoma. *Arch Phys Med Rehabil* 1985; 66:468.

118. Pitkeathly DA, Coomes EN. Polymyositis in rheumatoid arthritis. *Ann Rheum Dis* 1966; 25:127.

119. Pool KD, Feit H, Kirkpatrick J. Penicillamine-induced neuropathy in rheumatoid arthritis. *Ann Int Med* 1981; 95:457.

120. Rask MR. Superior gluteal nerve entrapment syndrome. *Muscle Nerve* 1980; 3:304.

121. Rayan GM, Foster DE. Handcuff compression neuropathy. *Orthop Rev* 1984; 13:81.

122. Reddy MP. Ulnar nerve entrapment syndrome at the elbow. *Orthop Rev* 1983; 12:69.

123. Reinstein L, Alevizatos A, Twardzik GH, DeMarco SJ. Femoral nerve dysfunction after retroperitoneal hemmorrhage: Pathophysiology revealed by computed tomography. *Arch Phys Med Rehabil* 1984; 65:37.

124. Reynolds G. Restless leg syndrome and rheumatoid arthritis. *Br J Med* 1986; 292:1203.

125. Reynolds G, Blake DR, Spall HS, Williams A. Restless leg syndrome and rheumatoid arthritis. *Br Med J* 1986; 292:659.

126. Sabin TD. Classification of peripheral neuropathy: The long and the short of it. *Muscle Nerve* 1986; 9:711.

127. Saeed MA, Gatens PF. Anterior interosseous nerve syndrome: Unusual etiologies. *Arch Phys Med Rehabil* 1983; 64:182.

128. Sander JE, Sharp FR. Lumbosacral plexus neuritis. *Neurology* 1981; 31:470.

129. Shabas D, Scheiber M. Suprascapular neuropathy related to the use of crutches. *Am J Phys Med* 1986; 65:298.

130. Siliski JM, Scott RD. Obturator nerve palsy resulting from intrapelvic extrusion of cement during total hip replacement. *J Bone Joint Surg* 1985; 67A:1225.

131. Silverstein A. Neuropathy in hemophilia. *JAMA* 1964; 190:554.

132. Simon KJ, Goodrich A. Femoral Neuropathy. *Orthopedics* 1984; 7:1841.

133. Somatosensory potentials and cervical cord disease. *Lancet* 1987; March 7: 546.

134. Sridhara CR, Izzo KL. Terminal sensory branches of the superficial peroneal nerve: An entrapment syndrome. *Arch Phys Med Rehabil* 1985; 66:789.

135. Stalberg E. Clinical electrophysiology in myasthenia gravis. *J Neurol Neurosurg Psychiatr* 1980; 43:622.

136. Stein B. Lumbar disk diagnosis (editorial). *Ann Neurol* 1984; 41:593.

137. Steinberg VL, Wynn Parry CB. Electromyographic changes in rheumatoid arthritis. *Br Med J* 1961; 630.

138. Stevens JC. The electrodiagnosis of carpal tunnel syndrome. *Muscle Nerve* 1987; 10:99.

139. Stevens JM, Kendall BE, Crockard HA. The spinal cord in rheumatoid arthritis with clinical myelopathy: A computed myelographic study. *J Neurol Neurosurg Psychiatr* 1986; 49:140.

140. Strandberg B. Certain muscular diseases in patients with rheumatoid arthritis. *Acta Rheum Scand* 1968; 12:1.

141. Suranyi L. Median nerve compression by Struthers ligament. *J Neurol Neurosurg Psychiat* 1983; 46:1047.

142. Susens GP, Hendrickson CG, Mulder MJ, Sams B. Femoral nerve entrapment secondary to a heparin hematoma. *Ann Int Med* 1968; 69:575.

143. Swift TR. Disorders of neuromuscular transmission other than myasthenia gravis. *Muscle Nerve* 1981; 4:334.

144. Synek VM. Somatosensory evoked potentials from musculocutaneous nerve in the diagnosis of brachial plexus injuries. *J Neurol Sci* 1983; 61:443.

145. Synek VM. Validity of median nerve somatosensory evoked potentials in the diagnosis of supraclavicular brachial plexus lesions. *EEG Clin Neurophys* 1986; 65:27.

146. Taylor AR. Ulnar nerve compression at the wrist in rheumatoid arthritis. *J Bone Joint Surg* 1974; 56B:142.

147. Thomas JE, Cascino TL, Earle JD. Differential diagnosis between radiation and tumor plexopathy of the pelvis. *Neurology* 1985; 35:1.

148. Tonzola RF, Ackil AA, Shahani BT, Young RR. Usefulness of electrophysiological studies in the diagnosis of lumbosacral root disease. *Ann Neurol* 1981; 9:305.

149. Trojaborg W. Electrodiagnosis in rheumatic diseases. *Clin Rheum Dis* 1981; 7:349.

150. Upton ARM, McComas AJ. The double crush in nerve entrapment syndromes. *Lancet* 1973; 2:359.

151. Vaisrub S. Autonomic neuropathy complicating rheumatoid arthritis (editoral). *JAMA* 1980; 243:152.

152. Van Lis JM, Jennekens FG. Immunofluoresence studies in a case of rheumatoid neuropathy. *J Neurol Sci* 1977; 33:313.

153. Vemireddi NK, Redford JB, Pombejara CN. Serial nerve conduction studies in carpal tunnel syndrome secondary to rheumatoid arthritis, preliminary study. *Arch Phys Med Rehabil* 1979; 60:393.

154. Walsh NE, Dumitru D, Kalantri A, Roman AM. Brachial neuritis involving the bilateral phrenic nerves. *Arch Phys Med Rehabil* 1987; 68:46.

155. Warmolts JR. Electrodiagnosis in neuromuscular disorders. *Ann Intern Med* 1981; 95:599.

156. Wasserman RR, Oester YT, Oryshkevich RS, et al: Electromyographic, electrodiagnostic and motor nerve conduction observations in patients with rheumatoid arthritis. *Arch Phys Med Rehab* 1968; 49:90.

157. Wertsch JJ, Melvin J. Median nerve anatomy and entrapment syndromes: A review. *Arch Phys Med Rehabil* 1982; 63:623.

158. Wilbourn AJ. Common peroneal mononeuropathy at the fibular head. *Muscle Nerve* 1986; 9:825.

159. Wilbourn AJ. Electrodiagnosis of plexopathies. *Neurol Clin* 1985; 3:511.

160. Wilbourn AJ. *True Neurogenic Thoracic Outlet Syndrome. Case Report '47*. American Association of Electromyography and Electrodiagnosis. Rochester, Minnesota, 1981.

161. Witt I, Vestergaard A, Rosenklint A. A comparative analysis of x-ray findings of the lumbar spine in patients with and without lumbar pain. *Spine* 1984; 3:298.

162. Wolf AM, Gonen B. Electrodiagnosic investigation of the neuromuscular lesions in rheumatoid arthritis. *Acta Rheum Scand* 1970; 16:280.

163. Wu JS, Morris JD, Hogan GR. Ulnar neuropathy at the wrist. *Arch Phys Med Rehabil* 1985; 66:685.

164. Yates DAH. Muscular changes in rheumatoid arthritis. *Ann Rheum Dis* 1963; 22:342.

165. Yiannikas C, Shahani BT, Young RR. The investigation of traumatic lesions of the brachial plexus by electromyography and short latency somatosensory potentials evoked by stimulation of multiple peripheral nerves. *J Neurol Neurosurg Psychiat* 1983; 46:1014.

166. Yiannikas C, Walsh JC. Somatosensory evoked responses in the diagnosis of thoracic outlet syndrome. *J Neurol Neurosurg Psychiat* 1983a; 46:234.

167. Yu YL, Jones SJ. Somatosensory evoked potentials in cervical spondylosis. *Brain* 1985; 103:273.

13
Electrotherapy

Joseph Kahn

Although electrotherapy may be thought of as a new discipline, a brief perusal of the literature indicates the use of electrical methods of treatment dating back to Benjamin Franklin in 1757. What is new, however, is the broad array of equipment now available to the clinician. Many of the modalities traditionally relegated to the office or clinic are now compact, portable, and available to the physical therapist regardless of treatment locale. With the emphasis today on home treatment, this facilitates continuity of quality care.

Rather than delve into the complexities of the field, this chapter offers an overview of the modalities now utilized in the physical therapy management of a wide selection of medical and surgical conditions. Note that all medical specialties appear on the roster of conditions referred for electrotherapy, including obstetrics, gynecology, urology, dermatology, otolaryngology, allergy, podiatry, and dentistry, in addition to the traditional trio of orthopedics, neurology, and surgery. For specific techniques, several excellent texts and manuals are available.[9,11]

THERMOTHERAPY

Shortwave Diathermy

Theory. The rapid oscillations of a high frequency current produce heat within tissues, dependent upon the density of the tissue. Wavelengths permitted in the United States for this purpose are 7, 11, and 22 (approximately 45, 27, and 13 megacycles, respectively). Heat could be produced by induction or by a condenser-like technique. Shape and intensity of the heat field produced are determined by position, distance, and type of electrodes utilized. A gentle warmth is obtained by correct "tuning" of the patient circuit to the diathermy's operating frequency.[21]

Indications. Probably one of the most underestimated and underutilized modalities in the field of electrotherapy is diathermy. Often found in the back of the clinic, housing running shoes and lunches, covered with dust and rarely used, this modality offers the clinician effective means of reaching deeper tissues for mild heating, bringing all of the benefits attributable to increased temperatures. This modality is effective in treating chronic obstructive pulmonary disease, sinusitis, bronchitis, pelvic inflammatory conditions, prostatic inflammation, otitis media, in addition to the arthritic joint, deep myositis, and other conditions for which heating procedures are commonly used.

Contraindications. Metallic implants, shrapnel, and fixation hardware contraindicate high frequency heating. Hemorrhage and tumors understandably require caution; hyposensitive areas offer the danger of unfelt burns, and acute inflammation generally serves as a warning away from additional heat. Diathermy should not be applied with a pacemaker present.

Rationale. The effects of heat are known (vasodilation, increased phagocytosis, relaxation of muscle spasm, increased circulation, sedation to nerve endings), but the depth of absorption with diathermy, dependent upon the density of the tissues heated, offers clinicians the opportunity to reach anatomic areas not readily treated otherwise from the surface.[6] Internal capsules of joints, deep cavities, and areas insulated from heat by bony and muscular overlay are prime targets for the astute physical therapist. The well-known physics principle of gas expanding when heated presents an excellent rationale for the diathermic approach to chronic obstructive pulmonary disease, bronchitis, sinusitis, otitis, etc., since the air trapped within these cavities, especially the alveoli, can be heated and thus expanded, allowing increased passage of air through them without heating the overlying tissues.

Microwave Heating (Microthermy)

Theory. Differing from diathermy in theory, microthermy offers a concentrated form of heat from extremely short wavelengths, 10–12.5 cm, or 2450 megacycles. This high frequency radiation is produced by a device, the magnetron, essentially different from the transistors and other components of diathermy apparatus. With the magnetron, a flow of electrons is impelled through a perforated iron disc, shaped like a doughnut, which tremendously increases the speed of the electron flow, therefore increasing the pitch, or frequency. At 2450 megacycles, the beam may be "focused" and directed and needs no "tuning" as with diathermy. At present, microthermy units are strictly controlled by the FDA; sales and manufacturing are regulated and restricted because of the inherent radiation dangers.

Indications. Microthermy apparently heats fatty tissues more readily than diathermy,[6] in which the heat seems to be absorbed in muscles. Where microthermy is absorbed, it is generally of a higher intensity than diathermy and more concentrated. It is advised to use microthermy for areas not overlaid with fatty tissues.

Contraindications. These are essentially the same as for diathermy, with additional caution advised in the areas of the eyes and genitals. The presence of pacemakers is a definite contraindication because of the high frequency involved.

Rationale. The effects of heat are the same as in diathermy, but since the area heated is more concentrated, target tissues are generally smaller and more specific.

ULTRASOUND AND PHONOPHORESIS

Ultrasound

Theory. High frequency soundwaves at 1 megacycle have favorable physiologic effects. These are chemical (increased reactions and metabolic processes), biologic (increased extensibility of tendons and cell membrane permeability), thermal (increased heating, mainly at interfaces), and mechanical ("micromassage" and oscillation of crystalline substances).[23] These beneficial sound waves are produced by imposing a high frequency alternating current (AC) on a piezoelectric crystal, usually a compound of lead-zirconium-titanate (PZT) or quartz, which oscillates at a sympathetic frequency, vibrating a metallic or plastic faceplate to produce the ultrasonic wave. Transmitted through a gel or water medium, this soundwave front is directed to the patient's skin and thence into the tissues beneath. Penetration is said to be in the range of 3–4 cm. However, absorption generally distorts these figures since sound is absorbed differentially by varying densities of the intervening tissues.

Indications. Controversy still exists about the value of ultrasound. I have always regarded it primarily as a mechanical modality, with the advantages of the oscillations for breaking up scar tissue and collagenous material, and as a chemobiologic assist for increased exchanges within the tissues and relaxation of tendon spasm. I have not considered it a prime heating mechanism, since the heat produced is highly localized at interfaces and is dependent upon the location of a moving soundhead at the surface. Some units are designed with stationary transducer heads, utilizing pulsed ultrasound to prevent overheating. (It must be remembered, however, that with pulsed ultrasound the sound is off half the time, so dosages must be regulated and revised constantly.) Some consider ultrasound a heating modality based upon the heat produced within the tissues. There is no question, at this point, about the effectiveness of ultrasound in treatment, but whether thermal or mechanical effects account for the apparent successes of this modality remains to be seen. I have not found ultrasound effective for pain control unless used with phonophoresis, as described below. Ultrasound has proven effective with musculotendinous spasm and as a sclerolytic agent with scar tissue. It is apparently effective in athletic injuries and is utilized to a great degree in sports medicine. Podiatric conditions (e.g., neuromas, warts) also respond well to ultrasound.

Contraindications. Ultrasound should not be applied to the eyes, over the growing epiphyses of children, or in the presence of pacemakers. Testicular function could be impaired by exposure to ultrasound. Caution should be used over bony prominences, because of the danger of periosteal burns, and in areas of sensory loss. Ultrasound in areas of metallic implants, hardware, etc., is also questionable because of the interface heating phenomenon at the metal/tissue boundary.

Rationale. The rapid oscillations tend to soften (disrupt) collagenous fibers, increase circulation and fluid transport across membranes, relax tendons, and speed up favorable metabolic processes, i.e., healing. Ultrasound is sometimes administered under water so that hard to reach anatomic regions and bony prominences may be treated safely and comfortably.

Phonophoresis

Theory. The ability of ultrasonic wavefronts to "push" molecules of substances through the skin is an accepted fact. Phonophoresis differs from iontophoresis, discussed later in the chapter, in that entire molecules rather than just ions are transported to the tissues through the skin. The physiologic effect of phonophoresis depends upon which molecule is introduced, since the mechanical procedure is exactly the same as a routine ultrasound treatment (except for massaging the therapeutic substance into the target skin at the beginning of the treatment). Substances used include anesthetics (lidocaine), antiinflammatory agents (hydrocortisone, salicylate), vasodilators (mecholyl), sclerolytic agents (iodine, salicylate), and metabolic stimulants (zinc). These are usually available to the clinician in ointment form, since phonophoresis is not administered under water or with solutions.[3]

Indications. Phonophoresis should be considered whenever the effects of specific molecules are indicated. Ultrasound becomes the mechanism for applying treatment rather than the treatment modality itself.

Contraindications. Specific sensitivities and allergies to the substances used serve as guidelines with phonophoresis. For example, if a patient's history shows allergic reactions to an intravenous pyelogram or seafood, a molecule other than iodine should be selected.

Rationale. Phonophoresis offers the clinician a noninvasive technique of introducing beneficial substances into the body. I have found it to be effective in pain control, (using lidocaine and hydrocortisone).

IONTOPHORESIS

Long abused by those who don't understand ionic transfer or whose techniques have produced nothing but poor results, this stands as one the most effective, useful, and exciting modalities in the entire spectrum of physical therapeutic approaches. The literature is full of reports of iontophoresis being relegated to the scrap heap because of poor results or burns. Close investigation of these writings, however, uncovers improper procedures, techniques ignorant of basic physics, and just plain poor clinical judgement. Those who utilize this modality properly, with sufficient knowledge of bio- and electrophysics as well as basic physical chemistry, have nothing but praise for this procedure. Thanks to improved equipment, newly available chemicals, and many articles written by clinicians in the past few years, this modality is once again achieving favorable status among practitioners and is being used much more than in the past 50 years. Several manufacturers have already begun to offer adjuncts to adapt existing equipment for iontophoresis (e.g., Electrostim Corp., Joliet, Ill.).

Theory. Briefly, ions of therapeutic substances are introduced into the skin by the electrical repulsion of similarly charged electrodes. A continuous direct current (DC, galvanic current) is mandatory for this, eliminating the use of high voltage DC equipment because of the intermittent nature of the current and the extremely short "on" duration. A 9-volt battery-operated unit is available to clinicians (Phoresor, Motion Control Corp., Salt Lake City). Currents of 5 mA are usually sufficient to provide ionic transfer. Low percentage solutions and ointments, i.e., 1%–5%, are generally adequate, bringing to mind the old Arndt-Schultz law in physiology about smaller stimuli producing the larger responses. Size of the electrodes, ion selection, electrode site placement, duration of treatment, intensity, and polarity all contribute to the success or failure of iontophoresis. These parameters must be studied carefully and selected prudently, based upon the physiologic needs of the patient and not what is "on hand." With iontophoresis technique is essential.[9]

Indications. The target condition determines the proper ion to be used. Anesthetic, anti-inflammatory, vasodilative, sclerolytic, antifungal, metabolic stimulant, and other ointments and solutions are available. Unlike phonophoresis, ion transfer may be administered with solutions, increasing greatly the available sources of substances.

Contraindications. Only individual sensitivities to substances limit the selection of ions to be used. Suggested procedures in the event of rare untoward reactions are found in most texts and manuals covering iontophoresis. Continuous DC is safe to use with pacemakers and metallic implants. The introduced ions penetrate less than 1 mm deep, are absorbed slowly over a variable period of time, and are then swept up into the bloodstream and transported systemically.

Rationale. A painless, noninvasive technique of introducing substances into the body for therapeutic purposes, depending upon the physiologic need of the patient.

ELECTRICAL STIMULATION

All electrical stimulation procedures have in common the purpose of eliciting responses from neuromuscular components, either contraction, relaxation, or analgesia. Muscle stimulation is really nerve stimulation, since the transmission of impulses to the muscle fibers follows the tracks of local neurofibrils and trunks to the end-plate and motor points. Whether one type of stimulation is superior to another seems to be a matter of personal preference. They all are designed to do the same thing—in most instances, to make a muscle contract. Manufacturers, dealers, and salespersons endeavor to promote their products in many ways, including claims of superiority of stimulating qualities and responses. The prudent clinician will filter and assess these claims based upon clinical results and procedural aspects with each model and brand. Again, all electrical stimulation uses the same electrons (perhaps in a different manner) to do the same thing: produce contraction of muscular tissue.

Low Volt DC/AC

Theory. Normal muscles respond to both AC and DC stimulation. With reaction of degeneration (RD) to the associated nerve, the ability to respond to AC is lost. Continuous DC will not elicit a response, since the body accommodates to a continuous flow of current and no response follows. Although the accommodation factor is lost to a great degree with RD, continuous DC will not elicit a response. An interrupted DC is necessitated by RD. With most non–RD conditions, AC is utilized for routine stimulation. A continuous or tetanizing AC, 50–100 Hz, is generally administered to obtain relaxation of a muscle in spasm, while the surged mode, 6–10 surges per minute, is used to exercise the muscle or muscle group. With RD present, an interrupted DC is utilized, usually at 1 pulse per second, traditionally with a hand-held interruptor key electrode. Low volt equipment is primarily designed to use less than 100 volts, at frequencies less than 2000 Hz.

Indications. AC stimulation is suggested in all cases when exercise is indicated for weakness, dysfunction, spasm, and paralysis, regardless of the special area or anatomic location of the muscle.

Contraindications. Care is necessary with recent fractures, hemorrhagic areas, phlebitis, and severe cardiac involvement. With pregnant patients, caution is suggested regarding stimulation in specific regions that can trigger premature contractions.

Rationale. The benefits of this technique include relaxation of spasmodic musculature, exercising of weakened or atrophic musculature, increasing of circulation because of the pumping phenomenon, and analgesia, probably due to the increased endorphin production resulting from electrical stimulation.

High Volt DC

Theory. The fact that higher frequencies (or shorter current durations) are known to offer greater penetration gives the high volt DC apparatus credibility. In an effort to get deeper penetration for muscle stimulation, high volt DC equipment provides an extremely short pulse duration, measured in microseconds, as the operating current. The strength/duration curve phenomenon in electrophysics, simply stated, says that the shorter the pulse duration, the higher the voltage must be in order to elicit a contraction from the stimulated muscle. With such a short pulse, the high volt DC units must use considerably higher voltages than with traditional low volt, usually in the range of 300–500 V. These short duration/high voltage stimuli are usually pulsed between 1 and 100 times per second for effective application, depending upon the clinician's techniques and treatment plans (e.g., tetanizing for relaxation or pulsed for contractions). The somewhat increased penetration may be advantageous for deeper muscular stimulation. No contribution to the recent literature indicates specific advantages to the DC polarity, since the pulses are too short to stimulate muscles with damaged nerve components, and the intermittent nature, plus the short pulse duration and low microamperage, negate the use of high volt DC apparatus for iontophoresis.

Indications. Same as for low volt. Many manufacturers claim favorable results with edema using the high volt DC.[2] Whether it is more effective than traditional low volt remains to be seen.

Contraindications. Same as for low volt, except the higher voltages and short pulse widths may be a problem with pacemakers.

Rationale. Same as for low volt. Additional advantages of polarity have yet to be proven clinically.

Interferential Current

Theory. Interferential currents use high frequency carrier circuits to afford deeper penetration. Frequencies of 4000–4100 Hz do not produce muscle contractions, since they are too rapid for the nerves to respond to. Among clinicians, a generally accepted range of optimal frequencies for stimulation is 80–100 Hz. This range, however, does not penetrate deeply. The physical phenomenon of reinforcement/cancellation of wave forms provides an ideal marriage of the two factors: depth of penetration with 4000–4100 Hz and an optimal frequency for stimulation, 100 Hz. With two separate circuits placed at 90 degrees to each other, (i.e., electrodes in a crossed pattern over the target area), the two higher frequency wave forms cancel each other except for the difference of 100 Hz, which remains to stimulate local musculature at a greater depth than if it had been applied from the surface. In most current apparatus, the clinician can sweep across several frequencies with one of the circuits (i.e., 4000 Hz through 4099 Hz), automatically obtaining stimulation at a variety of frequencies, sequentially or selectively.[8]

Indications. Same as for low volt and high volt DC, except perhaps for the targeting of deeper muscle groups.

Contraindications. Same as for low volt and high volt DC.

Rationale. Interferential current is similar to low and high volt techniques, except for the 80–100 Hz difference in most equipment and the crossed pattern of electrode placement. The differential frequency response offers the clinician the opportunity to stimulate increased numbers of fibers, all of which do not respond to the same frequency. (In our facility, we find that patients usually report the greatest sensation-response at 60 Hz, coincidentally close to the accepted normal tetanizing frequency of 50 Hz, which may indicate that the most fibers are responding. When patients report the greatest sensation-response at lower frequencies, e.g., 1–20 Hz, there is usually RD present, suggesting a better response at lower frequencies due to the increased chronaxie or duration necessitated by the RD. This interesting phenomenon is currently being studied by one of the manufacturers (Heterodyne, Birtcher Corp., El Monte, Calif.) as a research project, since the potential for interferential current as a diagnostic tool is evident.

TRANSCUTANEOUS ELECTRONIC NERVE STIMULATION (TENS)

Theory. This modality is based on two theories. One is the concept of controlling pain by electricity that emerged from the "gate theory" of Wall and Melzack.[15] In this theory, the gating mechanism in the substantia gelatinosa of the spinal cord selectively allows transmission to the higher centers. A fiber (proprioceptive) transmission apparently blocks C fiber (pain) transmission effectively enough to serve as an analgesic mechanism. By designing an apparatus—the TENS unit—that stimulates primarily A fibers, a system was developed to block pain transmission to the thalamus. By utilizing the wave forms, monophasic or biphasic, ideally suited for A fiber transmission, the pain gate is blocked and analgesia results. Proper parameter selection of pulse rate (frequency), pulse width (duration), amplitude (intensity), modulation, special wave form variations (spike, rectangular, sinewave, etc.) is an absolute necessity, as is correct electrode placement.

The second theory involves the enhanced production of endorphins, the endogenous painkiller, by electrical stimulation to the surface of the body.[18] Recent evidence directly attributes increased production of this natural morphine-like chemical to electrical stimulation. The stimulation does not have to be painful. Low frequencies, 1–10 Hz, apparently have a more stimulating effect than higher frequencies.

It may assumed, therefore, that the gate theory explains the immediate effect of TENS, while the delayed effects may be due, in part, to the production of endorphins. All the evidence is not yet in on this matter.

Prescription writing for TENS has as many variants as there are brands, models, and physical therapists. Most, if not all, clinicians report success. Therefore, it is quite probable that as with other modalities, it is still the clinician that makes the difference, not the brand.

Indications. Originally thought of as a control for chronic pain, clinicians have found TENS equally effective with acute pain. TENS does not cure or eradicate the cause of the pain, and proper diagnosis is essential before applying this pain control method so that symptoms are not masked. Current literature carries reports of TENS as an effective modality for treating circulation and blood pressure problems and paresthesia. The use of TENS as a stimulant to non-united fracture ossification is being studied in research and clinical practices throughout the world.[12] (The use of electricity to enhance fracture healing is not new; professional literature on the topic is plentiful, citing use of sophisticated, expensive, and widely different parameters.[5,22] With TENS, the parameters are set by the clinician, the unit is rented for a reasonable cost, and the patient is not immobilized or restricted since the units are worn or carried by the patient and used 4–5 times daily.)

Contraindications. Generally, there are few contraindications to TENS, other than areas to be avoided for electrode placements, severe cardiac disturbances, and the presence of certain pacemakers. The original proscription regarding pregnancy is questionable now, since TENS is being used successfully during labor and delivery.[4,19,20] Each manufacturer lists precautions in their patient manuals.

Rationale. TENS offers a noninvasive method of pain control as a viable alternative to narcotics. Used as prescribed and instructed, patients are not housebound, nor are they limited functionally. Use is widespread today, for a variety of conditions all involved with pain control. TENS is rapidly losing the honor of being the "last resort" and is taking its rightful place among the modalities of choice in painful conditions. As will the other modalities, all TENS units are designed to do the same thing—reduce pain and addiction to narcotics. No one manufacturer has the ideal unit, since that is a determination to be made by the clinician, based upon experience. Most do the job satisfactorily, a competent clinician will get good results from any of the units.

RADIATION

Infrared

Theory. Electromagnetic radiation from the sun or a therapeutic lamp in the wavelength range of 7700 AU to about 120,000 AU, is absorbed as heat at the 3 mm level of the body tissues.[17] When the radiation is absorbed locally, all the benefits of heat accrue and are distributed systemically via the circulation of the warmed blood. Radiant or nonradiant, the effects are generally the same, except for perhaps a slight difference in penetration depths.

Indications. Infrared radiation is administered when superficial, dry heat is indicated. Relaxing and analgesic to a high degree, this radiation has been used for decades in the form of bulbs, radiant "bakers," and other heat-producing devices. No permanent hyperemia is developed, and the effects are purely physical. Effects last up to an hour at most.

Contraindications. Infrared radiation, or heat, is generally not advised in acute inflammatory conditions. It should not be used to treat unusually heat-sensitive patients or those with sensory losses over the target area. The presence of potential or actual hemorrhage is a contraindication.

Rationale. This is a mild form of radiant heat, for superficial tissues, with all of the physiologic benefits of heat.

Ultraviolet

Theory. Ultraviolet radiation from the sun or therapeutic lamps, in the range of 3900 AU to 1500 AU, is absorbed by the body at depths less than 1–3 mm.[14] This is purely a chemical effect, with profound metabolic changes stimulated by the radiation, especially in the area of steroid production. The chemical effects are most pronounced with skin changes, i.e., "sunburn" and tanning, which are delayed in manifesting themselves (usually several hours after exposure) and last several days. The eyes are particularly sensitive to ultraviolet radiation and must be protected from direct exposure to radiation at all times.

Indications. Ultraviolet radiation has traditionally been utilized in the treatment and management of skin diseases such as alopecia, acne, decubiti, and especially psoriasis.

Contraindications. Ultraviolet radiation is not recommended for certain dermatologic conditions e.g. acute eczema, lupus erythematosus and herpes simplex, certain pulmonary tuberculosis, severe diabetic or advanced cardiac conditions, or for photosensitive individuals.[14]

Rationales. Ultraviolet radiation is a natural phenomenon useful in the management of various dermatologic and systemic diseases. Recently, a technique called PUVA has been adopted, although not without reservations.[1] One particular band in the ultraviolet spectrum, the 3600 AU band, is normally not readily absorbed by the body. A photosensitizing medication, a psoralen, is administered prior to exposure to lamps designed to emit ultraviolet radiation in the "black light" band of 3600 AU. Short (measured in seconds instead of minutes) exposures are given for psoriasis with favorable results. The medication, however, has been found to be problematic, affecting platelet count in blood chemistries, resulting in the discontinuance of the PUVA procedures in many facilities.

COLD LASER

The most recent addition to the physical therapist's armamentarium is the cold laser. Laser is literally *l*ight *a*mplification by *s*timulated *e*mission of *r*adiation. Although cold lasers are available in the 904 NM band (infrared), this discussion concerns only the helium-neon cold laser in the 632.8 NM band.

A mixture of helium and neon gases is stimulated electrically to emit radiation in the 632.8 NM band. This radiation is transmitted along an optic fiber to a probe tip where it is administered to the target tissues. This mechanical phase is the practical demonstration of the physical phenomenon of incandescent substances emitting characteristic, identifiable radiation. This phenomenon is useful in astronomy and astrophysics, enabling scientists to identify chemical properties of materials at great distances and at submicroscopic levels.

Three characteristic qualities differentiate laser light from ordinary light:[13]

1. *Monochromaticity*: All radiation emitted from the helium-neon mixture is in the same spectral band of 632.8 NM. Ordinary light is composed of all wavelengths from 390 NM through 770 NM—the visible spectrum of red, orange, yellow, green, blue, indigo, and violet.
2. *Coherence*: All radiation from a single laser source has identical wave forms. This phenomenon leads to reinforcement and amplification because of the conformity of troughs and crests. Ordinary light consists of a conglomeration of multiple wave forms of the various wavelengths in the visible spectrum.

3. *Nondivergence*: Laser emissions are unidirectional, with negligible divergence even at great distances, whereas an ordinary light source emits radiation in all directions.

It is not usually accepted that light has power, but is does, although minimal. The laser, because of the amplification, is considerably more intense than ordinary light. Current models of the cold laser are rated at 1 milliwatt (mW), which is about the same intensity as a 60 W household light bulb held several inches from the skin. When administered in the pulsed mode, the laser loses some of the available power and operates at approximately 50% efficiency, or 0.5 mW. Units with 5 mW of available power are being considered for clinical use.

Research indicates that the 632.8 NU wavelength is ideal for therapeutic purposes because of its unique quality of penetrating the cell membrane.[7] Stimulation is to the intracellular components (e.g., mitochondria) responsible for metabolic changes.[16] RNA, DNA, and other substances vital to growth and repair have been cited in the literature as affected by laser radiation at that level.

The physiologic effects of the helium-neon 632.8 NU cold laser are minimal heating and local dehydration, both of which are reversible. Overdosage may lead to protein coagulation, thermolysis, and evaporation, none of which should occur with prescribed dosages and recommended techniques.

Currently the primary use of the cold laser is to enhance healing of open lesions. Administration of the cold laser to open wounds, ulcerations, decubiti, and incisions has proven successful.[10] References are appearing in the literature even at this early stage in the development of this modality.

Exposure of only 90 sec/cm^2 of wound surface is apparently an optimal dosage. Administration is three times weekly, or even daily if indicated, with the probe tip 1–2 mm from the surface. A new lesion requires the beam in a continuous mode, while chronic wounds may respond better to a pulsed mode of 10–20 pps. A possible justification of this protocol is the same as with TENS: production of endorphins and other chemicals is enhanced in the slow rate mode rather than the high rate modes. My clinical experience with this procedure has shown moderate to excellent success with most lesions, except when active infection is present. Pus and drainage apparently preclude laser intervention.

When used for pain management, the cold laser simply becomes another stimulating modality at precise points on the body: trigger points, nerve roots, acupuncture points, and pain sites.

For pain control techniques, the probe tip is in direct contact with the target point(s). Recommended dosage is 15–30 sec per point. For acute pain, the laser should be in the continuous or high rate mode (80 pps). For chronic conditions, a lower rate is suggested (10–20 pps). Treating 6–8 points usually is sufficient to provide rapid relief. The duration of analgesia varies with individuals from minutes to hours.

Since light does have power and can stimulate tissues, it becomes another noninvasive alternative to narcotic dependency for analgesia. Additional physiologic benefits from the cold laser are being studied, including paresthesia modification, circulatory enhancement, increasing ranges of motion, and dental applications.

No reported side-effects or untoward reactions to the cold laser have been reported.

Direct lasing of the eye is contraindicated. The traditional precautions with pregnancy, cardiac problems, and small children apply also to the cold laser. In addition, clinicians must ascertain that patients are not ingesting photosensitizing medication for any condition, related to the target condition or not, since this could seriously affect the results and possibly lead to overabsorption.

COMMENTARY

- With any electrotherapy modalities, advertising and marketing methods by manufacturers, dealers, and salespersons must be carefully analyzed. Personal experience and FDA regulations must take precedence over commercial interests.
- Rote button-pushing without competency and background should be avoided at all costs.
- Treat **physiologically,** not according to rote procedures.
- There are advantages to the multiple modality approach, even to using several modalities in the same treatment session. Obtain good results then determine which modality is most effective by eliminating them one at a time.
- Change modalities if treatment is not successful after a reasonable period.
- Introduce physicians to your expertise in this field—they usually have little training in physical therapy and none in electrotherapy.
- These modalities should be used in conjunction with traditional physical therapy procedures such as hydrotherapy, exercises, mobilization, and massage.
- Proper utilization of electrotherapeutic modalities enhances clinical effectiveness and adds to the clinician's professional image. Most important, patients reap untold benefits.

REFERENCES

1. Adrian RM: Photochemotherapy in psoriasis and other skin diseases. *AFP* 1981; May:95.
2. Alon G: *High Voltage Stimulation*. Chattanooga, Tenn. Chattanooga Corp., 1984.
3. Antich TJ: Phonophoresis. *JOSPT* 1982; 4:2.
4. Augustinsson LE, Bohlin P, Bundsen P, et al: Pain relief during delivery by TENS. *Pain* 1977; 4:59.
5. Basset A, Pilla A, Pawluk R: A non-operative salvage of surgically resistant psuedoarthrosis and non-unions by pulsating electromagnetic fields. *Clin Orthop* 1977; 124:76.
6. *Choosing the Best Heat Therapy Method*. Anaheim, CA, Mettler Corp.,
7. Goldman JA, et al: Laser therapy of rheumatic arthritis. *Lasers Surg Med* 1980; 1:93.
8. Goulet MJ: Interferential current now being introduced as a modality. *P T Forum* 1984; 3:37.
9. Kahn J: *Clinical Electrotherapy,* 4th ed. Syosset, NY, J. Kahn, 1985.
10. Kahn J: Open wound management with the HeNe 632.8 AU cold laser: Case reports. *JOSPT* 1984; 6:203.
11. Kahn J: *Principles and Practice of Electrotherapy*. New York, Churchill Livingstone, 1987.
12. Kahn J: TENS for non-united fractures. *JAPTA* 1982; 62:840.
13. Kleinkort J, Foley RA: The cold laser. *Clin*

Management. 1982; 2:30.

14. Licht S: *Therapeutic Electricity and Ultraviolet Radiation*. Licht, New Haven, CT., 1959.

15. Melzack R, Wall PD: Pain mechanisms: A new theory. *Science* 1965; 150:971.

16. Mester E: The effect of laser rays on wound healing. *Am J Surg* 1971; 122:97.

17. Schriber WJ: *A Manual of Electrotherapy*, 4th ed. Philadelphia, Lea & Febiger, 1981.

18. Smith CW, Aarholt E: Effects of endogenous opiates from electricity. Kohn H: Commentary: *Chem Eng News* 1983; :59.

19. Tannenbaum J: TENS for labor and delivery: Protocols. Ein Karem, Israel, Dept. of Electronics, Hadassah Hospital, 1980.

20. TENS in the Management of Labor Pain: Annotated Bibliography. St. Paul, Minn., 3M Corp, Medical Products Division, 1985.

21. Therapeutic Microwave and Shortwave Diathermy. Rockville, MD, U.S. Dept. of Health and Human Services, 1984.

22. Weiss A, Parsons J, Alexander H: DC electrical stimulation of bone growth: Review and current status. *J Med Soc NJ* 1980; 77:7.

23. Williams AR: *Ultrasound: Biophysical Effects*. New York, Acadia Press, 1983.

14

Guidelines for Orthotic Management

Gary M. Yarkony

An orthosis is an orthopedic device used to support and align weak musculature, to prevent or correct deformities, or to improve the function of movable parts of the body. Orthotic devices are commonly referred to as braces or splints. The term orthosis is derived from the Greek word for "straightening," and the use of orthoses probably dates back to early Egyptian times.[15] Excavations from the Nechian Desert provide the first evidence of fracture bracing in mummies from the Fifth Dynasty (2750–2625 BC).[3] Although the term orthotics is now in common usage, the term orthetics has been proposed as more suitable.[3,15]

This chapter presents the general principles and various components of orthotics and reviews their use in specific conditions. The physician should prescribe an orthosis after determining the treatment goals and incorporating the orthosis into the overall plan of management. Since the orthotic device is usually an adjunct to a comprehensive program to correct a deformity, restore a lost function, or improve a physical capability, consultation with the orthotist and physical therapist is mandatory prior to final prescription. After the device is completed, it is the physician's responsibility to perform a final check of the orthosis before it is accepted for patient use. Collaboration between the physician, orthotist, and physical therapist should result in an appropriate device, which the physical therapist is comfortable in training the patient to use.

NOMENCLATURE

Orthotic devices are named for the joints they encompass.[1] For example, the orthosis called a long leg brace in the past is now called a knee-ankle-foot orthosis, or KAFO. A short leg brace is now an ankle-foot orthosis, or AFO. A resting hand splint is now a wrist-hand orthosis, or WHO. An orthosis that involves one joint is a knee orthosis or foot orthosis. This system was devised by the American Academy of Orthopaedic Surgeons (AAOS) to standardize terminology and eliminate confusion caused by the numerous eponyms of the past.[1]

After the joints are specified, the method of control is specified using AAOS nomen-clature.[1]

1. Free (F)—free motion at the joint
2. Assist (A)—external force is applied to increase the range, velocity, or force of a motion
3. Stop (S)—a static unit prevents motion in one direction
4. Resist (R)—external force is applied to decrease a given motion, velocity, or force
5. Variable (V)—allows for adjustment without requiring a structural change in the orthosis
6. Hold (H)—all motion in a desired plane is eliminated
7. Lock (L)—a lock is included as an option that may or may not be activated

This system is outlined in the AAOS publication *Atlas of Orthotics,* with specific examples given.[1]

UPPER EXTREMITY ORTHOTICS

Upper extremity orthotics are static or dynamic.[16] A static orthosis supports a joint in a desired position; a dynamic orthosis allows mobility of a joint. Devices may be as simple as rubber bands or as complex as external power sources using electricity or compressed air. Advanced dynamic systems may use myoelectric controls.

Static systems are generally used to prevent deformity or support tissues in a desired position after surgery, trauma, fractures, burns, or injuries to the nervous system. They may be readjusted or replaced frequently to move a joint into a more functional position. This is one means of treating joint contractures.

Static orthoses for the upper extremity are probably the most commonly used orthotic devices. Extremities paralyzed after stroke, spinal cord injury, or similar conditions are provided with resting hand splints to maintain the wrist and hand in a functional position and prevent flexion deformity. Commonly made of Orthoplast, these are applied to the palmar surface and fastened with velcro straps.

There are numerous variations of orthotic devices for the fingers and hands. They can be designed for stabilization of one or several interphalangeal joints. The opponens orthosis stabilizes the thumb in opposition and stops extension and adduction of the carpometacarpal joint; interphalangeal extension assists may be added. It is attached with a velcro or leather strap. The rigid thumb orthosis holds the thumb in extension for prehension by stabilizing the distal phalanx. To assist in finger extension, an external force, usually a rubber band or wire, is applied. Motion at the metacarpophalangeal joints may be limited by a lumbrical bar. The lumbrical bar or limited metacarpophalangeal extension bar may be fixed or variable.

A commonly used orthosis is the long opponens splint (Fig. 14-1). It stabilizes the wrist and thumb in a position of function. The version of this splint designed at Rancho Los Amigos Hospital (Downey, Calif.) can be modified and used as a support for dynamic splinting and can hold numerous assistive devices for feeding, writing, or typing.

The wrist-driven flexor hinge orthosis[17] (Fig. 14-2) is a device that converts finger

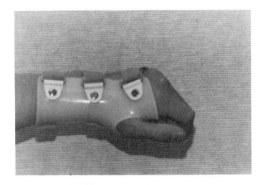

Figure 14-1. Long opponens splint.

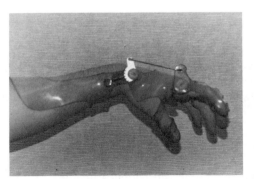

Figure 14-2. Wrist-driven flexor hinge orthosis.

flexion and extension at the metacarpophalangeal joints to provide for prehension. The thumb and index and middle fingers are stabilized to provide a three jaw-chuck. Significant skill and experience are required to fabricate this device. The patient must have a functional wrist extension, and the hand should be free from major deformity. There are numerous designs that incorporate external power, but they tend to be cumbersome and are not frequently used. The Rancho Los Amigos Hospital design is one of the more commonly used of these wrist-driven devices, particularly in C6 quadriplegic patients, who use their extensor carpi radialis longus to power the orthosis.

Support for the shoulder generally involves simple dressings or casts. However, many more complex devices are available. The airplane splint or orthosis holds the shoulder at 90 degrees. It may be used after burns of the axilla or after surgery. Scapular supports may be applied after injury to the long thoracic nerve (winged scapula).[16,18] They may be unilateral or bilateral and are attached via a breastplate.

The balance forearm orthosis[2,16] is also known as the ball bearing forearm orthosis or ball bearing feeder. It is designed for attachment to a wheelchair or other stationary device to assist the patient with loss of shoulder and elbow function. It again requires a skilled therapist because of the complex adjustments needed. It has a trough in which the forearm rests with a rocker arm assembly beneath, attached with swivel arms and ball bearings to the chair. The patient uses proximal movement to power the orthosis and position the hand for activities such as feeding and typing. It may be used initially to train patients in these activities as their proximal musculature improves or for long-term use in well-motivated patients.

Elbow orthoses may be rigid devices to hold the elbow in a desired position. They can have variable hinges for the treatment of contractures or dynamic elbow joints using gravity for extension and the patient's own power or external power for flexion.

SPINAL ORTHOTICS

Cervical

Spinal orthotic devices are used to correct deformity, protect the spine from injury, help decrease pain, and provide support to assist weakened or paralyzed musculature.[4,8] Again, their use should be part of a total treatment program. Inappropriate use can result in muscle weakness and contracture. This section reviews commonly available spinal orthoses.

The soft collar (Fig. 14-3) is probably the most ubiquitous cervical orthosis. It is typically made of foam rubber and does not significantly limit motion of the cervical spine. The major benefit is that of a feedback device reminding the patient to control cervical motion.[5,8]

The Philadelphia collar (Fig. 14-4) is firmer than the cervical collar. It is made of plastizote material reinforced with rigid plastic supports placed both anteriorly and posteriorly, which supports the cervical spine far more than the soft collar but still allows motion.

The four-poster brace (Fig. 14-5) has two rigid uprights anteriorly and posteriorly. They attach to mandibular and occipital supports and to thoracic and chest plates. This

Figure 14-3. Soft cervical collar.

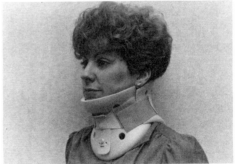

Figure 14-4. Philadelphia collar.

Figure 14-5. Four-poster brace.

Figure 14-6. Sternal occipital mandibular immobilizer.

brace primarily restricts flexion, extension, and lateral bending; rotation is less effectively controlled.[5]

The sternal occipital mandibular immobilizer (Fig. 14-6) is probably the best orthosis for limiting motion of the cervical spine.[5] This lightweight brace has an anterior adjustable bar to hold the chin piece and two metal rods to hold the occipital pad. It is easy to adjust, which is useful since the chin piece is set in different positions for sitting and lying in bed.

The halo brace[1,6] (Fig. 14-7) is a method of skeletal skull fixation that provides three-dimensional control of the head. The brace is secured to the head with four pins and allows for positioning of the cervical spine as desired. A halo vest is connected by metal support posts to the halo ring. Indications include cervical instability or deformity. The halo brace is considered the most effective means of cervical spine immobilization.

Thoracic, Lumbar, and Sacral Orthosis

Orthotic devices for these areas of the spine limit motion. They cannot provide total control. A large part of their effectiveness is in their ability to increase intra-abdominal pressure. This decreases the axial loading of the spine and pressure on the intervertebral discs.

Corsets supply abdominal compression. They can be sacroiliac, lumbosacral, thoracolumbar, or thoracolumbar sacral. They can be made of numerous fabrics including cotton, nylon, and rayon. They may have rigid back stays only or rigid lateral stays as well. These stays help the patient maintain adequate posture.

The chairback brace (LSO) limits trunk flexion and trunk extension in the lumbar spine. Two posterior uprights are attached to a pelvic band and a thoracic band. The Knight brace (Fig. 14-8) is a lumbosacral flexion extension and lateral control orthosis. It has lumbosacral posterior uprights and lateral uprights. The Williams brace (Fig. 14-9) is

Figure 14-7. Halo brace.

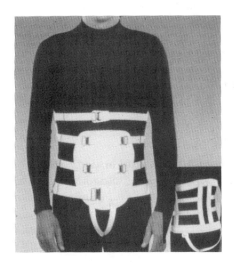

Figure 14-8. Knight brace.

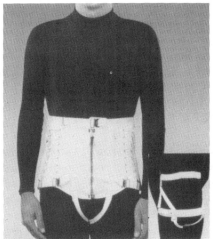

Figure 14-9. Williams brace.

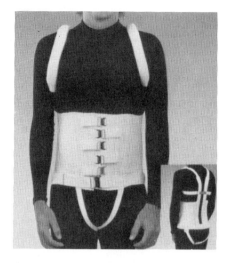

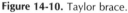

Figure 14-10. Taylor brace.

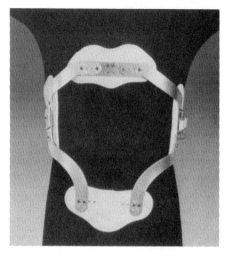

Figure 14-11. Hyperextension thoracolumbar orthosis.

a lumbosacral extension and lateral control orthosis. It allows for free flexion. The Taylor brace (Fig. 14-10) is a thoracolumbosacral flexion-extension control orthosis. To control thoracolumbosacral flexion and extension and provide lateral control, a Knight-Taylor brace is used. Control of rotary motion can be achieved by the addition of pectoral horns or cowhorns. All these braces have the common feature of a corset anteriorly.[1]

The Jewett anterior hyperextension or thoracolumosacral orthosis[1] (Fig. 14-11) is a three-point system with a sternal pad, suprapubic pad, and a thoracolumbar pad. It creates hyperextension and increased lumbar lordosis. It does not have a fabric or provide abdominal support. It can be used to prevent flexion after a compression fracture but should be avoided in elderly osteoporotic individuals.[4]

LOWER EXTREMITY ORTHOTICS

Orthotic devices for the lower extremity may be used to control or limit motion, correct or prevent deformity, or compensate for weakness.[3] Prevention of motion often involves a three-point system to immobilize a joint after trauma or surgery. An example of orthotic usage to treat deformity is the correction of torsional deformities in childhood. Orthotic methods to reduce weight bearing include quadrilateral cuffs in ischial weight-bearing orthoses and patellar tendon-bearing orthoses. The following discussion is limited to orthoses for weakness of the lower extremity.

Knee-ankle-foot orthoses are used by patients with the inability to stabilize their knees, along with weakness about the foot and ankle. Generally, the braces consist of bilateral uprights, upper and lower thigh bands, a knee joint, an ankle joint, calf bands, shoes, stirrups, and a strap to control the knees.[10]

Figure 14-12. Free, Klenzak (single-stop), and double-stop ankle joints.

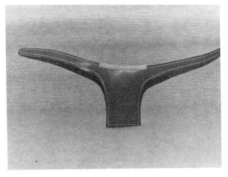

Figure 14-13. T-Strap.

Figure 14-14. Standard metal double upright ankle-foot orthosis.

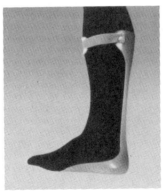

Figure 14-15. Posterior leaf spring orthosis.

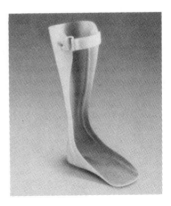

Figure 14-16. Rigid plastic ankle-foot orthosis.

The knee joint[13] can be free and allow extension and flexion but hold against varus and vulgus and can be modified to hold against extension at any angle. Various types of locking knee joints are available. Drop locks, which can be spring loaded, and bail locks, which provide for easy flexion when sitting, are the most commonly used knee joints.

Ankle joints[7] can be free, have a single or doubly stopped ankle joint, or have limited motion (Fig. 14-12). The single stop or Klenzak ankle joint provides a dorsiflexion assist and resists plantar flexion. The double-stopped or double-action ankle adds an anterior dorsiflexion stop.

Shoes are generally Blucher oxfords with a wide-tongued stirrup and an extended steel shank to the metatarsal heads. T-straps (Fig. 14-13) may be added to control varus or valgus deformity. The T-strap to control valgus is attached to the medial malleolus and strapped around the lateral upright. The varus T-strap is attached to the lateral aspect of the shoe and strapped to the medial malleolus.

The Scott-Craig orthosis[11,14] eliminates the lower thigh band, uses a bail lock at the knee and stabilizes the knee with a pretibial pad, and has a double-action ankle. This design is easier to donn and doff. It reduces energy requirements during ambulation because it decreases the oscillation of the center of gravity, the primary determinant of energy consumption during ambulation with KAFOs. This brace was designed specifically for use in paraplegia. Patients generally use a swing-to or swing-through gait.

Ankle foot orthoses[11] are available in numerous designs, including the standard metal double upright (Fig. 14-14), with components as described above, and various plastic orthoses. The plastic orthoses offer the advantage of light weight and improved cosmesis. However, Lehmann[9] has shown that it is the biomechanical function of the orthoses, not the weight, that is of primary importance. If biomechanical functions are the same, ambulation and energy consumption will be similar. Weight of the orthosis has only a minor effect since it does not influence the path of the center of gravity.

There are three commonly available designs of plastic ankle foot orthoses.[12] The Engen orthosis is vacuum-formed polypropylene trimmed posterior to the malleoli. The Teufel orthosis is a posterior leaf spring orthosis (Fig. 14-15) of thermoplastic polycarbon. The Seattle orthosis is laminated plastic trimmed anterior to the malleoli (Fig. 14-16). It provides a moderate force for toe pickup but little knee stability during stance. The Seattle orthosis provides the equivalent of an anterior and posterior stop. The predominant use of the Engen orthosis is to provide top pickup during swing phase.

SUMMARY

Prescription of an orthosis is an individualized process. Each patient should be evaluated in terms of the biomechanical problems that exist and the orthosis prescribed to control these problems. Although certain materials have advantages in terms of cosmesis, weight, or availability, one system should not be used routinely. Strict adherence to any orthotic system only deprives the patient of the potential benefits of each method. Finally, it must be stressed that the best orthoses are obtained only through close collaboration with the therapist, orthotist, and patient, who must accept its ultimate design and function. Without this collaboration the best-designed orthotic device will be one that fits easily in the closet.

REFERENCES

1. American Academy of Orthopaedic Surgeons: *Atlas of Orthotics*, 2nd ed. St. Louis, C.V. Mosby Co., 1975.
2. Bender LF: Upper extremity orthotics. In Kottke FJ, Stillwell GK, Lehmann JF (editors): *Krusen's Handbook of Physical Medicine and Rehabilitation*, 3rd ed. Philadelphia, W.B. Saunders, 1982; p 518.
3. Bunch WH, Keagy RD: *Principles of orthotic treatment*. St. Louis, C.V. Mosby Co., 1975.
4. Fisher JF (editor): *Krusen's Handbook of Physical Medicine and Rehabilitation*, 3rd ed. Philadelphia, W.B. Saunders, 1982, p 530.
5. Fisher SV, Bowa JF, Awad EA, Gullickson Jr G: Cervical orthoses' effect on cervical spine motion: Roentgenographic and goniometric methods of study. *Arch Phys Med Rehabil* 1977; 58:109.
6. Garfin SR, Botte MJ, Waters RK, Nickel VL: Complications in the use of the halo fixation device. *JBJS* 1986; 68A:320.
7. Heizer D: Short leg brace design for hemiplegia. In Perry J, Hislop H (editors): *Principles of Lower Extremity Bracing*. New York, American Physical Therapy Association, 1967, p 76.
8. Johnson RM, Owen JR, Hart DL, Callahan RA: Cervical orthoses: A guide to their selection and use. *Clin Orthop* 1981; 154:34.
9. Lehmann JF: Biomechanics of ankle-foot orthoses: Prescription and design. *Arch Phys Med Rehabil* 1979; 60:200.
10. Lehmann JF: Lower extremity orthotics. In Kottke FJ, Stillwell GK, Lehmann JF (editors): *Krusen's Handbook of Physical Medicine and Rehabilitation*, 3rd ed. Philadelphia, W.B. Saunders, 1982; p 539.
11. Lehmann JF, DeLateur BJ, Warren CG, Simons BC, Guy AW: Biomechanical evaluation of braces for paraplegics. *Arch Phys Med Rehabil* 1969; 50:179.
12. Lehmann JF, Esselman PC, Ko MJ, et al: Plastic ankle-foot orthoses: Evaluation of function. *Arch Phys Med Rehabil* 1983; 64:402.
13. Lehmann JF, Warren CG: Restraining forces in various designs of knee-ankle orthosis: Their placement and effect on the anatomical knee joint. *Arch Phys Med Rehabil* 1976; 57:970.
14. Lehmann JF, Warren CG, Hertling D, et al: Craig-Scott orthosis: A biomechanical functional evaluation. *Arch Phys Med Rehabil* 1976; 57:438.
15. Licht SL: Preface. In Licht S (editor): *Orthotics Etcetera*. New Haven, Elizabeth Licht, 1966; pp xi-xvii.
16. Long C: Upper limb orthotics. In Redford JB (editor): *Orthotics Etcetera*, 2nd ed. Baltimore, Williams & Wilkins 1980, p 190.
17. Nichols PJR, Peach SL, Haworth RT, Ennis J: The value of flexor hinge hand splints. *Prosthet Orthot Int* 1978; 2:86.
18. Troung XT, Rippel DV: Orthotic devices for serratus anterior palsy: Some biomechanical considerations. *Arch Phys Med Rehabil* 1979; 60:66.

15

Musculoskeletal Disorders in the Elderly

Saroj Shah and
Ellen D. Tanner

All demographic projections point to an increasing number of elderly people in the United States over the next few decades.[38] By the year 2030, people 65 years old and older will number almost 60 million, or 18% of our population—compared with today's 23 million, or 11% of the population.[24]

Musculoskeletal pain, stiffness, and reduced vigor are not uncommon problems in the aging population. Many elderly individuals accept their symptoms as an inevitable part of aging and are reluctant to seek medical help.[7] The concept, however, that aging necessarily implies a steady deterioration of physical capacities is no longer tenable in addressing such problems. Physicians should distinguish between rheumatic complaints requiring thorough investigations.

OSTEOARTHRITIS

Numerous epidemiologic studies show an increase in osteoarthritis (degenerative joint disease) with age. Among the risk factors for primary (idiopathic) osteoarthritis, age is the strongest. Furthermore, in people with grade 3 and 4 radiologic changes (those most likely to have symptoms), the increase with age is exponential.[25] Thus, between 45 and 64 years of age, 25% of men and 30% of women were found to have grade 3 and 4 osteoarthritis; in those aged 65 or older, comparable changes were seen in 58% of men and 65% of women.[26]

Elderly patients with osteoarthritis often present with joint pain, stiffness, loss of motion, decreased function, and deformity. These symptoms are associated with inflammatory signs of tenderness, swelling, warmth, and, rarely, erythema. Taking the patient's history is directed toward identifying secondary forms of osteoarthritis, especially trauma, overuse, or previous joint disease. Inflammatory symptoms suggest trauma, crystal deposits, disease, or in rare instances, sepsis.

317

Initially, pain occurs with motion; later, it can also occur at rest. Examinations reveal tenderness, crepitus, diminished range of motion, often bony enlargement, and occasionally inflammatory signs.

Laboratory findings are usually normal. Complete blood count and serum characteristics, although normal, provide indications of secondary causes. The sedimentation rate is normal or, rarely, minimally elevated (<30 mm/hr), mainly in inflammatory osteoarthritis. Arthrocentesis reveals clear, slightly yellow, viscous synovial fluid with <1000 white blood cells per mm^3, with predominance of mononuclear cells. Radiographic examinations allow assessment of severity of cartilage loss and demonstrate loose bodies. They also allow evaluation of adjacent bone density, detection of chondrocalcinosis, and determination of the presence of other joint disease (e.g., rheumatoid arthritis). Indeed radiologic evidence of osteoarthritis does not exclude other disorders as primary causes of the patient's symptoms.

INFLAMMATORY ARTICULAR DISEASE: RHEUMATOID ARTHRITIS

The most common inflammatory articular disease in the elderly is rheumatoid arthritis.[12] Although the disease usually appears in a younger age group, the average age being 35 years, rheumatoid arthritis frequently begins in the sixth or seventh decade of life.[8] Rheumatoid arthritis, a systemic disease, has constitutional symptoms, such as low grade fever, malaise, weight loss, and excessive fatigue. In the elderly it can be categorized into three subsets.

Subset 1 is classic rheumatoid arthritis. The small joints of both hands and both feet are predominantly involved. Rheumatoid factor is present, and rheumatoid nodules are frequent. The synovitis is persistent and tends to produce erosions detectable on radiographs. Disease in this subset can lead to progressive joint damage.

Subset 2 includes patients who have significant titers of rheumatoid factor but a limited and nonprogressive form of arthritis. They also have dry eyes, dry mouth, and labial salivary gland biopsy characteristic of Sjogren's syndrome. These patients meet the American Rheumatologic Association criteria for rheumatoid arthritis because they have persistent, symmetric inflammatory synovitis, frequently involving the wrist and metacarpophalangeal joints. Additionally, symptoms of carpal tunnel syndrome may be present.

Subset 3. The synovitis affects primarily the proximal joints, shoulders, and hips. The onset is sudden, and morning stiffness is often marked. The patient's wrist is often involved, with swelling, tenderness, impaired grip strength, and carpal tunnel syndrome. Rheumatoid factor is not present. This subset represents a self-limiting illness that is well-controlled by low doses of prednisone or anti-inflammatory agents and has a good prognosis.[9]

There is a strong association between polymyalgia rheumatica and the seronegative rheumatoid arthritis[17-20] but none between polymyalgia and seropositive or classic rheumatoid arthritis.[8] Extra-articular manifestations of rheumatoid disease, particularly vasculitis, peripheral neuropathy, purpura, and skin infarction, are rare in patients who developed rheumatoid arthritis after age 60, but they might appear in a patient of the same age who has suffered rheumatoid synovitis for many years.

Treatment of rheumatoid arthritis depends upon the stage of the disease when the diagnosis is made and the subset into which it fits. For pain relief using anti-inflammatory agents, the drug of choice remains aspirin. Aspirin's anti-inflammatory effect, however, is not present in doses of less than 12 tablets per day. At that level, side effects, especially gastrointestinal and central nervous system symptoms, are common. Nonsteroidal anti-inflammatory agents are useful, but manufacturer's recommended anti-inflammatory dosage is advised, and careful monitoring for side effects is necessary. If the patient is unresponsive, especially in seropositive rheumatoid arthritis, potent anti-inflammatory agents, such as gold, should be considered.

Steroids in low dosage are very useful in subsets 2 and 3. The synovitis is well controlled by this well-tolerated regimen. The arthritis commonly subsides in 6 to 24 months. General measures are similar to those described for the treatment of osteoarthritis.

GOUT AND PSEUDOGOUT.

Gout and pseudogout, both inflammatory arthropathies, are noted in older age groups. They are provoked by the formation of microcrystals of monosodium waste and calcium pyrophosphate in joints and surrounding tissue. Crystal deposition is strongly associated with aging.

The onset of crystal arthropathies is abrupt. Classic gout presents with acute onset of pain in the great toe (first metatarsophalangeal joint), swelling, redness, fever, and leukocytes. Gout has a strong preponderance in the middle-aged male and is associated with obesity, hypertension, hyperlipidemia, high IQ, and a high alcohol intake. In the elderly population, diuretics are considered to be responsible for precipitating gout.

In elderly patients, women appear more susceptible than men, and the presentation is different. For example, tophus formation occurs in osteoarthritic joints of the hand with or without inflammation. Acute attacks are less common. These differences have not been explained, but combinations of age, diuretic-induced hyperuricemia, and focal osteoarthritis seem to predispose to tophus formation in the osteoarthritic tissues.

In pseudogout crystal deposition in articular cartilage is common. Pseudogout is a predominantly age-related phenomenon, showing radiographic features that include chondrocalcinosis. Pseudogout is, in fact, the major cause of acute monarthritis in the elderly patient. The knee is the most common site, although wrist, shoulder, and small joints of the hand and temporomandibular joints can all be involved. More than one joint is affected in about 10% of cases. Attacks occur spontaneously or follow several days after a minor injury, intercurrent illness, or surgery. At times, sepsis coexists with pseudogout.

The typical elderly patient with gout or pseudogout is a woman with chronic pain or stiffness of her knees, which often show characteristically marked patellofemoral crepitation. Occasionally, severe destructive disease develops that is similar to Charcot joints.

Less typical presentation includes a self-limiting meningitic syndrome associated with chondrocalcinosis and cervical myelopathy due to deposits in ligamentum flavum.[13]

Aspiration of the affected joint is important in elderly individuals since septic arthritis in this population is always a concern. Indeed, mortality is significantly increased in elderly patients with septic arthritis. The fluid should be cultured and examined by Gram stain. The most likely organism to cause pyoarthrosis in the elderly is *Staphylococcus aureus*.

Treatment

No specific treatment is available except for antibiotic therapy of the septic arthritis. Acute attacks are frequently responsive to aspiration alone. In difficult cases, nonsteroidal anti-inflammatory drugs (NSAID) and intra-articular injection of steroids are useful. For chronic pseudogout, management is similar to that used in osteoarthritis. Analgesics, NSAID, physical therapy, and weight reduction are helpful in relieving symptoms. Oral magnesium carbonate enhances solubility, inhibits precipitation of the crystal, and gives good symptomatic relief.[10,11]

DERMATOMYOSITIS AND POLYMYOSITIS

Polymyositis is a diffuse inflammatory disorder of striated skeletal muscles. It causes symmetric weakness and atrophy, mainly of central groups of muscles—shoulder girdle, pelvic girdle, neck, and pharynx—with involvement of one or more internal organs. This disease has been observed in infancy as well as late in life. The terms dermatomyositis and polymyositis are frequently used synonymously. The clinical findings are similar except that dermatitis is present in dermatomyositis. This disorder can be divided into primary idiopathic polymyositis with or without skin lesions and multisystem connective tissue disorder. In adults, associated neoplasms are noted without any preferred organ site. In 1977, a large number of patients with polymyositis and dermatomyositis were reported. Specific diagnostic criteria were cited and are generally accepted.[4,29]

Dermatomyositis and polymyositis with connective tissue disorders are found mostly in women. Nevertheless, polymyositis associated with malignancy has been observed in men. Moreover, tumor-related polymyositis emerges especially in older age groups.[6] Presenting symptoms include proximal limb weakness, which is more often insidious at onset. Initially, patients are far better judges of muscle strength related to their baselines than any outside observer. Patients usually seek medical advice for muscle weakness noted by difficulty in standing up after sitting, getting out of the bathtub, climbing stairs, and combing their hair. At times patients complain of easy fatigability, malaise, and low grade fever. Weight loss is often rapid and concomitant with swallowing difficulties. Proximal weakness occurs in as many as 95% of patients and dysphasia in 50%. Weakness of neck muscles occurs in a smaller percentage.

Dermatologic findings can be used in determining the diagnosis. The characteristic skin rash is dusky red and most evident about the eyes, with edema of the eyelids. The rash can be seen on the dorsum of interphalangeal and metaphalangeal joints. In the elderly patient, polymyositis must be approached and followed with the possibility of occult malignancy in mind.[6] Malignancy may be detected before, after, or during polymyositis. Raynaud's phenomenon has been observed with polymyositis and dermato-myositis but not usually with malignancy. In addition to skeletal muscle dysfunction, the involvement of pharyngeal muscles, with dysphasia, and respiratory muscles, including intercostal and myocardium, presents major management problems and is associated with a poor prognosis.

Serial muscle enzyme levels are elevated. This assortment of enzymes includes tran-skinase, creative phosphokinase, creatinine, phosphokinase, and the hydrogenous and aldolase indicators of muscle injury. Myoglobinemia and myoglobinuria of mild degree are frequently observed. Electromyographic findings are helpful in establishing the diag-nosis as well as a site for biopsy. Nonetheless, an entirely normal EMG can be present in 10% of the cases with otherwise classic symptoms.[4,29] Muscle biopsy might indicate the presence of inflammatory myositis with cellular infiltration and necrosis of muscle fibers. Diffuse infiltrates of chronic inflammatory cells are sometimes seen near blood vessels or between individual muscle fibers.

Treatments consist of supportive therapy with steroids and in some cases immuno-suppressive or cytotoxic agents. Initial starting steroid dosage for adults is usually high. Analysis of serum enzyme levels is a guide in determining effectiveness of this therapy. The prednisone dosage should be reduced slowly while monitoring enzyme levels. Con-tinued maintenance steroid therapy is commonly necessary. Immunosuppressive agents have been used to supplement steroids.

GIANT CELL ARTERITIS

Giant cell arteritis is commonly seen in association with polymyalgia rheumatica (PMR) (39% in one large series).[19] The age-adjusted incidence rates increase after the sixth decade. About 25% of giant cell arteritis patients present with symptoms characteristic of PMR.[17,19,30] Therefore, patients with PMR should be questioned about visual symptoms, headache, and other symptoms suggestive of giant cell arteritis.

Patients in this subgroup complain of temporal headaches, transient visual loss, jaw dysfunction, or pain in the tongue when chewing. Giant cell arteritis is a serious disease since it affects most of the arterial branches of the aortic arch, including brainstem and ophthalmic arteries. Sudden blindness or brainstem infarction is not uncommon in un-treated temporal arteritis.

Initially, cephalgia (32%), PMR (25%), and unexplained fever (15%) are leading man-ifestations.[5] On examination, decreased temporal artery pulsation and focal arterial ten-derness support diagnosis of giant cell arteritis. Most rheumatologists still recommend temporal artery biopsy in patients with PMR. Characteristic granulomatous arteritis is sometimes found in the sampled tissue. The ratio of positive to negative biopsies can be improved by techniques that include adjacent tissue and bilateral biopsy sites. If arteritis is present in the biopsy, or if the patient has symptoms described above, (even with a negative biopsy) high dosage steroid therapy is often indicated. The level of steroid med-ication is kept at a high level until all symptoms have abated and the erythrocyte sedi-mentation rate has returned to normal. The dose is then gradually reduced.

SJOGREN'S SYNDROME

Sjogren's syndrome is a chronic inflammatory process with diminished or absent secretion of lacrimal and salivary glands. Dryness of the mouth, or xerostomia, and dry eyes, keratoconjunctivitis sicca, are associated with lymphocytic and plasma cell infiltration of glandular tissue (the sicca complex). Sjogren's syndrome occurs alone (primary) or emerges in the setting of another disorder (secondary) like rheumatoid arthritis, systemic lupus, or systemic sclerosis. Sjogren's syndrome is associated with classic rheumatoid arthritis in 50% of the total cases.[16] Primary Sjogren's syndrome is predominantly a disorder of older women.

The sicca complex appears in patients who have had rheumatoid arthritis for a long time and in otherwise normal persons. Eye symptoms can begin with a gritty or sandy sensation. Thick, ropy strands of mucous at the inner canthus of the eye upon awakening can cause distress. Persistent dry mouth, causing difficulty in chewing and swallowing, recurrent epistaxis, otitis media, and bronchitis occur. Some patients have parotid gland enlargement. Extraglandular manifestations affecting many organ systems have been reported, including interstitial pulmonary infiltration, glomerularnephritis, and biliary cirrhosis. This syndrome is probably underdiagnosed, and a high index of suspicion is valuable in examining elderly patients.

Moreover, with these elderly patients, alertness to the possibility of related vasculitis as cause of neurologic events is therapeutically important. This combination produces myelitis, neuritis, and radiculitis. The Schirmer filter paper test (Whatman H-41) provides a practical measure of tear formation. More reliable diagnostic tests include application of rose bengal dye and biomicroscopy techniques to evaluation of these problems.

Treatment is symptomatic. Artificial tears are helpful as is frequent ingestion of fluids. Meticulous oral hygiene is imperative. The management of associated rheumatoid arthritis is similar to that of uncomplicated rheumatoid arthritis. After confirming the diagnosis, treatment effort should be directed to other complications such as intensive chemotherapy for lymphoma. Steroid or immunosuppressive drugs are indicated for severe functional disability or life-threatening complications.

OSTEOPOROSIS

Osteoporosis is the most common metabolic bone disease in older patients. It is defined as an absolute reduction in bone mass. In osteoporosis there is great loss of trabecular bone, accounting for the primary features of the disease, which predisposes the crush fractures of vertebrae and fractures of the neck of the femur and the distal end of the radius. Fractures associated with osteoporosis are considered among the main causes of morbidity and disability in old age.[33,34]

Osteoporosis is more common in women than men (5:1). About 30% of women over 60 years of age have clinical osteoporosis. It is more common in Caucasians than in Blacks. Bone mass declines by about 10% per decade in women after menopause and about 5% per decade in men after age 50.[36,37] Several risk factors have been identified in osteoporosis. These include the female sex, Caucasian or Asian ethnicity, positive family history, nulliparity, early menopause, sedentary lifestyle, alcohol abuse, low calcium intake, high caffeine or sodium intake. To oversimplify, elderly white women do not eat enough dairy foods.

Immobilization plays an important role in bone loss. Patients lose calcium if confined to bed rest, and this continues until they become ambulatory.[2] In response to immobilization, calcium output increases for about 5–6 weeks and then continues at a high level. Loss of bone density in motor paralysis is commonly observed in paraplegia and also in hemiplegia. Weight bearing, however, should be combined with strengthening exercises in limiting disuse osteoporosis. Muscular contraction can be an effective stress upon bone for prevention of disuse osteoporosis. High levels of physical activity during youth provide individuals with higher skeletal mass during the third and fourth decades. This reservoir of bony mass at midlife can delay the clinical manifestations of involutional osteoporosis.[36,37] Exercise can modify involutional bone loss.[2] The most important exercise is the one *not* done with osteoporosis—stretching.

Most patients with osteoporosis are asymptomatic until fractures occur. Fractures of the distal radius and vertebrae are common. Minimal trauma, such as bending or turning in bed, can precipitate fractures. Inappropriately vigorous range of motion exercise precipitates fractures. These fractures usually take the form of wedge and crush fractures of the lower thoracic and lumbar vertebral bodies. Furthermore, one spinal fracture predisposes the patient to others as a result of further alteration in spinal biomechanics. Stress fractures can occur through bony shafts or through joints.

Knowing the nature of the pain is helpful in determining whether an osteoporotic fracture has occurred. Back pain, local tenderness, and muscle spasm are primary symptoms of compression fractures of the spine. Sometimes radiating pain to the flank occurs. Nerve root and spinal cord compression are rare. The pain is usually aggravated by activity, but rest might not relieve it completely. Dull, aching pain of long duration is common, usually in the lower thoracic or lumbar region, and it is made worse by sitting or standing.

Other causes of osteopenia should be ruled out, which could be treatable. Exclusion of osteolytic vertebral lesions should be considered a differential diagnosis. Survey of the skeleton by radiograph bone scan, and CT scan techniques aids in establishing a differential diagnosis.

Current methods available for quantitation of skeletal mass include quantitative CT for measurement of the spine, single energy photon absorptiometry for measuring peripheral bones, and dual energy photon absorptiometry for measuring the spine, hip, and total body calcium. These methods offer dramatic improvement over standard radiographs. Each method provides different, complementary information.

Treatment

The management of osteoporosis after fracture is difficult. The most effective approach is prevention, based on (1) reduction of risk factors, (2) adequate calcium intake, (3) adequate nutrition, (4) physical activity, and (5) low dose estrogen therapy.

Reduction of Risk Factors. Advising the patient about prevention is generally simple—removal of risk factors, i.e., reducing alcohol and caffeine consumption, and increasing calcium intake. Yet, even with educated patients, compliance is often impossible to obtain.

Adequate Calcium Intake. An adequate calcium intake is important for all osteoporosis patients. The usual recommended dose is about 1 gm/day for premenopausal women, increasing to 1.5 gm/day for postmenopausal women. Controlled studies, however, evaluating the effects of supplemental calcium on bone loss have generally, though not uniformly, been negative. Commonly, physical stress determines skeletal status.

Physical Activity. Gravity-resistive exercises probably stimulate new growth formation. Commitment should be made to a regular exercise program; 3–4 hours per week of weight-bearing exercise are beneficial. Physical exercise increases peak bone mass at maturity and reduces bone loss thereafter. Physical exercise is the safest, most economic method for maintaining bone mass. People with compression fractures should be instructed to try pectoral stretching, deep breathing, and back extension exercises after 2–3 weeks of bed rest.[33] In the management of the osteoporotic patient, flexion exercises are contraindicated since they predispose the spine to compression fractures.[34]

Estrogen Replacement Therapy. Estrogen replacement therapy is generally believed to prevent postmenopausal acceleration of bone loss and reduce the incidence of hip fracture in aging women. Low dose estrogen with calcium has been found to be a sufficient prophylactic dose against bone loss in women having spontaneous menopause.[14] The beneficial effects are greater if estrogen is given early in the postmenopausal period of life. Estrogen has serious side effects, such as endometrial hyperplasia, cancer, and vaginal bleeding and should be used only when there are no relative or absolute contraindications and in the smallest effective dosage. Epidemiologic evidence indicates reduction of fracture frequency in estrogen-treated women.

Calcitonin. Calcitonin is a useful drug in the prevention of bone resorption. At present, calcitonin's main disadvantage is that it is expensive and requires parenteral administration. Calcitonin should be considered in high risk women who are not candidates for estrogen, in men with osteoporosis, and in patients with secondary osteoporosis.

Sodium Fluoride. Sodium fluoride is a potent stimulant of new bone formation that is not currently approved by FDA for use in osteoporosis. A combination of fluoride, estrogen, and calcium has been suggested to reduce fracture recurrence. Unfortunately, while fluoride makes the bones sclerotic and radiologically dense, they also become more brittle. Sufficient amounts of calcium are necessary to maintain chemical composition of bone, but exogenous deposition of calcium occurs with a fluoride excess.

PAGET'S DISEASE

Paget's disease is a chronic multifocal disorder of bone, characterized by accelerated bone formation and resorption, that occurs in about 3% of people older than 40 years. Although this disorder is common in its subclinical form, widespread, deforming, or disabling disease is uncommon.[3] Most patients remain asymptomatic, but those who become symptomatic complain of pain or bony deformity. The most common areas affected include the back, pelvis, skull, and lower extremities.

The disease has male predominance and occurs almost exclusively in Caucasians. In more than 80% of cases, Paget's disease is detected after the age of 50. Often the diagnosis is made by an incidental finding of an elevated alkaline phosphatase or a radiograph for other purposes. The diagnosis is frequently made when a patient presents with fracture through a pagetic site.

Paget's disease is usually readily distinguishable from other disorders in its advanced stages. Serum alkaline phosphatase is elevated in active disease. Paget's disease must be differentiated from metastatic prostatic carcinoma. When serum acid phosphatase is elevated, radioimmunoassay for prosthetic acid phosphatase is helpful.[35] Occasionally, bone biopsy is necessary to differentiate Paget's disease from other disease.

Radiologic findings are very typical. The initial radiologic lesion is a radiolucent zone. This lytic zone is replaced by a coarse, heavy, abnormal-appearing trabecular pattern, which is frequently heavily calcified. The bone scan is useful in delineating the extent of disease since the active Paget sites accumulate intense amounts of nuclide—even more than in most malignancies.

Calcitonin, a polypeptide hormone of the C-cells of thyroid, reduces both the osteoclast-mediated bone resorption and the serum calcium concentration. Parenteral administration of calcitonin produces significant suppressive influence on the bone disease. Synthetic salmon calcitonin, administered subcutaneously at intervals ranging from daily to once a week, is probably the most effective form of therapy. Nonetheless, it is also associated with allergic phenomena and is very expensive. Diphosphonates and methramycin have been used with questionable symptomatic relief but are not recommended because of toxic side effects.

MYASTHENIA GRAVIS

Myasthenia gravis is not highly prevalent in the older population and is treatable. When initial symptoms appear late in life, the patient's sole complaint is often weakness, "tiredness," or both. The disease can present in three different forms: (1) mild generalized dysfunction, (2) severe generalized weakness, (3) a subacute progressive course that evolves into respiratory crisis.

In a mild form of myasthenia gravis, generalized symptoms are vague and nonspecific. Significant complaints include difficulty in climbing stairs, grooming, and walking long distances. Bulbar paresis presents initially with problems of chewing and swallowing. Progressive mastication problems are often insidious at onset, becoming evident in time. Difficulty with swallowing first thin, then thick liquids is often reported.

The initial ocular manifestations can be subtle. The patient relates a history of intermittent ptosis that is usually not present in the morning but becomes more obvious as the day progresses.[25] Another frequent ocular symptom, diplopia, is often variable and fluctuating. Routine physical examination includes determination of blood pressure, pulse, respiratory rate, and vital signs. Eyelid closure can be weak since orbicular oculi muscle weakness is present in many patients. This weakness is subtle but can usually be detected with a simple test. In normal patients, no eyelashes are seen with firm eye closure, but the lower eyelashes of myasthenia patients often remain visible. This phenomenon becomes more obvious when repeated several times.

Muscle function is assessed by having the patient repetitively perform simple exercises such as lifting the head from a pillow, holding the arm over the head, extending the fingers, or rising from a low chair. The weakness is not associated with muscle atrophy, fasciculation, loss of reflexes, or sensory loss.

Weakness and fatigability in the elderly can be produced from a variety of disorders affecting nerves, muscles, bones, or joints. Differential diagnostic tests for myasthenia gravis can become complex.[21,22] Polymyositis, for example, can be present with progressive weakness. EMG and muscle biopsy are necessary to confirm the diagnosis before treatment is started.

In myasthenia gravis, circulating antibodies directed against the postsynaptic acetylcholine (ACH) receptors are present and can be detected.[15,31] The antibody blocks and degrades receptors. This process leaves fewer functional receptor sites.[15,24] Symptoms occur because of impaired transmission.

Several readily available tests can help confirm the diagnosis of myasthenia gravis. One is the Tensilon test. The most frequently used pharmacologic aid is edrophonium chloride.[38]

Traditionally, the electromyographer has used repetitive nerve stimulation whereby the peripheral nerve is stimulated supramaximally at the rate of 3 pulses per second. The decremental response of compound muscle action potential amplitude greater than 10% strongly suggests dysfunction of neuromuscular transmissions. Overall, 60%–80% of generalized myasthenia patients can be identified with this method. The accuracy of this test depends on the diligence of the electromyographer and the choice of muscle study. The yield is significantly lower in patients with primary ocular manifestations. The most recent electrophysiologic advance to significantly improve diagnosis yield in patients with myasthenia gravis is single-fiber EMG.[12]

Another highly sensitive test is to assay the titer of circulating ACH receptor antibody. If elevated, it is diagnostic of myasthenia gravis. Although at present the titer does not correlate well with severity of the disease, the test is still useful.

Treatment

Major advances in the definitive treatment of myasthenia gravis have been made in the past decade.

The usual treatment is ACH drugs, which provide symptomatic relief. These are no longer recommended as the only therapy for generalized myasthenia gravis in older patients because more deaths have been reported in this group when they were treated with ACH alone (37.5%). Treatment results are better with a combination of ACH and steroids, steroid medication alone, or plasma exchange.

In many patients, surgical removal of the thymus gland leads to an eventual cure.[26,32] Thymectomy can often induce remission and relieve the possibility of medication becoming a lifelong requirement. Immunosuppressive therapy along with steroid administration can result in long-lasting remission and is recommended[10,11] especially in patients who are older or who are medically unable to tolerate surgical treatment.

Patients who present in myasthenia gravis crisis and who are selected for plasmaphoresis and immunosuppressive therapy usually require years of treatment with immunosuppressive drugs to obtain complete remission. Although the medication is well tolerated, careful medical attention and close supervision are indicated throughout the course because of the potential side effects.

The major embarrassment of treating elderly patients is that they are thought a homogeneous group rather than patients with diverse musculoskeletal medical problems. Consequently, previously mentioned disorders are not even considered, much less organized into a differential diagnosis. Much of the geriatric literature is weak on this matter, emphasizing the control of symptomatology rather than the treatment of disease.

REHABILITATION OF THE ELDERLY PATIENT

Specific therapeutic interventions for many of the diagnostic categories above are discussed in earlier chapters. We shall not, therefore, review the treatments for the arthritides, nor describe general principles of rehabilitation. Numerous factors, however, can complicate the rehabilitation process in elderly patients. These complications may be present in patients with relatively simple musculoskeletal dysfunction but are more likely in patients with complex conditions that require extensive rehabilitation.

Of the special needs and problems of the elderly—about which volumes have been written—two points require special mention. First, the longer a person lives the more prone to physiologic changes he or she becomes. Thus, older patients are more likely to have multiple diagnoses that chronically affect several organ systems. It is, therefore, necessary to take these pre-existing problems into account when rehabilitation programs are begun. For example, it would be expected that a patient with a history of heart diseases must be carefully monitored, both for heart rate and for blood pressure response, for signs of cardiovascular distress during any strenuous exercise. In the case of a patient with Parkinson's disease who has lost the ability to ambulate, certain assistive devices, such as wheeled walkers, make the process of regaining ambulation easier and safer. A wheeled walker eliminates the difficult sequencing process necessary for using a standard walker and might even make the difference between safe ambulation and no ambulation for certain patients.

Second, sensory changes are common in older persons and can cause serious problems. Therefore, with elderly patients extra care should be taken by the rehabilitation team member to be sure that (1) communication is effective, (2) unperceived dangers are eliminated, and (3) no injury is caused iatrogenically.

Vision

Vision is commonly compromised in the elderly, being affected by presbyopia ("old sight"), cataracts, glaucoma, retinitis pigmentosum, macular degeneration, and other problems. The more severe the visual loss, the more the patient loses contact with the environment, and the greater the fear becomes. Especially when out of familiar settings, patients can become confused and disoriented, fearful of injuring themselves on unseen obstacles; somtimes they even withdraw, refusing to participate in the rehabilitation process. Patients with severe vision losses are no longer able to watch television, read, do handwork, play cards, or busy themselves with other enjoyable pastimes.

Many things can be done to assist patients with poor vision, some of which are quite simple. Providing adequate lighting, reducing glare, and using bright colors to identify and locate important objects are helpful. Some people with cataracts are caused discomfort by fluorescent lights. Strong color contrasts should be used on stair treads, and handrails are essential. Throwrugs and other environmental barriers should be eliminated from the home, since they are a serious danger in causing falls. Low vision aids, such as magnifying glasses and large print playing cards and books, should be employed to help these people to improve their quality of life by restoring the ability to read or do handwork.

Hearing

Presbycusis, or "old hearing," involves reduced ability to discriminate sound and a loss of sensitivity to the higher pitched sounds. When hearing is impaired, effective communication is more difficult. Older people often cannot understand what is said, although sound is heard. Since it is embarrassing to ask repeatedly, "What did you say?", some people simply nod, pretending they heard.

Communication with a hearing-impaired person can be improved in several ways. Reducing the speed of speech rather than increasing the loudness, facing the person being addressed, avoiding obstacles in front of the speaker's mouth, and enunciating carefully make it easier for the hearing-impaired to understand. Hearing aids may or may not improve hearing, depending upon what part of the auditory system is involved. Careful audiologic evaluation is indicated to determine what the specific problem entails and whether auditory aids will be of assistance. Patience is required in determining whether inappropriate responses in an older person are due to confusion or to hearing loss.

Sensation

Because of decreased sensation, heat and cold are not well perceived in some elderly patients. Burns or ischemia can be inadvertently caused by carelessly applied modalities. Regular skin checks are necessary during heat or cold application. Reduced intensity of heating modalities, including more toweling on the hot packs, lower intensity ultrasound, etc., are indicated because of the reduced ability of the older person's circulatory system to disperse the heat and thus prevent burns. Further, there is less water content of the tissues, which changes the quality of the heat exchange of ultrasound and diathermy modalities.

Pain

Pain is often perceived differently by the elderly; they have difficulty localizing pain, and the decreased secretion of enkephalins—the body's natural pain killers—causes increased perception of pain. As a result, extra care is required in evaluating and treating pain in the elderly. Special attention must be paid to determining the cause of pain, since older people are sometimes hesitant to complain of pain at all. Complicating the picture are stimuli that might modify the pain, making it either more disabling or less distressing. For example, patients who are depressed and who are not being stimulated sense the pain more acutely, creating a vicious cycle that is difficult to break. While therapeutic modalities are often effective in lessening discomfort, psychological and social counseling can also be helpful.[23] Visual imagery, transcutaneous electric nerve stimulation (TENS), relaxation, and biofeedback are often effective in treatment of chronic pain in the elderly.[27]

Drugs

In elderly patients, as many as 7–12 medications may be prescribed simultaneously for underlying disease processes such as heart conditions, hypertension, visual impairments, diabetes, Parkinson's disease, and pain from one or more sources, as well as psychological manifestations. Ideally, one physician should be in charge of all medication for a patient. All too frequently, however, an older person acquires medications from more than one physician and more than one pharmacy. This increases the likelihood of drug interactions and negative side effects. In addition to the prescription medications, any consumption of over the counter drugs further complicates matters. Some common side effects are

- Sedation
- Depression
- Confusion
- Anxiety
- Postural hypotension
- Blurred vision
- Involuntary movements
- Skin rash
- Parkinson-like symptoms
- Fatigue
- Weakness
- Ataxia
- Incontinence
- Dizziness
- Constipation
- Urinary retention
- Diarrhea
- Tremor

The rehabilitation process can be seriously impaired by these symptoms. Problems like temporary incontinence are sometimes perceived by an older person as a sign of no longer being able to function appropriately, which causes great anguish and fear and can destroy motivation. Without the desire to improve, rehabilitation teams can do little to help a patient to recover lost functional abilities.

Secondary Problems

Secondary medical complications, such as joint contracture, decubitus ulcer, and muscular atrophy from prolonged bed rest are common results of the treatment of a primary problem. These disabilities are difficult to treat and should be prevented; once present, they can greatly prolong the rehabilitation process.

Deconditioning is a frequent result of protracted bed rest. Younger people are affected by bed rest, but the degree of deconditioning is often much worse in the elderly, regardless of the premorbid activity level.[23] During the process of rehabilitation, therefore, it is essential to concentrate on restoring endurance.

Patience on the part of the rehabilitation team is also necessary, since progress can be slowed by both psychological and physical factors. For example, the patient might be terrified of falling, preventing attempts to stand or walk. Shame of the fear compounds the problem, causing depression and withdrawal.

Reduced strength and flexibility, which can create posture and gait abnormalities, may be caused by pathologic entities such as arthritis or by hypokinesis (decreased activity).[27] The less active a person is, the greater the chance that strength and flexibility will be impaired. Sitting in a chair all day encourages muscle shortening, especially of the flexor muscles. While all joints might be affected by these changes to a certain extent, loss of range of motion is not necessarily an indication of loss of function. For example, many older women show a loss of some of the range of motion in flexion, abduction, and external rotation of the shoulder, but often with no noticeable change in the ability to complete functional tasks such as combing the hair. If crepitus is present, however, arthritis could cause the limitation of range of motion: motion is limited to avoid the pain, which in turn causes weakness and further limits range of motion.

Posture

Posture is often affected by loss of strength and flexibility, resulting in flexion at the hips and knees, kyphosis, and a forward head (Fig. 15-1). Tightness of the Achilles tendon is often found in women who wear shoes with heels 1 in or higher. Postural changes are successfully treatable as long as there is no bony restriction. Correction of flexion contractures in one area should not be attempted without correcting the shortened muscles both proximally and distally. In a patient with tight hip flexors, for example, stretching the

Figure 15-1. Common posture changes with age. Note forward head, round shoulders, thoracic kyphosis, reduced lumbar lordosis, and flexed hips and knees.

knee flexion contractures alone cannot significantly improve the posture. Commonly, the neck, shoulders, upper and lower back, hips, knees, and ankles require stretching of tight structures and strengthening of weak muscles in order to restore erect posture.

Patients who return to the same sedentary lifestyle will probably gain little functional advantage, since in time the shortened condition of the muscles will return. Thus, achieving the clinician's goal, i.e., restoration of "normal" posture, offers sustained benefit only to patients who are willing to maintain the gains by making some changes in lifestyle. Patients whose only goal of treatment is to increase the ability to perform a simple activity, such as rising from a chair, may not be interested in a "makeover." Investigation of the specific needs and goals of each patient and insight into his or her lifestyle are invaluable in formulating rehabilitation goals.

Gait

Gait is affected by changes in strength and flexibility. Careful evaluation helps determine the causes of gait deviations and what is required to correct them. Usually the causes are a combination of tight muscles and weak muscles (in the absence of neurologic causes) for which normal balance must be restored. Ambulation devices are invaluable in permitting the restoration of mobility, but are frequently misused. Walkers and canes are often used at inappropriate heights—usually too high, thus reducing the mechanical advantage they afford when used correctly. Once poor habit patterns have been acquired in using a device, the patient must be retrained.

Mental Confusion

While it is beyond the scope of this discussion to describe the numerous conditions of which confusion is a symptom, suffice it to say that there are many causes of confusion in the elderly that can present enormous barriers to the goals of rehabilitation. A confused patient usually cannot learn nor carry over learning from one day to another. Mental confusion is frightening to the family and causes frustration for health care providers.

Finding and reversing the cause of acute confusion is essential if the patient is to progress. Confusion might diappear quickly, or it might linger for long periods. The cause could be as simple as fecal impaction or urinary tract infection or less readily discernible, as in metabolic imbalances. Medications are frequently suspect, especially if the patient takes four or five medications that can interact. "Drug holidays"—periods of taking no drugs—are vital to determine if the side effects of multiple agents are at fault. Confusion after major surgery, often as a result of the anesthesia, is common, and usually clears in time. The cause of confusion can sometimes require complete neurologic investigation to rule out stroke or other brain trauma. These measures are necessary if the patient is to be able to again function with any degree of independence.

Social and Emotional Considerations

Last, but far from least, are the social and emotional considerations that play an enormous part in the lives of the elderly. Older people who have been self-sufficient and independent for 60 or 70 years or more, prefer to live in their own homes and make their own decisions. They like to have the option to socialize or to be alone when the mood strikes. When physical impairments present themselves, the elderly require family or community support systems in order to maintain their lifestyles. Dependence upon others furthers the already perceived loss of identity, productivity, or selfworth. Societal stereotypes of the elderly as useless, ugly, or a burden further the older person's feelings of desperation and helplessness. Most people "need to be needed," as valuable, contributing members of society. Family and friends furnish an "anchor," providing stability and encouragement during times of crisis.

Substantial emotional support from the providers of health care sometimes makes the difference between success and failure of the rehabilitation process and returning the patient to an independent lifestyle. Understanding and making provisions for the unique problems of the elderly makes this process rewarding for the practitioner and the patient alike. Kindness, patience, and understanding will improve the most impaired patient, while their lack will undermine an otherwise effective rehabilitation program for any patient. A smile is worth a thousand words.

REFERENCES

1. Abramson AS, Delagi EF: Influence of weight bearing and muscle contraction on disuse osteoporosis. *Arch Phys Med Rehabil* 1961; 42:147.
2. Aloia JF, et al: Prevention of involutional bone loss by exercise. *Ann Intern Med* 1978; 89:356.
3. Barry HC: *Paget's Disease of Bone*. Baltimore, Williams & Wilkins, 1970.
4. Bohan A, et al: A computer-assisted analysis of 153 patients with polymyositis and dermatomyositis. *Medicine* 1977; 56:255.
5. Calamia KJ, Hunter G: Clinical manifestations of giant cell (temporal) arteritis. *Clin Rheum Dis* 1980; 6:389.
6. Callen JP: Myositis and malignancy clinics. *Ann Rheum Dis* 1984; 10:117.
7. Cape RL: *Aging, Its Complex Management*. Hagerstown, MD, Harper & Row, Inc., 1978.
8. Chuang T, et al: Polymyalgia rheumatica: A 10-year epidemiologic and clinical study. *Ann Intern Med* 1982; 97:672.
9. Corrigan AB, et al: Benign rheumatoid arthritis of the aged. *Br Med J* 1974; 1:444.
10. Dieppe PA, Doherty M: *Crystals and Joint Disease*. London, Chapman & Hall, 1982.
11. Dieppe PA, et al: Crystal-related arthropathies. *Ann Rheum* 1983; 42(Suppl 1):13.
12. Ehrlich GE, et al: Rheumatoid arthritis in the aged. *Geriatrics* 1970; 25:103.
13. Ellman MH, et al: Calcium pyrophosphate deposition in ligamentum flavum. *Arth Rheum* 1978; 21:611.
14. Ettinger B, et al: Postmenopausal bone loss is prevented by treatment with low dosage estrogen with calcium. *Ann Intern Med* 1987; 106:40.
15. Fries JF: Aging, natural death and comparison of morbidity. *New Engl J Med* 1950; 303:130.
16. Hahn B: *Clinical Immunology*, Vol 1. Philadelphia, W.B. Saunders, 1980.
17. Hall S, et al: The coexistence of rheumatoid arthritis and giant cell arthritis. *J Rheum* 1983; 10:995
18. Healey LA: Long term follow-up of polymyalgia rheumatica: Evidence for synovitis. *Semin Arthr Rheum* 1983-84; 13:322.
19. Healey LA: Polymyalgia rheumatica and giant cell arteritis. In McCarty DJ: *Arthritis and Allied Conditions*, 10th ed. Philadelphia, Lea & Febiger, 1985, p 902.
20. Healey LA: Rheumatoid arthritis in elderly. *Clin Rheum Dis* 1986; 12:173.
21. Howard FM, Jr.: Myasthenia gravis with onset after 65 years of age: A preliminary report. In Dau PC (editor): *Plasmapheresis and the Immunobiology of Myasthenia Gravis*. Boston, Houghton Mifflin, 1979, p 281.
22. Howard JF Jr, Sanders DB: The manage-

ment of patients with myasthenia gravis. In Albuquerque EX, Evefrawi AT (editors): *Myasthenia Gravis*. London, Chapman and Hall, 1983, p 457.

23. Jackson O: *Physical Therapy of the Geriatric Patient*. New York, Churchill Livingston, 1983.

24. Kane RL, Kane RA: *Values and Long-Term Care*. Lexington, Mass., Lexington Books, 1982, p 10.

25. Lawrence JS: Surveys of rheumatic complaints in the population. In Dixon ASJ (editor): *Progress in Clinical Rheumatology*. London, J & A Churchill, 1966; p 1.

26. Lawrence JS, et al: Osteoarthrosis. *Ann Rheum Dis* 1966; 25:1.

27. Lewis CB (editor): *Aging: The Health Care Challenge*. Philadelphia, F.A. Davis Co, 1985.

28. Oosterhuis HJ: Observation of the natural history of myasthenia gravis and the effect of thymectomy. *Ann NY Acad Sci* 1981; 377:678.

29. Pearson CM, Bohan A: The spectrum of polymyositis and dermatomyositis. *Med Clin North Am* 1977; 61:439.

30. Pereira M., Kaine L: Polymyalgia rheumatica and temporal arteritis: Managing older patients. *Geriatrics* 1986; 41:54.

31. Perkins LC, Kaiser HL: Results of short term isotonic and isometric exercise program in persons over 60. *Phys Ther Rev* 1961; 41:633.

32. Perlo VP, et al: The role of thymectomy in the treatment of myasthenia gravis. *Ann NY Acad Sci* 1971; 183:308.

33. Sinaki M: Postmenopausal spinal osteoporosis: Physical therapy and rehabilitation principles. *Mayo Clin Proc* 1982; 57:699.

34. Sinaki M, Mikkelsen BA: Postmenopausal spinal osteoporosis flexion vs extension exercises. *Arch Phys Med Rehabil* 1984; 65:583.

35. Singer ER, et al: Salmon calcitonin therapy for Paget's disease of bone. *Arthr Rheum* 1980; 23:1148.

36. Smith DM, et al: Age and activity effects on rate of bone mineral loss. *J Clin Invest* 1976; 58:716.

37. Smith EL, Reddan W: Physical activity—a modality for bone accretion in aged. *AJR* 1976; 126:1297.

38. U. S. Senate Special Committee on Aging, in conjunction with the American Association of Retired Persons: *Aging America: Trends and Projections*. Washington, DC, 1984, p 19.

Index